AQUATIC MICROBIOLOGY

Third edition

Aquatic Microbiology

Third Edition

G. Rheinheimer

Institut für Meereskunde,
University of Kiel, West Germany

A Wiley-Interscience Publication

JOHN WILEY & SONS

Chichester · New York · Brisbane · Toronto

Library of Congress Cataloging in Publication Data:

Rheinheimer, G. (Gerhard), 1927 –
  'A Wiley–Interscience publication.'
Aquatic microbiology. – 3rd ed.

  Translation of: Mikrobiologie der Gewässer.
  Bibliography: p.
  Includes index.
  1. Aquatic microbiology. I. Title.

QR105.R49513   1985   576'.192   84–25816

  ISBN 0 471 90657 3

Composition and Printing by the German Democratic Republic

Dedicated to Professor Dr. HORST ENGEL

Preface to the Third English Edition

In nearly all countries in recent years, high priority has been given to measures for water protection. In view of the great importance of micro-organisms in pollution and self-purification of waters, numerous new microbiological investigations, devoted both to applied problems as well as to fundamental research, have been conducted on very varied waters. Our knowledge of these matters, therefore, has again increased greatly since the appearance of the previous edition of this book.

Consequently, this present third edition has had to be thoroughly revised and in many parts extended. The illustrations have also been brought up to date and to some extent replaced. Although about 180 published papers — for the most part, from the last 5 years — have been reviewed and quoted, consideration could not be given to all the new publications. The selection has been so made, therefore, that with the help of literature data the remaining work could also become accessible to the reader.

I would like to express my warm thanks to all colleagues and collaborators who have afforded me help and advice. I am specially grateful to Dr. Norman Walker who, with much understanding, has made the translation.

Gerhard Rheimheimer

Preface to the Second English Edition

Lively interest in all branches of hydrography has been maintained undiminished in recent years and has led to intensified research effort. The activity of micro-organisms in aquatic ecosystems and their role in the transformations of matter in waters and sediments have received special attention. The application of improved radiobiological methods as well as fluorescence and scanning electron microscopy has provided new knowledge which, in some cases, has resulted in a revision of previously held views.

The second English edition of *Aquatic Microbiology*, therefore, has received a thorough and extensive revision. For the sake of completeness, the Cyanophytes or blue-green algae have been included because they stand closer to the bacteria, which likewise are procaryotic, than to the true algae. For the rest, the arrangement on the whole has been retained — most of the chapters, however, have been extended and here and there supplemented by new sections. The illustrations also have been suitably improved and increased in number. Thus, the book has been enlarged by about 50 pages. Nevertheless, in its new edition, it still remains an introduction to aquatic microbiology which it is hoped may provide a survey of the present position of this branch of study for as wide a circle as possible of readers from science, teaching and practice.

I would like to thank all my colleagues who have contributed to the new edition by advice, by drawing my attention to various points or by supplying photographs. Particularly I am indebted to Dr. Norman Walker who, with his great experience and knowledge of the subject, has made the translation.

Gerhard Rheinheimer

Contents

CHAPTER 1

1. Introduction

Aquatic microbiology deals with the structure and life of the micro-organisms in springs, rivers, lakes and seas, and with their role in the cycling of elements in water and in sediment. It includes the relationships between micro-organisms and aquatic plants and animals and the behaviour of terrestrial forms in aquatic environments.

Strictly speaking, all microscopically small plants and animals should be regarded as micro-organisms, i.e. bacteria, fungi, unicellular algae, protozoa and the smallest metazoa, like rotifers and similar organisms. To these must be added also the viruses. For technical reasons, however, the subject microbiology was often limited formerly to bacteria, fungi and viruses (Rippel-Baldes, 1955; ZoBell, 1946a). In more recent times the Cyanophyta also are usually included because of their close relationships to the likewise procaryotic bacteria (Schlegel, 1976). Therefore the present edition of *Aquatic Microbiology* also deals with bacteria, cyanophytes, fungi and viruses. These micro-organisms play a dominant role in the living activities in waters. The contribution of bacteria and fungi consists in destroying organic material and ensuring the quickest possible return of the most important inorganic nutrients into the cycle of matter so that green plants may produce new organic substances. In addition to their function as primary producers, Cyanophyta are often capable of fixing molecular nitrogen and are also able to colonize extreme aquatic sites. A comprehensive limnology or oceanography is, therefore, no longer possible without taking into consideration microbiology, i.e. the study of the activity of bacteria, cyanophytes and fungi. In the last two decades, most larger research institutions have recognized this and have established microbiological units or departments.

Aquatic microbiology is a branch of ecology which deals with the relations of organisms to their environment, i.e. to their living space and to the other living members of the community within it. The microbiologist working in this field will, therefore, gain his knowledge mainly through work on and in an aquatic environment and will endeavour to supplement this by appropriate laboratory experiments. The adoption of continuous measuring instruments, modern biochemical and serological methods together with transmission and scanning electron microscopy should quickly extend our knowledge of the manifold functions of micro-organisms in the living events in waters. It should not be overlooked, however, that many methodological difficulties still remain. Thus, the problem of taking sterile water samples, particularly from the surface of sediments, has not yet been solved satisfactorily. It is also a very great drawback that water samples undergo physicochemical changes as soon as they are put into containers. Adsorption on the walls of the containers plays a particularly important role. ZoBell (1946a) calls this phenomenon 'solid surface effect'. It rapidly changes the bacterial population of a water sample after collection. The number of types of organisms decreases, but the number of individuals may reach a multiple of the original within 24 hours. A further cause of change is the great sensitivity of many aquatic micro-organisms to a relatively small rise in temperature. If possible, therefore, water samples should be dealt with immediately after collection. Microbiological investigation of large bodies of water can be carried out satisfactorily only if a ship with a suitable laboratory is available. With most research ships, the laboratories have been accordingly so designed that they can be used also to microbiological work. Likewise, research submarines have repeatedly been involved for microbiological purposes. In recent years, marine microbiology in particular has thereby taken a powerful upturn and, as an example, quite unexpected results have been disclosed in connection with investigations of thermal vents in the deep ocean.

As a branch of ecology, aquatic microbiology is dependent on close links with other branches of aquatic science, particularly hydrochemistry, hydrobotany and hydrozoology. Only close collaboration between microbiologists, chemists, botanists and zoologists can lead to the elucidation of the complex interplay between micro-organism and the rest of the living world of the waters. Above all, the implication of microbiology is indispensable to investigations of ecosystems and for the present-day very important research on water pollution and its control. Therefore a knowledge of the most important microbiological processes in waters is an essential prerequisite for the success of the work of limnologists, oceanographers and specialists in the fields of fisheries and public health. This is also valid for some geologists and hydraulic engineers.

With regard to the prescribed size of this book, information on methods has had to be kept to a minimum. This seems to be justifiable since various books have appeared which deal in more detail with the methods of aquatic microbiology (Daubner, 1972; Reichardt, 1978; Rodina, 1972; Schlieper, 1968).

Reference may also be made to the books *Modern Methods in the Study of Microbial Ecology* (Rosswall, 1972) and *Microbial Ecology of a Brackish Water Environment* (Rheinheimer, 1977c), in which, among other things, various methods that were recently modified or newly developed for aquatic microbiology are described. Numerous technical details of methods may be also found in the special literature quoted.

2. *Environments of Aquatic Micro-organisms*

In almost all accumulations of water — on as well as under the surface of the earth — in the smallest puddle as well as in lakes, in rivers and in the sea, micro-organisms are able to live. Geographers use the terms 'subterranean waters' and 'surface waters', and distinguish between *inland waters*, for example springs, rivers and lakes, and the *seas*, i.e. the oceans and the regions bordering them. They all provide a suitable environment for aquatic micro-organisms. Individual waters are dominated by very different physical and chemical conditions, and on these the composition of the living community — including the microflora — depends. Subterranean waters (groundwater) and springs are poor in nutrients and are therefore, colonized as a rule by a microflora of only a few types, higher plants and animals being almost completely absent. Rivers, lakes and seas, on the contrary, may have a flora and fauna of great diversity including, besides bacteria and fungi, representatives of all important groups of plants and animals. The waters thus offer a variety of niches for life, and knowledge of them is important to an understanding of aquatic microbiology.

2.1. Subterranean Waters

Everywhere on earth water evaporates constantly, condenses as clouds in the atmosphere and returns to the surface of the earth as dew, rain or snow. Some of these precipitations ooze through loose soil and form *groundwater*, which is dammed up above impervious layers. Subterranean waters may also originate when the water of rivers or lakes penetrates through the banks. The layer of the soil which carries the groundwater is called the groundwater carrier; the layer underneath, which is impervious to water, is the groundwater barrier. The groundwater level is often subject to seasonal fluctuations caused by variations in the precipitation of water. In river valleys there is usually some exchange between river water and groundwater. Many lowland rivers keep receiving groundwater, which ensures that they carry water even in times of little rainfall. The level of the groundwater table is of great importance for the agricultural utilization of the soil. There are groundwater currents with more or less steep gradients, and corresponding rates of flow. They may reach considerable lengths. Thus, for example, groundwater currents run from Scandinavia under the Baltic to northern Germany. The mineral content of the groundwater varies according to geological conditions, and determines whether the water is 'hard' or 'soft'. Normally the groundwater is very poor in nutrients because of the filter effect of the soil layers is has trickled through. Consequently also there are few bacteria; hence the important role of groundwater for the supply of drinking water.

In the capillaries of permeable rocks we find the so-called *bed water*. It occurs mainly in coarse sandstone and some limestone rocks. Under as well as over the water-carrying layers are impermeable rocks. Where the water-carrying layer of a high-lying area reaches lowlands, as for example in Australia, artesian wells occur on a large scale.

In rocky areas water often reaches great depths by way of crevices and fissures, and collects in caves and chambers to form subterranean streams and lakes. This so-called *joint water* is found mainly in calcareous mountains and is responsible for the characteristic features of Karst scenery. The caves, with their fantastic variety of stalactites and stalagmites, are due to joint water. Joint water is much richer in nutrients than groundwater or bed water, and has often therefore a richer microflora. Subterranean lakes may even be inhabited by higher animals like the olm (*Proteus anguinus*) and certain fish.

2.2. Surface Waters

Springs form wherever subterranean water reaches the surface. According to the specific geological conditions, springs differ greatly with regard to their volume and the temperature and chemical conditions of the water. In the loose soils of the plains and the boulder clay areas of mountains, there are mainly groundwater springs. Scree springs fed from melted snow or ice occur in various mountain regions. The joint water of calcareous mountains may emerge as a spring. Where it emerges, bed water is referred to as a strata spring. Karst springs often yield much water, but there are some which run dry during droughts.

Besides the cold springs, there are also warm and hot, so-called thermal, springs whose water originates in volcanic areas or comes from great depths. The well-known geysers are found in volcanic areas — for example in Iceland, in the American Yellowstone Park, in Japan, and in New Zealand. These gush forth periodically; at definite intervals, jets of boiling hot water are thrown high into the air.

Depending on geological conditions, some springs carry distinctive mineral waters and many of these have given rise to spas. There are, for instance, salt springs, springs with magnesium salts, iron and sulphur springs. Acidulous waters which contain carbonic acid occur widely. In many places radioactive springs are found; these, too, have led to the establishment of spas.

Calcareous spring water often forms extensive sinter terraces. This happens particularly with hot springs — for example in Yellowstone Park in the USA. Silicic terraces are also formed there.

Springs are colonized almost exclusively by micro-organisms, predominantly algae and bacteria. The exact composition of the microflora is always determined by the chemical composition and the temperature of the spring water.

Rivers and *streams* are troughs following a more or less winding course which is determined by geological as well as climatic conditions. A river bed is bordered on both sides by banks which often show a characteristic vegetation. Its width changes with that of the valley. The depth is generally relatively small, a few metres, but large rivers may be very deep. The greatest depths are found in straits; they arise as a result of a large local increase in the erosive activity of the flowing water and are called pot-holes. In the 'Iron Gate' of the Danube, for instance, they attain a depth of about 54 m. Deep pools alternate with shallow fjords. The river water always moves towards lower-lying regions and finally runs into the sea or into lakes with no outflow.

Mountain rivers usually have steep gradients which, in turn, means strong currents. For instance, the gradient for the upper course of the Rhine as far as Chur is 23 in 1000, from there to Lake Constance 2 in 1000, from Rheinau to Basle 1 in 1000 and further downsteam less than 1 in 1000. In the lowlands it varies between 0.15 and 0.3 in 1000. The Danube has a gradient of only 0.05 in 1000 in the Hungarian plain, i.e. over 1000 m the river only falls by 5 cm. The speed of flow depends on the gradient as well as on the volume of water and the cross-section of the valley; in turn, the speed

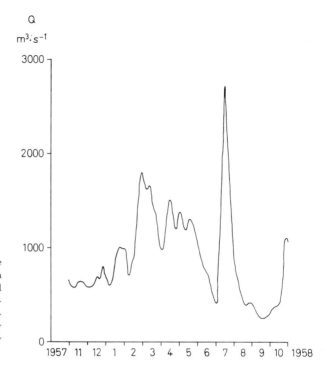

Figure 1   Water flow of the Elbe at Neudarchau (km 534) in 1958 (Annual Report of the Wasser- und Schiffahrtsdirektion Hamburg). Heavy summer floods in July

of flow determines the extent of transport of rock fragments. In the case of mountain rivers, the speed of flow amounts to several metres per second; with lowland rivers of central Europe, it is still 0.3–0.5 m s^{-1} (Czaya, 1981). Many rivers carry large quantities of suspended matter which is finally deposited in their estuarial regions. The Nile and Mississippi deltas, for instance, developed in this way. The suspended particles may carry on their surfaces a diverse flora which may be of great importance to the microbiology of the particular river.

The volume of water carried, in all rivers, is subject to fluctuations, often of considerable magnitude. It depends on meteorological conditions, i.e. the amount of precipitation and the temperature. Rainy periods and thaw lead to a large volume of water being carried, often resulting in floods. In dry periods, on the other hand, the volume of water decreases and some rivers may turn into poor rivulets. Figure 1 shows the fluctuations of the water flow of the river Elbe at the water gauge Neudarchau (river kilometre mark 534). The fluctuations, in many parts of the world, are seasonal, i.e. periodic. The rivers of the temperate climatic zone carry their greatest volume of water in early spring, after the thaw, and their smallest often in September. However, non-periodic fluctuations may occur; thus, after heavy summer rains rivers may cause devastating floods.

In subtropical regions, many rivers carry water only during the rainy season. In large rivers which have their origin in mountains — as, for example, the Euphrates — the amount of water carried may, curiously, decrease towards the estuary.

The mineral content of rivers varies and is dependent on geological conditions. Rivers of the plains may contain large quantities of humic acids which come from bogs. In densely populated areas the river water is often much polluted by industrial and domestic sewage. The Elbe and the Rhine, for instance, show considerable anthropogenic salination. Moreover, most of the larger rivers are loaded, more or less heavily, with organic waste substances. Such pollution, of course, affects the biology of the rivers decisively. According to the degree of pollution, we distinguish oligosaprobic, mesosaprobic and polysaprobic rivers.

In temperate and cold climatic zones the seasonal fluctuations of the water temperature are considerable and rivers may carry an ice cover for some time in winter.

Depending on the physical and chemical conditions, a variety of communities of living organisms may occur. Thus, plankton can develop only if the rate of flow is not too high. This holds also for the benthos vegetation. Rapidly flowing waters, therefore, have poor productivity and are inhabited by only a few kinds of animals and plants, all adapted to the extreme conditions of rapid flow. Algae can develop only as crustlike coatings on the surface of stones. Insects display flat shapes so as to offer the least possible resistance to the current. However, at rates of flow below $1\,\mathrm{m\ s^{-1}}$ there are more organisms of more varied kinds, and the conditions gradually approach those of lakes (Gessner, 1955).

In slowly flowing waters (particularly the larger rivers) plankton is found and can even multiply in this habitat, but river plankton (potamoplankton) originates without exception in lakes and abandoned channels and must be replaced from there all the time. Beds and banks also have a multifarious flora and fauna.

Lakes, whatever their size, are water-filled hollows of the mainland. In addition to lakes which are too deep to allow the vegetation from the banks to spread into them, there are shallow ponds which can be colonized completely by submerged plants, and swamps whose vegetation reaches the surface everywhere and rises above it here and there (Nussbaum, 1933). Lakes vary a good deal, according to the geographic, geological and climatic conditions. Most common are ponds and small lakes; in America and Asia, however, lakes occur which cover areas of 30000 to 80000 square kilometres. There are shallow lakes with a depth of only a few metres, and lakes which may be more than 1000 m deep — for example Lake Tanganyika in East Africa and the Siberian Lake Baikal, whose greatest depth is around 1650 m.

River lakes obtain their water from the rivers which flow through them — Lake Constance from the Rhine and Lake Geneva from the Rhone.

Groundwater lakes, on the other hand, are fed from groundwater. These are found mainly in calcareous mountain regions and in regions of boulder clay. A lake of this kind is the Lake of Lunz in Austria, well known to all limnologists. The lakes of Carinthia obtain water from warm springs, hence their temperature is relatively high. In calcareous mountain regions there are numerous lakes with subterranean outlets.

By building retaining dams, reservoirs are formed which, as a rule, serve various water or energy purposes such as, for example, irrigation, drinking-water supplies, water control and the generation of electricity. The largest reservoirs have a content of more than a hundred thousand million cubic metres and are situated in the Soviet Union (Bratskaja) and in Egypt (Sadd-el-Ali). The character of whole river systems has been completely changed in recent decades by such extensive retention dams and, in place of rivers, chains of artificial lakes have now appeared. A good example of this is the Volga.

Even though lakes are regarded as still waters, they are not without water movements. Inflow and outflow cause currents, and wind may raise considerable waves, particularly in large lakes. These water movements, in turn, influence the shaping of the banks.

In the temperate climatic zone, the temperature at the beginning of spring is usually the same (3 °C) throughout the body of water and the wind may achieve complete mixing of the water (spring circulation). Later, as the surface layers of the water warm up quickly, layers become stabilized with a warm zone above a cold one (summer stagnation). At the boundary between the two zones the temperature drops rapidly and this boundary zone is known as the thermocline (or metalimnion). This stratification of many lakes is, of course, extraordinarily important biologically.

The warm, well-lit so-called epilimnion becomes a productive zone due to the assimilation activity of the phytoplankton which develops copiously, whereas in the cold, poorly lit hypolimnion breakdown of organic matter takes place and is due predominantly to heterotrophic micro-organisms. With the lower autumn temperatures, the layers become less stable until the autumn circulation is reached, which may be followed by a winter stagnation. In the hypolimnion of eutrophic lakes so much oxygen is consumed that it often disappears completely and this creates the right conditions for the production of hydrogen sulphide (H_2S) (Ruttner, 1962). An environment of this kind, which contains H_2S is called, according to Bass-Becking (1925), a *sulphuretum*. They are colonized almost exclusively by micro-organisms.

Of all waters, lakes are the most influenced by the living communities in them. The remains of dead plants and animals become deposited on the bottom and contribute to a gradual decrease in depth. Plants take part in the precipitation of $CaCO_3$. The vegetation of the banks encroaches all the time, until finally the area becomes dry land. This occurs particularly rapidly in lakes rich in nutrients, 'eutrophic' lakes. But 'oligotrophic' lakes, poor in nutrients and in humus, and 'dystrophic' lakes, which are rich in humus, are also subject to this process. First of all swamps become established, and finally marshy woods. The pollution of many lakes by inorganic fertilizers (from agriculture) and by sewage causes massive eutrophication in the densely populated areas of the world, and thereby an acceleration of the processes leading to the water areas becoming land (Ohle, 1965).

Salt lakes are distinguished by a high salt content. They are either fed from salt springs or they are lakes without an outlet in arid regions where evaporation causes an increase in salt concentration. The largest salt lake is the Caspian Sea, whose salt concentration is, however, very low at 1.1–1.3 per cent. It is very much higher in the Dead Sea, and in the Great Salt Lake of Utah (USA), where the salt concentration may go up to 28 per cent. In other salt lakes it may rise to saturation. Shallow salt lakes which, at times, dry up more or less completely turn into salt bogs.

The salt composition may be very different in individual lakes. Besides NaCl, they contain different amounts of $MgCl_2$, $MgSO_4$ and $NaHCO_3$. There are the salt lakes proper which contain predominantly NaCl, and soda lakes where $NaHCO_3$ predominates; an example of the latter is Wadi Natrum in Egypt. Bitter lakes are rich in $MgSO_4$. There are also borax lakes with $Na_2B_4O_7$ at the bottom; these are found mainly in California and Nevada (USA).

Lakes with high salt contents represent extreme environments with living communities consisting of only a few types and composed mainly of micro-organisms (bacteria, blue-green algae and flagellates), whereas lakes with lesser salt contents are, in some biological regards, more like the sea.

The *sea* covers 70.8 per cent of the surface of the earth and thus represents the largest continuous water surface. It is already distinguished by its size from all inland waters, and also from most of them by its salt content which, on an average, is 3.5 per cent. Sea water has a higher density, a higher osmotic value and a lower freezing point than fresh water. The salinity of the oceans does not vary much; it lies between 3.2 and 3.8 per cent. In coastal waters, however, it may sink to less than 0.1 per cent, or rise to 4.4 per cent. The composition of the salts dissolved in sea water is approximately constant. The NaCl part is 77.8 per cent. The most important constituents of sea water (S 3.5 per cent) are:

Sodium	10.752 g kg^{-1}	Chlorine	19.345 g kg^{-1}
Potassium	0.390 g kg^{-1}	Bromine	0.066 g kg^{-1}
Magnesium	1.295 g kg^{-1}	Fluorine	0.0013 g kg^{-1}
Strontium	0.013 g kg^{-1}	Sulphate	2.701 g kg^{-1}
		Bicarbonate	0.145 g kg^{-1}
		Boric acid	0.027 g kg^{-1}

In addition, it contains traces of about half of all the known elements. By reason of its salinity, the sea is a biological environment *sui generis*, and of the marine plants and animals most, by far, can live only in the sea, while most limnic kinds cannot penetrate into the marine region. Hence, in brackish water like the Baltic the number of different kinds decreases with the decreasing salt content — without a corresponding rise in the number of fresh water forms. Relatively few organisms are adapted specifically to brackish water.

The shallow regions of the sea extending along the coasts are regarded, down to a depth of 200 m, as belonging to the mainland and are called *shelves*. The topography of the sea bottom is as complex as that of the land. There are underwater mountain ranges some of whose peaks may rise as islands above the level of the sea; there are flat basins, deep trenches and dome-shaped hill tops. Large parts of the sea are deeper than 2000 m and regarded as *deep sea*. The greatest depths, more than 10000 m, are reached in the Pacific trenches, along the east Asiatic groups of islands.

The hydrostatic pressure increases with the depth of the sea, by about 1 atm every 10 m. In the deep sea there is thus a pressure of several hundred atm, which only some marine creatures can endure.

In the deep sea there is permanent darkness as the light diminishes rapidly in the water. Down to about 200 m green plants are still able to assimilate. This range is therefore called the productive zone. In very turbid coastal waters the range is, however, much shallower; in the German Bight, for example, not deeper than 15 m (Gessner, 1957).

The temperature of the surface water depends to a large extent on climatic conditions. Thus in the temperate and subtropical zone it may be subject to considerable fluctuations, whereas in the tropics it remains steady throughout the year. In depths of more than 1000 m the temperature is always between $+5$ and $-1.5\,°C$. This holds for about 90 per cent of the sea (ZoBell, 1946a). For this reason, psychrophilic, i.e. cold-loving, creatures play a large role here.

In the mixing zones of deep sea thermal vents — for example, in the vicinity of the Galapagos Islands in the Pacific, where hot water of 300–400 °C occurs — a completely different microflora is found in which, however, chemoautotrophic sulphur bacteria are abundantly represented (Jannasch and Wirsen, 1981). In the top layer of 10–100 m thickness the temperature is usually fairly homogenous, but then falls quickly over a narrow zone. The region of rapid temperature drop is known as the thermocline. Not infrequently it is associated with a sudden discontinuity in salt content as well; consequently, there may be differences in density which influence the vertical distribution of organisms.

The oxygen content of the water decreases many times in the aphotic zone; in the Arabian Sea, for example, it decreases to less than 1 ml O_2 l^{-1} at 1000 m but may increase again slowly in still greater depths, where the supply of nutrients is so small that no more oxygen is used up (Dietrich *et al.*, 1966). In parts of the sea which are partially land-blocked so that the water exchange is small, oxygen may disappear completely and hydrogen sulphide accumulates. This happens, for instance, in the Black Sea where, below 200 m, up to 7.3 mg H_2S l^{-1} are dissolved in the water.

The pH lies, as a rule, between 7.5 and 8.5, but in some coastal waters may show a greater divergence.

The plant nutrients, nitrate and phosphate, are generally bound by the phytoplankton in the photic zone and gradually released in the aphotic zone, mainly through the agency of bacteria. Thus, with increasing depth the concentration of these substances goes up while the organic matter decreases. Areas with upwelling water movements therefore permit vigorous development of phytoplankton and are, consequently, rich fishing grounds.

In many places, the sea water is constantly moving. Currents like the warm Gulf Stream affect our climate, and the tides shape many coastlines.

In contrast to the *oceans* there are the *enclosed seas* (e.g. the Baltic and the Mediterranean), and the coastal waters are often much cut up. Creeks and fjords penetrate far into the mainland, and at estuaries the tides may continue into the limnic areas. Regions of brackish water form in these border areas (Caspers, 1959) which differ biologically both from the inland waters and from the sea.

The coastal waters represent a variety of habitats due to their geographical and climatic peculiarities. Usually they are richer in nutrients than the open sea, so that they are much more colonized by plants and animals. Along the coasts there are extensive meadows of sea grass and forests of algae which offer protection and food to numerous animals. Here also are the spawning grounds of numerous economically useful fish, and a rich microflora proliferates.

The oceans, however, are not so uniform as they may appear. There are regions which are relatively rich in nutrients with a multifarious animal and plant population, and others of extreme nutritional poverty, with phosphates and nitrates present in scarlecy demonstrable concentrations where only few creatures can develop. The dark deep sea is a domain of microbes and of bizarre sea animals. There is life not only in the water, but also on the sea bed. Even deep in the Black Sea, full of hydrogen sulphide, micro-organisms flourish. Because of its size, the sea is still the least well-known habitat on earth — but oceanography has undergone an enormous boom in the last 15 years, so that our knowledge and the possibilities of sea utilization are growing year by year.

As large environments in waters, open water or the *pelagic zone* and the water bottom or *benthic zone* may be distinguished.

Organisms suspended in the water form *plankton*, which is subdivided into phyto- and zoo-plankton. Phytoplankters are the most important primary producers, of which cyanophytes and unicellular algae are the main representatives. However, heterotrophic bacteria and fungi should also be included. Zooplankters are, above all, primary consumers — namely, protozoa, small metazoa and the larvae of larger animals. For practical reasons, plankton is frequently classified according to size (Götting *et al.*, 1982). Actively swimming animals belong to the *nekton*. By far the largest group is the fishes, to which must be added cephalopods (ink fishes) as well as some reptiles, birds and mammals.

Organisms that live on or in the water bed are included in the *benthos* and, again, phyto- and zoo-benthos may be distinguished accordingly. Classification is often made, according to the size of the organisms, into micro-, meio- and macro-benthos. Micro- and macro-algae belong to the phytobenthos, as do also higher plants and, in addition, bacteria and fungi. The zoobenthos embraces a large number of animals of most varied forms, some of which are sessile, i.e. attached to the substrate and, for example, form mussel banks and coral reefs; others move about as free-living organisms on or in the sediment.

3. *Bacteria*

Soon after the invention of the microscope, *bacteria* were discovered in water, as were other micro-organisms. Antonie van Leeuwenhoek in 1683 not only provided the first illustrations of cocci, rods and spirilla, but also studied their movements in drops of water. But it was only in the nineteenth century that any progress was made in aquatic microbiology. Stimulated by the successful work on microbes done by Louis Pasteur and Robert Koch, a variety of waters were searched for bacteria, and it was soon found that they were present there just as in other environments. It was also shown that the bacterial inhabitants of water are not by any means uniform, but show extraordinary variety, just like those of the soil. Besides the genuine (authochthonous) aquatic bacteria whose proper home is in water and which can develop optimally only there, a number of bacteria from other habitats are also found. Thus many soil bacteria are found in fresh water, as flowing waters in particular are in close contact with the soil. A constant rain of bacteria falls from the air on to surface waters. Other sources of contamination are plants, animals and humans. Some of these bacteria are ubiquitous and can proliferate in the most diverse, habitats, including water. Others can remain alive in water for a limited time only. Some bacteria found in the human intestines, both normal and pathogens, belong to this group. According to the kind of aquatic habitat, the composition of the bacterial flora differs widely — dependent not only on the water's content of organic and inorganic material, its pH, turbidity and temperature, but also on the sources from whence organisms can enter the water. The bacteria living in the sea are different from those in fresh water and, amongst the latter, those of the rivers are different from those in lakes. The

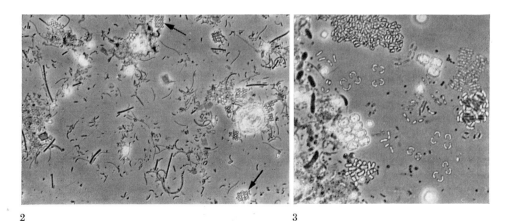

2   3

Figure 2   Bacterial flora of the anaerobic sediment surface of the shallow Forest Pond, Michigan (USA), in spring, with *Thiopedia* ($\times 360$). (P. Hirsch)

Figure 3   Bacterial flora of Lake Wintergreen (Michigan, USA) in summer at a depth of 4 m, with *Brachyarcus* ($\times 850$). (P. Hirsch)

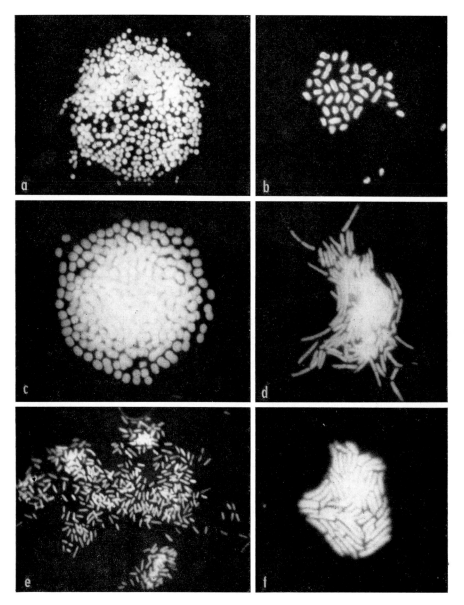

Figure 4   Colonies of cocci and rods from the Baltic Sea ($\times 2\,000$). (L. A. Meyer-Reil)

majority of aquatic bacteria are heterotrophic, i.e. they live on organic substances. By far the most numerous are saprophytes on dead material of plant or animal origin; the number of parasites is relatively small. In addition, there are in water photo- and chemoautotrophic bacteria needing only inorganic nutrients. They are either — just like green plants — able to carry out photosynthesis, or else they can, by means of chemical energy, reduce carbon dioxide and synthesize organic material. Chloro- and purple bacteria belong to the first group, nitrifying, sulphur and some iron bacteria to the second.

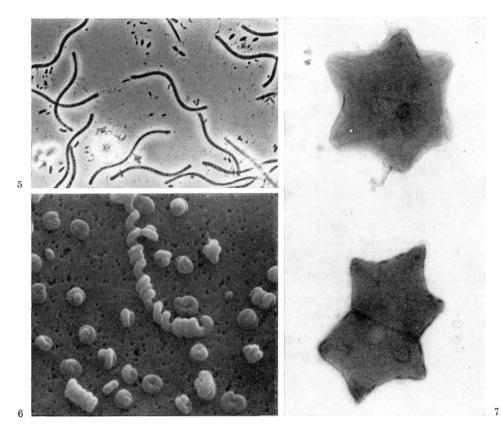

Figure 5   Spirilla from the Forest Pond ($\times 850$). (P. Hirsch)

Figure 6   *Microcyclus marinus* from the Kiel Fjord ($\times 5400$). Scanning electron micrograph by R. Schmaljohann

Figure 7   *Stella aquatica*, a star-shaped bacterium from the Kiel Fjord. Top: single cell. Bottom: cell shortly before division ($\times 21000$). Electron micrograph by H. Schlesner

Morphologically, most aquatic bacteria have their equivalents amongst the basic types of terrestrial bacteria, as already described by van Leeuwenhoek, i.e. the cells are spherical, rod-shaped, commas or spirals (Figures 2–5). In addition, filamentous and band-shaped organisms occur, and stalked forms are also found. The filaments may be branched or simple, and they may occur singly or in bundles. Various aquatic bacteria may form aggregates consisting of a few or many individual cells; the aggregates may be spherical or shaped like an egg, a star, a ribbon, a net or a sheet. Bacteria present in a body of water often show very considerable differences in size (Figures 6 and 7). In investigations in the Kiel Fjord and in the estuary of the Elbe the volumes differed approximately by the ratio of $2:1000$.

In most waters the predominant bacteria are gram-negative; that is, organisms which, after the so-called Gram staining method with crystal violet and subsequently treated with a solution of iodine in aqueous potassium iodide, are then decolorized by alcohol. This staining method, devised in 1884 by the Danish physician, H. C. Gram, represents an important taxonomic feature which is dependent on the cell wall structure and is also correlated with other characteristics.

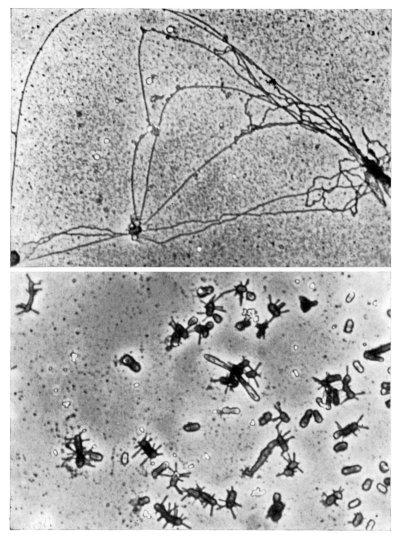

Figure 8   Bacteria with fimbriae, isolated from the Baltic Sea: *Agrobacterium stellulatum* (top) with very long fimbriae and *A. ferrugineum* (bottom) with very short ones (× about 3000). (R. Jeske)

The majority of aquatic bacteria are motile, as a rule by means of flagella, but some by creeping along solid surfaces. More recently, fimbriae have been found in addition to flagella (Figure 8). They are associated with the formation of star-shaped aggregates (Ahrens *et al.*, 1968). The fimbriae or pili are generally thinner than flagella but vary a great deal in length (Moll and Ahrens, 1970).

Genuine aquatic bacteria are distinguished by their ability to utilize very small concentrations of nutrients. According to Wright and Hobbie (1966), the water of some lakes contains acetate and glucose in concentrations of 1–10 μg l^{-1}; even in these low concentrations they can be absorbed by bacteria which thus have an advantage over heterotrophically growing algae.

The bacteria may live in the water free, or growing on some solid substratum (i.e. mainly on detritus). Most are capable of either of these modes of life; however, there are some which can live only one way or the other. Nevertheless, all are included in the great living community of the plankton (the floating organisms).

Systematically, aquatic bacteria are not a homogeneous group. Their representatives are found in almost all orders of the class of Bacteria.

Thus, obligate or facultative aquatic bacteria are to be found in 16 of 19 groups which are listed in the 8th edition of *Bergey's Manual of Determinative Bacteriology* (Buchanan and Gibbons, 1974). Of course it has to be borne in mind that any systematic classification of bacteria is still very problematical. In the absence of any other indication, all taxonomic details are based on the 8th edition (at present the most recent) of *Bergey's Manual*.

As the classification of bacteria in the 8th edition of *Bergey's Manual* deviates very much from the earlier one, used in the 7th edition, a compilation of the groups that are important for aquatic microbiology is given in Table 1. To permit a better survey, the simply constructed bacteria — which may be referred to three basic forms, cocci, straight rods and curved rods — are included as eubacteria and are separated from the complicated forms among which the phototrophic bacteria also belong (see Table 1).

The colourless sulphur bacteria of the genus *Beggiatoa* show great similarities to those of the genus *Oscillatoria* which belongs to the Cyanophyta. Therefore, formerly, they were also considered as colourless blue-green algae.

Recently, the so-called Archaebacteria have been differentiated from all others particularly with regard to the composition of their cell walls and membranes as well as the RNA. Thus, their cell walls contain no murein. Representatives are the methane producers (*Methanobacterium*, *Methanospirillum*, *Methanogenium*, *Methanosarcina* and *Methanococcus*), and possibly also the extremely halophilic *Halobacterium* and *Halococcus*, the acidophilic and thermophilic *Sulpholobus* and also *Thermoplasma* — these are all inhabitants of extreme habitats. Obviously, the Archaebacteria were already separated from the phylum of the eubacteria at the beginning of biological evolution.

As there are great biological differences between inland waters carrying fresh water and seas, the bacteria of these types of waters will be treated separately.

3.1. Bacteria of Inland Waters

There are relationships between the bacterial flora of inland waters and that of the soil. These are particularly close if one considers flowing waters, which are constantly exposed to infection from the soil. Thus, a sharp separation of soil bacteria and aquatic bacteria (i.e. terrestrial and limnic bacteria) is not easily possible. Most bacteria of groundwater, spring water and stream water also occur in the soil which, indeed, shares some of the features of waters. One need only think of the water in capillaries, or of badly drained soils holding water. Considering the small size of bacteria, there is no fundamental difference for them between soil which is well supplied with moisture and waters where a great part of the bacterial flora exists growing on a variety of solid surfaces. Usually different, however, is the supply of nutrients which is very much greater in the soil, particularly the top layers, than in most waters. Hence only soil bacteria with the most modest requirements will also be able to grow in water, for instance, of springs and streams, and these, therefore, have a very characteristic microflora.

Particularly poor, however, in micro-organisms and in nutrients are groundwater and spring water because of the filtration effect of the soil. In the groundwater examined

Table 1   Classification of aquatic bacteria, based on the 8th edition of *Bergey's Manual* (Buchanan and Gibbons, 1947), and their most important distinguishing characters

---

Simply constructed bacteria (eubacteria)

Cocci

Gram-positive	aerobic	*Micrococcus*
	anaerobic	*Sarcina*
Gram-negative	aerobic	*Paracoccus, Lampropedia*

Rods without spore formation

Gram-negative

heterotrophic

	aerobic	*Pseudomonas, Zoogloea, Azotobacter, Methanomonas, Alcaligenes, Photobacterium* etc.
	anaerobic	*Fusobacterium*

chemoautotrophic

	aerobic	*Nitrobacter, Thiobacillus*

Rods with spore formation

Gram-positive	aerobic	*Bacillus*
	anaerobic	*Clostridium, Desulphotomaculum*

Curved or spiral-shaped cells

Gram-negative	aerobic	*Vibrio, Bdellovibrio, Spirillum*
	anaerobic	*Desulphovibrio*

Bacteria with complicated structures

Phototrophic bacteria

Purple bacteria
(coloured red, pink, violet or brown)

Chromatiaceae (H-donor H_2S)	anaerobic	*Chromatium, Thiocystis, Thiospirillum, Thiocapsa, Thiodictyon, Thiopedia, Ectothiorhodospira*
Rhodospirillaceae (H-donor organic compounds)	microaerophilic	*Rhodospirillum, Rhodopseudomonas, Rhodomicrobium*

Chlorobacteria
(coloured green or brown)

Chlorobiaceae	anaerobic	*Chlorobium, Prosthecochloris, Pelodictyon, Chlathrochloris*
Sheathed bacteria	aerobic	*Sphaerotilus, Leptothrix, Crenothrix*
Stalked bacteria	aerobic	*Hyphomicrobium, Rhodomicrobium, Caulobacter, Gallionella, Nevskia*
Actinomycetes	aerobic	

Coryneform bacteria (rods with thickened ends)		*Corynebacterium, Arthrobacter*
True Actinomycetes (form mycelium and spores)		*Nocardia, Streptomyces, Actinoplanes*
Spirochaetes (spiral cells with axial threads)	aerobic, anaerobic	*Spirochaeta, Cristispira, Treponema, Leptospira*
Mycoplasmas (polymorphic bacteria without cell walls)		*Thermoplasma, Metallogenium?*

Table 1 (continued)

Gliding bacteria (creeping on solid substrates)	aerobic, anaerobic
Beggiatoaceae (threads of numerous cells, partly with sulphur granules)	*Beggiatoa, Thioploca*
Cytophagaceae (flexible rods or filaments)	*Cytophaga, Sporocytophaga, Flexibacter, Flexithrix*
Leucothrichaceae (filaments attached to solid substrates, forming rosettes)	*Leucothrix, Thiothrix*

from various sites immediately north of the Harz mountains (Germany), gram-negative non-sporing rods predominated (Wolters and Schwartz, 1956). Particularly numerous were organisms belonging to the genera *Achromobacter* and *Flavobacterium*; in the groundwater at Börsum, for example, out of 1905 colonies which were examined they comprised 79.1 per cent of the total flora. Small numbers of gram-positive rods were found and some members of the genera *Micrococcus, Nocardia* and *Cytophaga*. Vibrios, spirilla, corynebacteria and mycobacteria were completely absent. In groundwaters from geologically different layers the general character of the bacterial associations are similar but certain qualitative and quantitative differences in the distribution of species were detected.

Hirsch and Rades-Rohkohl (1983) found 72 different morphological types of bacteria (besides ten protozoa and eight fungi) in groundwater from the region of Segeberg in Holstein. Most of the bacterial cells possessed fimbriae, eight were stalked and seven spore-formers. They were identified as belonging to the genera *Microcyclus, Hyphomicrobium, Prosthecomicrobium, Planctomyces* (Figure 9), *Gallionella, Caulobacter, Agrobacterium, Clostridium* and *Nocardia*. The majority of the cultures grew best on poor nutrient media. They showed also considerable physiological differences which pointed to different microhabitats in the groundwater region. Bacteria isolated from the groundwater of the Marmot basin in Alberta (Canada) corresponded to the

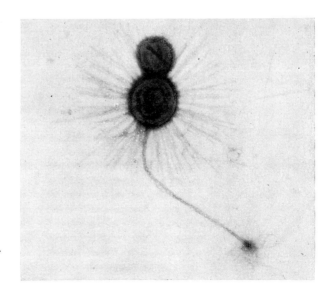

Figure 9   *Planctomyces* from groundwater (×13 000). Electron micrograph by P. Hirsch

species which occur there in soil. Total bacterial numbers and heterotrophic activity were higher than in an adjacent stream (Ladd *et al.*, 1982).

The groundwater of oil-bearing rocks contains large numbers of bacteria which decompose hydrocarbons. Not infrequently *Desulphovibrio desulphuricans* is also found here (Kusnezow *et al.*, 1963).

In springs, other bacteria are added to those mentioned above — there are more opportunities for infection from the surroundings as well as a greater supply of nutrients. In springs of iron-containing water, iron bacteria like *Gallionella ferruginea*, *Leptothrix ochracea* and *Crenothrix polyspora* are frequently found; in springs of sulphur-containing water we find various colourless sulphur bacteria and purple bacteria (Chromatiaceae) also if there is sufficient light. The bacterial flora of thermal springs consists of thermotolerant and thermophilic species as, for instance, *Sulpholobus acidocaldarius* and *Leptothrix thermalis*. Brock (1967a) found bacteria in hot springs, even at temperatures of more than 90 °C.

The bacterial flora of surface waters is, as a rule, much more diverse than that of subterranean waters. Its composition depends, above all, on the supply of nutrients in the water.

In streams, which are poor in nutrients, gram-negative non-sporing rods still predominate. In addition, there are often various stalked bacteria like *Hyphomicrobium*, *Caulobacter* and *Gallionella*, and also pseudomonads. According to Baker and Farr (1977), the genera *Pseudomonas*, *Flavobacterium*, *Acinetobacter* and *Moraxella* are dominant in relatively clean streams of the southern English chalky districts. With increasing eutrophication of the water, the proportion of *Flavobacterium* and *Achromobacter* species diminishes more and more, and representatives of the Pseudomonadaceae (particularly those of the *Pseudomonas fluorescens* group), Bacillaceae and the Enterobacteriaceae gain in importance (Wolters and Schwartz, 1956). The number of soil bacteria in flowing waters is generally still rather high. Thus, *Azotobacter chroococcum* and the nitrifying bacteria *Nitrosomonas europaea* and *Nitrobacter winogradskyi* have been found regularly in the water of the lower Elbe and its tributaries (Rheinheimer, 1965a); these organisms have their main habitat in arable soil and there play an important role in the bacterial nitrogen cycle. Besides *Azotobacter chroococcum*, *A. agile*, distinguished by the absence of the brown pigment, was found in the Elbe; this also can fix free nitrogen, and although it is regarded as an aquatic organism, it does occur in soil. Many other saprophytic bacteria of rivers behave similarly. Besides the organisms already mentioned, river water has regularly yielded vibrios, spirilla, thiobacilli, micrococci, sarcinae, nocardiae, streptomycetes, cytophagae and spirochaetes. In α- and β-mesosaprobic rivers, *Siderocapsa treubii* is found frequently (Drake, 1965). This organism has coccoid or rod-shaped cells, up to eight in a capsule, stained brown or blackish due to a deposit of iron or manganese hydroxide; it is usually attached to aquatic plants. It is a heterotrophic bacterium which utilizes organic iron compounds as nutrients. *Siderocapsa treubii* and the other species of this genus are probably autochthonous aquatic organisms, and thus limnic bacteria in the proper sense. The same holds for the majority of the family Siderocapsaceae.

Depending on the sewage load, rivers carry more or less numerous sewage bacteria. Of particular interest here are the intestinal bacteria *Escherichia coli*, the so-called coliform strains and the pathogenic salmonellae which can cause typhoid fever. Although intestinal bacteria can remain alive in water for only a limited time, they are often present in considerable numbers in polluted water, without loss of virulence as far as the pathogenic forms are concerned. Frequently, sewage-laden streams and rivers contain *Proteus vulgaris*. Clostridia are often present abundantly; and while in rapidly flowing waters with a correspondingly favourable oxygen supply they occur mainly as spores, in water zones poor in oxygen and also in sediments vegetative *Clostridium* cells may develop. Under these conditions the comma-shaped cells of *Desulphovibrio desulphuricans*, which can reduce sulphate, may be found. Flowing waters leavily loaded with organic garbage are the favourite habitat for *Sphaerotilus natans* and related sheathed bacteria. Particularly at sites where the sewage from cellulose factories and certain food industries enters the rivers, this bacterium can develop in such masses that the bottom and the banks are covered with dense lawns of sphaerotilus filaments. Characteristic also of such highly polluted flowing waters are the strangely branched colonies of *Zoogloea ramigera*.

Of all the surface waters, streams and rivers are those whose microflora is the most exposed to terrestrial influences. Hence the proportion of autochthonous aquatic bacteria is much smaller than in stagnant waters, where stable stratification reduces the influence of the surroundings, at least temporarily.

The groundwater lakes still contain many groundwater bacteria; added to these, however, are other kinds, depending on the particular chemical and physical conditions. The microflora of river lakes, similarly, is affected by that of the rivers which feed them. The smaller the lakes, the more significant this influence. The bacterial flora of a lake is, however, always quite distinct from that of its river.

According to Kusnezow (1959), non-sporing rods predominate, at least in the lakes of the temperate and boreal climatic zones. Their relative proportion is greatest in eutrophic lakes. Sporers rarely represent more than 10 per cent. In mesotrophic lakes, the number of sporers is often greater; they may be 20–25 per cent of all saprophytic bacteria. Their importance is greatest in dystrophic lakes, where they may sometimes even exceed the non-sporing bacteria in numbers. In the Russian Tshornoye Lake, which belongs to the dystrophic type, 85 per cent of the bacteria were sporers. In many lakes there is a preponderance of gram-negative bacteria. According to Taylor (1942), 95.5 per cent of bacteria isolated from some English lakes were gram-negative. In the neuston of north German lakes, Babenzien (1966) found 85.5 per cent gram-negative bacteria. By contrast, Jegorowa *et al.* (1952) found only 18 per cent in Siberian lakes.

Within the saprophyte flora of clean lakes the simply structured eubacteria seem to predominate, as they do in groundwater. In addition, there are, in some lakes, actinomycetes also. The latter occur particularly in the sediments, but they do not play so important a role here as they do in the soil.

Looking through the literature, it can be gathered that bacteria, particularly of the genera *Achromobacter*, *Flavobacterium*, *Brevibacterium*, *Vibrio*, *Spirillum*, *Micrococcus*, *Sarcina*, *Bacillus*, *Pseudomonas*, *Nocardia*, *Streptomyces*, *Micromonospora* and *Cytophaga*, occur widely in lake water. Also, *Sporocytophaga cauliformis* Knorr and Gräf seems to play an important role in the microflora of lakes (Ruschke and Rath, 1966). Numerous other genera may occur, depending on the type of lake and local conditions. Stalked bacteria like *Caulobacter* have frequently been found (Henrici and Johnson, 1935) (Figure 10), *Hyphomicrobium* (Figures 11 and 12) and *Planctomyces* (= *Blastocaulis*) (see Hirsch and Rheinheimer, 1968; Gorlenko *et al.*, 1977). In the hypolimnion of eutrophic lakes there is an abundance of unpigmented sulphur bacteria (e. g. *Thiospira*, *Thiothrix* and *Thioploca*), and methane oxidizers like *Pseudomonas methanica* (Anagnostidis and Overbeck, 1966). Here we also find numerous clostridia and *Desulphovibrio desulphuricans*. In the neuston — i.e. the living community which develops during calm weather on the surface of lakes at the water–air interface — a characteristic microflora develops (Babenzien, 1966), with representatives of the following bacterial genera: *Pseudomonas*, *Caulobacter*, *Nevskia*, *Hyphomicrobium*, *Achromobacter*, *Flavobacterium*, *Alcaligenes*, *Brevibacterium*, *Micrococcus*, *Sarcina* and *Leptothrix*. Particularly interesting here is *Nevskia ramosa* (Figure 13), which by unilateral mucous secretion forms curious stalks with dichotomous branching.

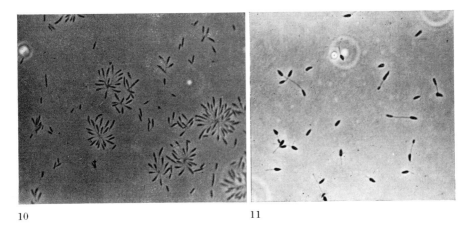

10   11

Figure 10   Rosette-shaped colonies of *Caulobacter* (×850). (P. Hirsch)

Figure 11   *Hyphomicrobium vulgare* (×900). (P. Hirsch)

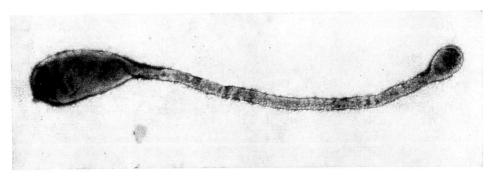

Figure 12   *Hyphomicrobium vulgare* with daughter cell (×16000). Electron micrograph by P. Hirsch)

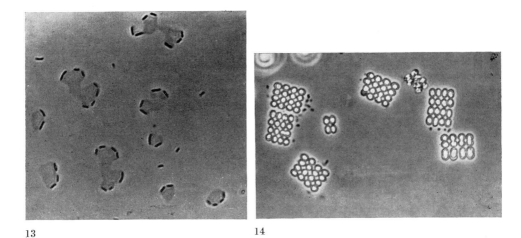

13   14

Figure 13   *Nevskia ramosa* with dichotomously branched slime stalks (×850). (P. Hirsch)

Figure 14   *Thiopedia* sp. (×900). (P. Hirsch)

An analysis of the bacterial population of Lake Pluss at different seasons, with the aid of numerical taxonomy, showed that, besides the basal population, bacterial groups specific for the season in question occur in spring, summer and autumn and they develop only under the particular ecological conditions of that season (Witzel et al., 1982).

Chemoautotrophic bacteria also play an important role in lakes, for example *Nitrosomonas europaea* and *Nitrobacter winogradskyi*, thiobacilli and iron bacteria.

Very important, particularly in eutrophic lakes, are photoautotrophic bacteria. They are particularly numerous in those anaerobic zones of the hypolimnion which still receive sufficient light. According to Skuja (1956), purple and chlorobacteria occur widely in Swedish lakes. This was confirmed for north German lakes by Anagnostidis and Overbeck (1966). For example, they found in the upper, still well-lit, hypolimnion of Lake Pluss in Holstein the following Chromatiaceae: *Thiopedia rosea* (Figure 14), *Lamprocystis roseopersicina*, *Rhodothece conspicua* and *nuda* Skuja; and further representatives of all the genera of the Chlorobiaceae, usually with large numbers of individuals.

Some of these bacteria contain numerous gas vacuoles, perhaps a floating mechanism (Figure 15).

The stalked photosynthetic bacterium *Rhodomicrobium vannielii* (Figure 16) also occurs in some waters.

It was established already some decades ago that in salt lakes, such as the Great Salt Lake in Utah (USA) and the Dead Sea, bacteria live in these waters in spite of their very high salt content. In some salt lakes, the salt concentration may reach saturation, and extensive salt efflorescences may precipitate out at their edges. A similarly high salt concentration also occurs in the spray water ponds on the coasts of arid regions and in the salt pans which serve to extract salt from the sea. The microflora of salt lakes — like that of other extreme environments — contains few kinds of organisms, but the numbers of individuals may be very great. Salt concentration and composition differ greatly between the different salt lakes and affect the composition of the microflora.

The overwhelming majority of bacteria living in salt lakes with a high salt concentration are halophilic forms; the proportion of salt tolerants is, as a rule, small (ZoBell, 1946a). Most interesting are the extreme halophils, whose optimal development takes place at salt concentrations of 20–30 per cent. They all have red pigment and belong to the genera *Halobacterium* and *Halococcus*. Since their cell walls do not contain any murein they are assigned to the Archaebacteria (p. 25). Such extreme halophilic bacteria occur also in salt pans. They remain alive even in the salt crystals. When food (fish, meat, vegetables) and skins are pickled in salt, the organisms may grow in the brine. Occasionally this leads to red stains on the infected articles — a fact which helped research into these organisms.

The greatest salt requirement is found in members of the genus *Halobacterium*, which can grow only in salt concentrations above 12 per cent. Their optimum lies at 25–30 per cent, and they develop even in saturated salt solutions (Larson, 1962). They are very pleomorphic, gram-negative rods. The motile species possess polar bundles of flagella. The cells are red-orange or brilliant red due to the presence of carotenoids. Frequently the cells have peculiar vacuoles, possibly filled with gas. Quite frequently, in the cytoplasmic membrane, there are dark red spots which are caused by the content of bacteriorhodopsin. This pigment, in the light, induces a proton gradient and hence an electrochemical membrane potential. With the equalization of the charges, regeneration of ATP can take place. The energy gained by this special form of photophosphorylation, however, may represent only a supplementation to the energy produced by oxidation (Schlegel, 1981). Halobacteria utilize chiefly

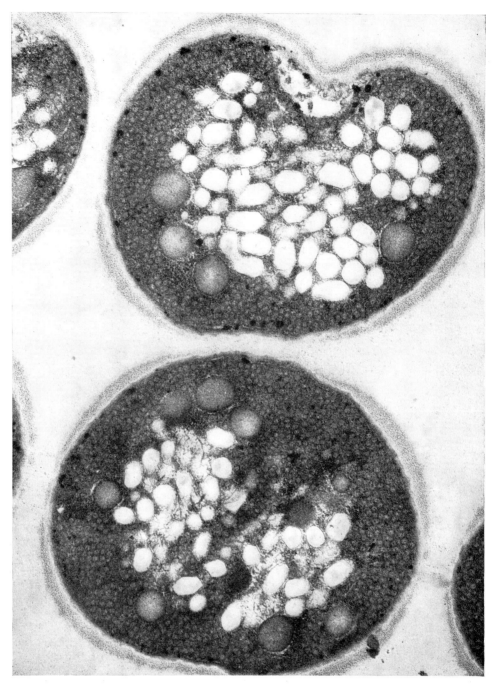

Figure 15 *Thiopedia* sp., two cells with gas vacuoles (×42 000). Electron micrograph by P. Hirsch

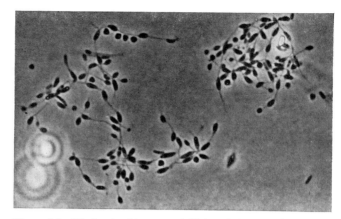

Figure 16   *Rhodomicrobium vannielii* ($\times 1\,200$). (P. Hirsch)

proteins and amino acids as food and carbohydrates only to a very small extent (Larsen, 1981).

Besides the '*Halobacterium* group', Larsen (1962) classes as extreme halophilic bacteria some coccoid organisms. Various strains of *Halococcus morrhuae* were isolated from the Dead Sea. These organisms also show red pigmentation. They grow best at a salt concentration of 20–25 per cent, and not at all at less than 10 per cent NaCl.

Besides the extreme halophils, salt lakes also contain some moderately halophilic bacteria with salt optima of 5–20 per cent. *Chromobacterium maris-mortui* (Rhizobiaceae), with a salt optimum of 12 per cent, belongs to this group. It is able to grow, however, at much lower salt concentrations (down to 0.5 per cent) and is therefore — probably wrongly — classified as halotolerant in the 7th edition of *Bergey's Manual* (Breed *et al.*, 1957). This organism has a bluish pigment and is motile by means of 4–6 peritrichously inserted flagella. The moderately halophilic bacteria of salt lakes are simply structured eubacteria.

In salt lakes with relatively low salt concentrations (i.e. less than 5 per cent), salt-tolerant bacteria are found in great numbers, in addition to weakly halophilic ones; extreme halophils are absent from this habitat.

In salt lakes containing hydrogen sulphide, large quantities of chloro- and purple bacteria can develop — for example, members of the genera *Chlorobium*, *Pelodictyon*, *Prosthecochloris*, *Chromatium*, *Ectothiorhodospira*, *Thiocapsa* and *Thiodictyon* (Butlin and Postgate, 1954; Trüper and Genovese, 1968; Cohen *et al.*, 1977; Gorlenko, 1977). *Ectothiorhodospira halophila* and *E. halochloris*, like the halobacteria, are extremely halophilic micro-organisms, which develop optimally at salt contents of 20–30 per cent and at 40 °C (Trüper and Imhoff, 1981).

3.2. Marine Bacteria

While the bacterial flora of inland waters shows close relationships to that of the soil, the open sea allows the development of an autochthonous marine flora. Most marine bacteria are halophilic, i.e. they need NaCl for optimal development. According to ZoBell and Upham (1944), they grow best at salt concentrations of 2.5–4.0 per cent; they either do not grow at all or grow very badly in fresh water media. The salt concentration of the sea, $\pm$ 3.5 per cent, represents the optimal salt concentration for genuine marine bacteria. According to MacLeod and Onofrey (1956), they need, as a rule, a certain minimum of Na$^+$ ions; some require, in addition, Cl$^-$ ions. The sodium is involved apparently in the transport of substances into the cell, possibly through the activation of proteases in the cell wall. Nevertheless, many marine bacteria do better in sea water media than in isotonic NaCl solutions. The ionic composition of sea water then, besides providing the necessary Na^+ ions and — for those organisms which require them — Cl^- ions, seems to have a favourable effect on growth generally.

Larsen (1962) classifies marine bacteria, with their salt optimum of 2–5 per cent, as weakly halophilic — in contrast to the moderately and extremely halophilic organisms, with optima of 5–20 per cent and 20–30 per cent, respectively. Besides halophilic marine bacteria there are also in the marine habitat others which are merely halotolerant, i.e. they are able to grow in fresh water media. Their importance in the open sea is, however, very small. They are found mainly near the coasts, particularly in creeks and estuaries; even there, their proportion of the total flora is not very significant (Rheinheimer, 1968a). In coastal waters there are, in addition — depending on the opportunities for infection from dry land — terrestrial bacteria which grow much better in fresh water media and are viable in the sea for only a limited time.

The majority of marine bacteria stain negatively with Gram's stain. Investigations of ZoBell and Upham (1944) on the coast of southern California yielded a proportion of 80 per cent gram-negative species. On agar plates inoculated with sea water, this proportion rose to 95 per cent. The author's own investigations of the North Sea and the Baltic, and also of the Arabian Sea, gave similar results.

In the sediment of a sea grass stand on the Australian coast, Moriarty and Hayward (1982) found 90 per cent of gram-negative bacteria in the aerobic surface zone (0–1 cm) and 70 per cent in the underlying anaerobic zone (20–21 cm). Amongst soil bacteria, on the other hand, the proportion of gram-negatives was found to be only 27–36 per cent (Taylor and Lochhead, 1938).

Most bacteria found in the sea are motile. According to ZoBell (1946a), 75—85 per cent of pure cultures examined for this characteristic possess flagella.

Sporers do not seem to occur widely in sea water. Their importance is greater in sediments. Thus, ZoBell and Upham (1944) isolated several species of the genus *Bacillus* from the marine mud around the Pacific coast. According to investigations by Rüger (1975) in the north Atlantic, such aerobic spore-formers make up a high proportion of the bacterial population of sediments at depths of 1000 m. Anaerobic clostridia may also be found in the mud of polluted coastal waters.

Most marine bacteria are facultative anaerobes, but grow better in the presence of oxygen. There are relatively few obligate aerobes and fewer still obligate anaerobes.

Marine bacteria generally grow more slowly than most soil bacteria (ZoBell, 1946a). Thus, on agar plates inoculated with soil and incubated at the optimal temperature, colony counts may be carried out after 2–7 days, whereas after inoculation with sea water or marine sediment the maximum number of visible colonies is only reached after 14–18 days.

Most marine bacteria can utilize nutrients present in minute concentrations. This ability is a prerequisite for their growth in sea water which, throughout, is very poor in nutrients. Even so, many marine bacteria can only find sufficient food while growing as Aufwuchs.

Adaptation to the very small concentrations of nutrients in sea water might cause the marked pleomorphism of many marine bacteria in culture. Pure cultures, both on solid and in liquid media, frequently show the most diverse cell shapes. Organisms which normally are rod-shaped may occur as cocci, vibrios, filaments and spirals. In older cultures, branching may be observed. Pleomorphism may, of course, occur also in bacteria from other habitats, but it does not seem to occur so frequently as in marine bacteria.

Many marine bacteria can grow at temperatures as low as between 0 and 4 °C. The temperature optimum frequently lies between 18 and 22 °C and the maximum only a few degrees higher. The proportion of psychrophilic, i.e. cold-loving, organisms in the marine domain is large. These are mainly facultative psychrophils (Stokes, 1962) which grow well at 0 °C, but their optimum is 20 °C or higher. Morita (1968) states that there are also obligate psychrophils in the sea, i.e. organisms whose growth optimum is below 20 °C. Considering the fact that about 90 per cent of the sea is permanently at temperatures below 5 °C, the importance of psychrophilic bacteria in this habitat must be much greater than it was thought to be some time ago (Morita, 1966; 1968).

Large parts of the seas must be classed as deep sea; hence barophilic (pressure-loving) and barotolerant bacteria are obviously important. However, in the surface zones there are also numerous barophobic bacteria which are inhibited by pressures above 100 atm.

The proportion of proteolytic (protein-decomposing) bacteria seems to be greater in the marine habitat than in most fresh water and soil environments. According to ZoBell (1946a), almost all heterotrophic marine bacteria can release ammonia from peptone, and nearly three-quarters of them liquefy gelatine. On the other hand, saccharolytic (sugar-decomposing) organisms play a much smaller role in the sea than in other habitats. Notwithstanding the predominance of proteolytic bacteria, there is hardly any natural organic compound which cannot be attacked by some marine bacteria and used as a nutrient. Besides organisms which decompose sugars and fats, there are others which decompose substances of high molecular weight like cellulose, agar, alginate, chitin, or hydrocarbons, phenols, etc.

Bacteria able to carry out denitrification (reduction of nitrate via nitrite to free nitrogen) occur, and others able to reduce sulphates to hydrogen sulphide. Both groups become enriched in anaerobic aquatic habitats and sediments; but while the number of denitrifying species is rather great, there are few sulphate reducers, of which the genus *Desulphovibrio* is, however, particularly important.

A great number of marine bacteria are capable of a wide variety of metabolic reactions and, consequently, can utilize a great variety of nutrients. But there are also distinct specialists, dependent on quite definite chemical compounds.

Morphologically, marine bacteria no more represent a group *per se* than other aquatic bacteria. In the overwhelming majority, the shape of cells corresponds to the four basic types: cocci, rods, vibrios and spirilla. For the greater part they are gram-negative, flagellated, non-sporing rods. Filamentous, branched or stalked forms may be found in small numbers. In the North Sea, micrococci also are important (Anderson, 1962). Curved bacteria evidently are strongly represented in the oceans (Sieburth, 1979). In addition to flagellated *Vibrio* and *Beneckea* species, often non-flagellate, strongly curved or even circular *Microcyclus marinus* are found in greater numbers.

Not infrequently, aggregates of varying numbers of cells are formed.

The occurrence of so-called ultramicrobacteria, which pass through membrane filters of 0.2 µm pore size, has also been repeatedly observed. Usually they are tiny short rods, some of which belong to the genera *Pseudomonas*, *Aeromonas*, *Vibrio* and *Alcaligenes*, according to investigations of an estuary on the American gulf coast by MacDonell and Hood (1982). In this case, it may be undersized or starvation forms that are involved, as already described by Jannasch (1955).

With regard to their systematic position, the marine bacteria are by no means a single group but belong to numerous families and genera, in which terrestrial kinds are also found. Frequently the only difference between the marine bacteria and the closely related terrestrial forms, with almost identical metabolic reactions, is the halophilic and facultatively psychrophilic character of the marine forms. Often the

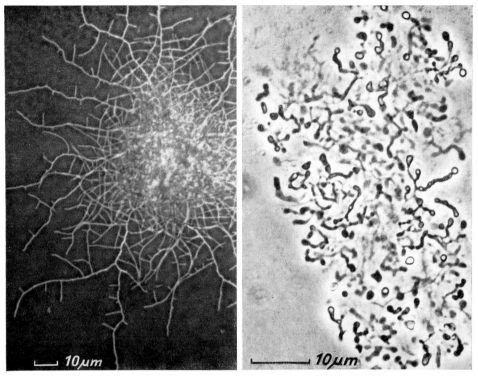

Figure 17 *Actinoplanes* sp. from North Sea sediment. Left: mycelium. Right: formation of sporangia. (H. Weyland)

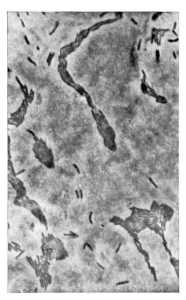

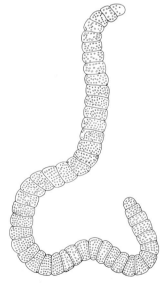

18   19

same genus contains marine organisms together with soil bacteria and fresh water bacteria.

According to Starr *et al.* (1981), members of the genera *Pseudomonas, Beneckea, Vibrio, Spirillum, Alcaligenes* and *Flavobacterium* (in sediments, *Bacillus* also) are widely distributed in marine regions. All these genera also contain numerous species of soil bacteria and fresh water bacteria; and they play very similar roles in clean inland waters as they do in the sea. The close relationship between the most common terrestrial and the marine bacteria has raised doubts as to the existence of genuine autochthonous marine bacteria (Scholes and Shewan, 1964); time and again it has been assumed that bacteria living in the sea are only terrestrial forms adapted to life in sea water. MacLeod (1965) believes that, although the ability to live in the sea is the only characteristic which clearly distinguishes them from other bacteria, this one characteristic is nevertheless sufficient to delimit them because under natural conditions — for example as a river enters the sea — adaptation of the fresh water bacteria which have been carried along does not occur. MacLeod (1965, 1968) assumes, on the other hand, that only a few mutational steps are needed to change a marine form into one which will survive in a non-marine environment. Both marine and terrestrial bacteria might have developed from original marine ancestors.

The bulk of marine bacteria belong to the Eubacteria (Brisou, 1955). But other forms are also found in the sea (Breed *et al.*, 1957; Hirsch and Rheinheimer, 1968) and of these the following groups and genera in particular play a role in the marine microflora:

Sheathed bacteria:	*Phragmidiothrix*
Stalked bacteria:	*Hyphomicrobium* (Figures 11 and 12)
Spirochaetes:	*Spirochaeta*
Actinomycetes:	*Nocardia, Streptomyces, Actinoplanes* (Figure 17)
Gliding bacteria:	*Cytophaga* (Figure 18), *Flexithrix, Beggiatoa* (Figure 19), *Thioploca, Thiothrix, Leucothrix*

Of Archaebacteria, methane producers of the genus *Methanogenium* occur in marine regions (Mah and Smith, 1981).

The sea is the focal point of distribution for the interesting luminous bacteria. These are able to transform chemical energy into light energy and so produce a rather bright greenish or bluish light (see Section 10.4). In the 7th edition of *Bergey's Manual*, some of the luminous bacteria have been grouped together as the separate genus *Photobacterium* (Pseudomonadaceae). Others have been put into the genus *Vibrio*. Spencer (1955), however, believes that the establishment of a genus *Photobacterium* solely on the grounds of luminescence is not justified, as pure cultures may lose this ability after a short time. If this were accepted, *Photobacterium* would have to be divided up between the genera *Aeromonas* (Pseudomonadaceae) and *Vibrio* (Spirillaceae). Based on a more recent investigation, Hendrie *et al.* (1970) suggest, on the contrary, a separation of the luminous bacteria into three groups, which would be placed in *Vibrio, Photobacterium* and a new genus *Lucibacterium*. According to Baumann and Baumann (1981) and Hastings and Nealson (1981), they should now be assigned to the genera *Photobacterium* and *Beneckea*.

Figure 18   *Cytophaga salmonicolor* from the North Sea. Single cells and characteristic cell associations (× 900). (B. Baeker)

Figure 19   *Beggiatoa mirabilis*. Filament with transverse walls and sulphur granules (× 1 200). (Based on Krassilnikow, 1959, modified)

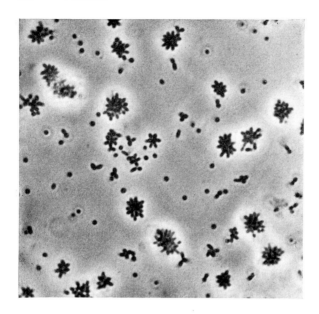

Figure 20 *Agrobacterium sanguineum*. Single cells and star-shaped cell aggregates (×3000). (R. Jeske)

There are free-living luminous bacteria and others that live as symbionts in the luminous organs of cephalopods (cuttlefish) and bonefish. Most of them are halophils which grow optimally at a salt concentration of 2–4 per cent. Up to now, only a single fresh water form (*Vibrio albensis*) has been described.

In sea water and in some marine sediments the proportion of pigmented organisms is strikingly large. ZoBell (1946a) reports that more than half of all bacteria living in sea water are pigmented. The author's own investigations in the North Sea, the Baltic and the Arabian Sea also indicated that a large number of marine bacteria are pigmented, but there were considerable differences according to area and season. On yeast extract peptone agar (ZoBell nutrient medium 2216 E), the proportion of pigmented bacteria varied between 10 and 95 per cent. Yellow, orange, brown and red forms are most frequent — much more scarce are those which are violet, blue, black or greenish. Bacteria which fluoresce green, blue or yellow also occur, but rarely more than just a few per cent.

Besides heterotrophic bacteria, photo-and chemoautotrophs also occur in the marine habitat. Photoautotrophic organisms are present wherever there is hydrogen sulphide and sufficient light for assimilation. These conditions are met above all in the immediate surroundings of coasts, such as secluded bays, beach lakes, sea-water pools within the area reached by the tide, on algae flung up by the water and in the so-called colour-band sand shoals.

Chemoautotrophic bacteria are found in coastal waters as well as in the open sea. Sulphur-oxidizing *Thiobacillus* species have been demonstrated particularly in those marine habitats where hydrogen sulphide is produced, for example in polluted coastal waters, and also in the depths of the Black Sea and in sulphide-containing sediments. Nitrifying bacteria (which oxidize ammonia to nitrite or nitrite to nitrate) have been demonstrated in the North Sea (Rheinheimer, 1967) and in the Atlantic (Watson, 1963). Watson, for the first time, found a marine nitrite bacterium, *Nitrosocystis oceanus*, in various depths of the Atlantic Ocean. Iron and manganese bacteria (oxidizing Fe^{2+} and Mn^{2+} to Fe^{3+} and Mn^{3+} respectively) are also present in the marine habitat.

In fringe areas of the sea which carry brackish water (i.e. water whose salt content lies below 3 per cent), not only genuine marine bacteria and salt-tolerant fresh water

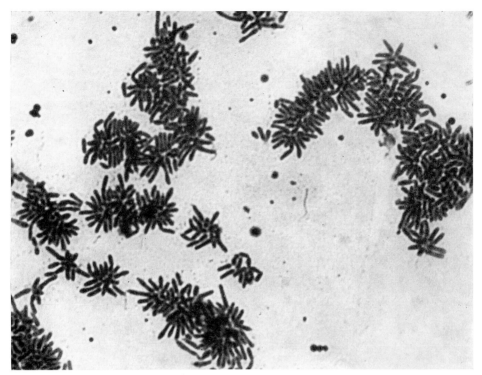

Figure 21   *Agrobacterium stellulatum*. Star-shaped and caterpillar-shaped cell aggregates
($\times$ 7 500). (R. Jeske)

bacteria are found, but also halophilic bacteria whose specific habitat is brackish
water (Rheinheimer, 1970a). Investigations on bacteria from the western and central
parts of the Baltic showed that some of them, with salt optima of 0.5-2 per cent,
are clearly distinct from the marine bacteria. These organisms from brackish water
show either scanty or no growth at all in fresh water media and are often, on the other
hand, inhibited by salt concentrations above 3 per cent. By contrast, many genuine
marine bacteria are markedly inhibited at salt concentrations of 1.5 per cent or below
(Larsen, 1962). Thus they do not find suitable living conditions in brackish water.

In the western Baltic, a group of brackish-water organisms which form characteristic star-shaped cell aggregates play an important role (Figures 20 and 21). They
are Rhizobiaceae, belonging to the genus *Agrobacterium* (Ahrens and Rheinheimer,
1967; Ahrens 1968). They seem to be particularly numerous in the Bay of Kiel and
the adjoining sea areas, where the salt concentration varies between 1 and 2.5 per cent.
Up to 90 per cent of the total number of bacteria isolated on yeast extract-peptone
agar from this region is made up of these organisms (Ahrens, 1969). In the eastern
Baltic where the salt concentration is less, and in the Kattegat where the salinity is
higher, these bacteria diminish considerably in number. Adaptation experiments
showed that at least some of the bacteria of brackish water have stable salt optima
of between 0.5 and 2 per cent.

Seen as a whole, the bacteria of brackish water represent no more of a systematic
group in their own right than the genuine marine bacteria. It is likely that, besides
heterotrophs, photoautotrophs also belong to this group. Thus, some purple sulphur
bacteria (Chromatiaceae) grow preferentially in brackish water, inland as well as along
the sea coasts.

4. Cyanophyta

Cyanophytes or blue-green algae are procaryotes, like the bacteria, and therefore have recently also been termed cyanobacteria. Their cell nucleus has no envelope, in contrast to the eucaryotic true algae, and they lack plastid membranes and mito-chondria. Chlorophyll a, β-carotene and phycobilin are their pigments for photo-synthesis. Many species owe their characteristic blue-green colour to phycocyanin. Some cyanophytes, however, are a more yellowish green and some are even red in colour due to phycoerythrin. At greater depths they undergo chromatic adaptation, i.e. adaptation to the changed light quality. Cyanophytes react to the greenish blue light prevailing there by the increased formation of phycoerythrin and by assuming the complementary colours, red or violet, can so achieve the highest levels of assi-milation.

Their pigments distinguish the cyanophytes both from the majority of true algae and also from the chloro- and purple bacteria as well. As a rule the true algae have even more pigments (chlorophyll b, other carotenoids) — the chloro- and purple bacteria for their part possess certain bacteriochlorophylls. Like the true algae, cyanophytes are capable of cleaving the water molecule, whereas the anaerobic chloro- and purple bacteria, on the contrary, need H_2S or organic compounds as hydrogen donors. That the photosynthetic pigments are not bound to chloroplasts, as is usual in green plants, but to often strongly coiled lamellae, is a common char-acteristic of all photoautotrophic procaryotes.

The morphology of cyanophytes assumes many forms. Thus, in addition to single-celled forms (Figure 22a), spherical, egg-shaped or plate-like colonies (Figure 22) occur which consist of more or less numerous cells and are often surrounded by slime coatings. These Chroococcales contrast with the filamentous hormogonal cyanophytes. The filaments represent cell associations and are termed trichomes (Figure 22 d, f). They are a physiological unit and frequently are surrounded by a sheath. Trichomes are able to creep on solid surfaces — like the closely related gliding bacteria (*Beggiatoa, Thiothrix, Leucothrix*, etc.; see Chapter 3). Multiplication of cyanophytes takes place by cell division. In the Hormogonales, multicellular trichome sections, the so-called hormogonia, can creep out of the sheath. In various species, non-motile resting cells (spores) Figure 22 e, f) and nodule-like thickened heterocysts occur (Figure 22 d). Nitrogen fixation takes place in the latter (Fogg *et al.*, 1973). At the sites where they are formed, false branching often occurs. The most highly developed forms even show true branchings and in the *Stigonema* a central row of cells is already surrounded by peripheral cells which are all linked together with pits. The classification of the cyanophytes and their most important distinguishing features are presented in Table 2.

Cyanophytes play an important role in the living events of surface waters. There are free-living and also attached forms. The latter are distributed as surface growths on water plants and animals as well as on dead substrates. Some species live in symbiosis with different lower plants and animals.

Although they are photosynthetic organisms, many hormogonal cyanophytes are also capable of a heterotrophic existence and can develop therefore not only in the

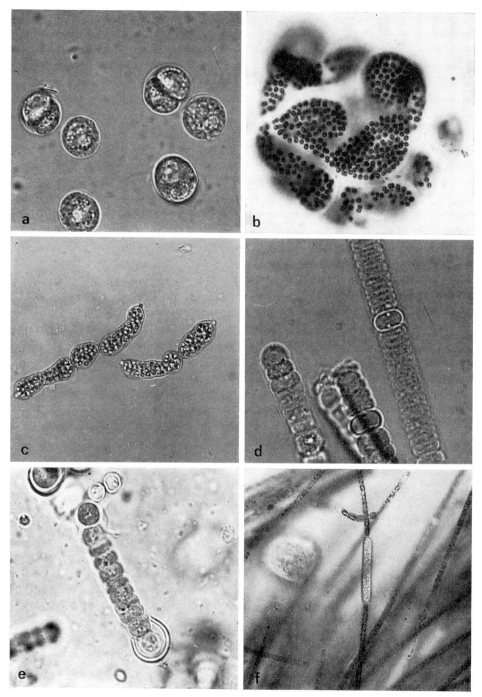

Figure 22   Cyanophytes from inland and coastal waters. (a) *Chroococcus*, single cells with division stages (×1000). (b) *Microcystis*, colony (×250). (c) *Nostoc*, colonies (×250). (d) *Nodularia spumigena*, trichome with two heterocysts (×450). (e) *Nodularia spumigena*, germinating resting cell (×450). (f) *Aphanizomenon flos-aquae*, trichome with resting cell (×250). (U. Horstmann)

Table 2   Classification of the Cyanophyta (blue-green algae), according to Golobic (1976)

Chroococcal

Single cells or colonies, reproduction by cell division and spores

Chroococcales

single cells, colonies
(*Aphanothece, Chroococcus, Merismopedia, Microcystis, Synechococcus*)

Chamaesiphonales

sessile single cells, exospores
(*Chamaesiphon*)

Pleurocapsales

filamentous cell aggregates, endospores
(*Hydrococcus, Pleurocapsa*)

Hormogonal

Cell filaments with cell differentiations (trichomes), reproduction by hormogonia

Oscillatoriales

filaments without heterocysts
(*Lyngbya, Oscillatoria, Phormidium, Spirulina, Trichodesmium*)

Nostocales

filaments with heterocysts
(*Anabaena, Aphanizomnon, Gloeotrichia, Nostoc, Rivularia*)

Stigonematales

filaments with true branchings
(*Mastigocladus, Stigonema*)

photic but also in the aphotic zone of waters. In any case in the dark they can only utilize carbohydrates as organic nutrients (Carr and Whittob, 1973).

The capacity to fix molecular nitrogen is widespread particularly among the hormogonal cyanophytes which form heterocysts (see Section 11.5). This property they share only with certain bacteria. It is therefore a character which is restricted to the procaryotes.

Some cyanophytes that attach themselves firmly to substrates, precipitate chalk and so contribute to sediment and rock formation.

Alkaline waters are colonized preferentially — acid ones on the contrary are avoided by many species. Planktonic forms frequently possess gas vacuoles, with the aid of which they can remain in suspension and rise to the surface. They occur in eutrophic waters often in massive quantities and so give rise to the formation of so-called water blooms. Most of these species exhibit a worldwide distribution.

4.1. Cyanophytes of Inland Waters

Inland waters are the chief habitat of the cyanophytes. Not only do most species occur here, but they play a very substantial part in the transformation of matter. Where water blooms are formed they can even be temporarily responsible for the total primary production.

Whilst cyanophytes, which are predominantly photosynthesizing organisms, have only a very trivial significance in subterranean waters, they occur with some frequency in springs. Their occurrence in thermal springs is particularly interesting. According to Castenholz (1973), in western North America there seem to be hardly any hot springs without cyanophytes provided there is sufficient light for photosynthesis, the pH is above 5 and the temperature does not exceed 74 °C. Along the edges they often form thick, mat-like layers which usually persist throughout the whole year. In thermal springs and their outflows are found members of the genera *Aphanocapsa*, *Pleurocapsa*, *Synechococcus*, *Calothrix*, *Mastigocladus*, *Oscillatoria*, *Phormidium* and *Spirulina* (Figure 23). Many of them have a worldwide distribution — as is also the case with the inhabitants of other extreme environments. They are not distinguished morphologocally from related species, which live in cooler waters.

Sites that are temporarily heated by strong solar radiation likewise are often colonized by cyanophytes. These include the blackish mats termed 'ink streaks' on steep slopes in the mountains which run in streaks in the direction of the run-off water, and shallow waters in hot deserts (see Castenholz, 1973). The ink streaks are produced especially on chalk and dolomite in addition to others by *Scytonema myochrous*.

In rivers the importance of the cyanophytes declines in favour of the true algae. However they are also widely distributed in flowing waters. In rapidly flowing streams *Pleurocapsa*, *Hydrococcus* and *Chamaesiphon* species are often found as surface growths on stones, mosses and algae. They are firmly attached to their substrates, so that they can resist the current. In mountain streams *Rivularia haematites* forms thick layers which become encrusted with chalk and coloured brown by iron. *Nostoc verrucosum* also lives in quickly flowing waters (Fott, 1970).

In large rivers and streams planktonic forms predominate. Schulz (1961) listed 19 species for the lower Elbe. Nearly all are found also in standing waters. From old river beds and lakes they continue to reach the rivers and here some of them are also able to multiply greatly. Thus, in the Elbe, mass occurrences of *Aphanizomenon flos-aquae*, *Microcystis aeruginosa* and *M. flos-aquae* have been observed again and again. In the bank regions various cyanophytes are found and also in surface growths on plants, stones and other substrates.

Figure 23   *Spirulina platensis* (× 2000). Scanning electron micrograph by R. Smaljohann

Several hundreds of cyanophyte species are known to occur in lakes (see Geitler, 1932, and Desikachary, 1959). With the great majority it is a question of planktonic forms. These include both chroococcal as well as hormogonal cyanophytes and altogether they seem to be much more widely distributed in warmer zones than in the temperate and cool regions.

A group of species which can form water blooms in nutrient-rich lakes is of particular interest. These forms all possess gas vacuoles. Altogether close on 20 bloom-forming species are known, of which the following occur most frequently:

Chroococcal	Hormogonal
Microcystis aeruginosa	Anabaena circinalis
Microcystis flos-aquae	Anabaena flos-aquae
Synechococcus plancticus	Anabaenopsis
Synechococcus parvula	Aphanizomenon flos-aquae
	Gloeotrichia echinulata
	Oscillatoria redekii
	Oscillatoria rubescens
	Spirulina

Most of these species have a worldwide distribution. Some few, like *Anabaenopsis* and *Spirulina*, both of which have coiled trichomes, occur much more in the tropics whilst *Aphanizomenon* and *Oscillatoria redekii* occur mainly in the temperature climatic zones (Carr and Whitton, 1973).

The numerous other planktonic cyanophyte species are more or less frequent components of the phytoplankton of lakes. The majority are limited to particular climatic zones or even to individual types of waters. Gas vacuoles are frequently present — but they can also be lacking.

Among the chroococcal cyanophytes, apart from those species which exist as single cells (*Synechococcus*), there are many colony-forming forms. Members of the widely distributed genus *Microcystis* produce relatively small spherical or elongated cells, many hundreds of which are enclosed by a structureless jelly and in this way are combined into mainly spherical or egg-shaped colonies (Figure 22 b). The differentiation of the individual species is very difficult so that the corresponding details in the literature deviate markedly from one another. *Coelosphaerium* forms hollow spherical colonies, *Merismopedia* regularly constructed tablet-shaped ones (Figure 24). In *Gomphosphaeria* the individual cells have small gelatinous stalks. *Aphanothece* has large gelatinous layers in which the cells lie relatively distant from each other. In the case of *A. stagnina* the colonies can attain the size of a pigeon's egg. Although they certainly develop in shallow waters at the surface of the mud, they are easily detached and float to the surface (Fott, 1970).

Numerous hormogonal cyanophytes are also found in the plankton of lakes. The members of the genus *Anabaena* have chain-like trichomes with heterocysts and resting cells. The vegetative cells contain gas vacuoles. *Nostoc* is constructed similarly, but in this case the trichomes occur in gelatinous layers (Figure 22 c). With *Gloeotrichia echinulata* the trichomes are combined into spherical colonies of 0.5 mm diameter. The individual filaments are arranged radially in the sphere and terminate in a colourless tip. *Aphanizomenon* forms macroscopic layers which contain numerous trichomes and at times resemble pine needles. *Oscillatoria* has neither heterocysts not spores. Reproduction takes place by hormogonia. *O. rubescens* is coloured red and when it occurs in large amounts can cause a corresponding coloration of the lakes involved. *Phormidium* also has a very similar structure. *Lyngbya* has filaments with firm sheaths.

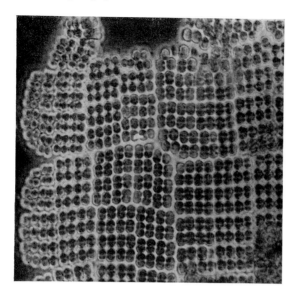

Figure 24   *Merismopedia* from the
Forest Pond (× 480).
(P. Hirsch)

In small ponds as well as in the bank regions of larger lakes, soil-dwelling and firmly attached cyanophytes also occur. In this respect the genus *Rivularia* is especially interesting. The individual trichomes become tapered from the base to the tip and run out into a hair. Quite often thick layers are formed which in part precipitate chalk. Water plants, stones, mussel shells and similar substrates are at times thickly overgrown by cyanophytes. For example, various *Gloeotrichia* species are found in such sites. *Tolypothrix lanata* forms dark, nearly black tufts which attain a length of 2 cm.

Different cyanophyte species also occur in bogs. High peat bogs have a characteristic cyanophyte flora. Here are found *Synechococcus aeruginosus*, *Chroococcus turgidus*, *Anabaena augstumalis*, *Haplosiphon hibernicus* and *Stigonema ocellatum* and others (Fott, 1970). Most of them are cosmopolitans.

Many cyanophytes have a relatively high salt tolerance and accordingly quite a number of species, whose main habitats are in inland waters, are found in salt lakes with moderate or low salt contents. This is true, for example, of the Caspian Sea, the phytoplankton of which contains a series of cyanophytes of worldwide distribution which are also found in inland waters and in brackish water regions like the Baltic Sea. Among these species are also some which form plankton blooms — as, for example, *Aphanizomenon flos-aquae*. Furthermore, the genera *Aphanothece*, *Coelosphaerium*, *Chroococcus*, *Gomphosphaeria*, *Microcystis*, *Merismopedia*, *Anabaena*, *Nodularia*, *Lyngbya* and *Oscillatoria* are represented in the Caspian Sea.

Chlorogloea sarcinoides, which occurs in large amounts in salt lakes of the Soviet Union, can be considered to be a halophilic form (Fott, 1970. In the Solar lake on the Sinai peninsula, with salt contents up to 18 per cent, cyanophytes both in the plankton and also in the benthos play an important role. Cohen *et al.* (1977) found in the water of this salt lake both chroococcal forms (*Aphanocapsa littoralis*, *Aphanothece halophytica*) as well as hormogonal ones (*Oscillatoria limnetica*, *O. salina*). Mats of *Microcoleus chtonoplastes* occur in the crystallization ponds of salts on the northern Adriatic. This organism can not only tolerate high salt contents but H_2S also (Schneider 1979).

4.2. Marine Cyanophytes

The distinction between most marine cyanophytes and related fresh water species is only trivial. Accordingly, there are only few purely marine genera, as, for example, *Trichodesmium*. *Dermocarpa* also includes predominantly species living in the sea. As autoecological investigations have been made in only few instances, practically nothing can be said about the occurrence and significance of obligate marine forms when considering the salt tolerance of many cyanophytes.

In marine regions the cyanophytes do not play so great a role as in inland lakes. Nevertheless they occur almost everywhere in the sea — only the arctic and antarctic parts of the oceans form an exception. According to the conditions in the remaining environments an increase in cyanophytes in the warm zones may be established in the sea also. But there are also many cosmopolitans among the marine forms as there are with those in inland waters.

In the plankton, the genera *Trichodesmium*, *Oscillatoria*, *Pelagothrix*, *Haliarachne* and *Katagnymene* are chiefly found (Fogg, 1973). *Trichodesmium* develops mainly in tropical waters. The members of this filamentous genus can form plankton blooms. The chroococcal cyanophytes are represented, in particular, by *Synechococcus* (Schmaljohann, 1984). In the deep waters of the Atlantic and Indian Oceans as well as in the Mediterranean Sea, *Nostoc* and *Dactyliococcopsis* were found by Bernard and Lecal (1960). Different species of fresh water plankton occur in the Baltic and similar regions of the sea which contain brackish water (see Pankow, 1976). These are particularly numerous, of course, in the areas poorest in salt — especially in strongly eutrophic bays and lagoons. *Anabaena flos-aquae*, *Aphanizomenon flos-aquae* and *Nodularia spumigena*, however, also frequently form plankton blooms in the open Baltic Sea. It is possible that there are also obligate brackish water species — but this is not yet established unequivocally.

In the coastal regions firmly attached cyanophytes live on plants and stones as well as on the sea bottom. This is equally true of the intertidal region, the sand and also of the mud. A very characteristic cyanophyte flora is found in the spray water region of rocky coasts, where it often forms a black band. Here is the habitat particularly of members of the genera *Calothrix*, *Phormidium*, *Gloeocapsa*, *Nodularia* and *Rivularia* and, quite often also, of *Plectonema*, *Oscillatoria* and *Lyngbya*. Apart from cyanophytes, almost only lichens occur as well, of which some likewise contain cyanophytes as , for example, *Lichina*, the phycobiont of which is *Calothrix* (see Fogg, 1973).

In the eulittoral of limestone coasts, endolithic cyanophytes are often found — namely, those that penetrate into the stome. These include members of the genera *Entophysalis*, *Hormatonema*, *Hyella* and probably also *Gloeocapsa* and *Scopulonema*.

5. Fungi

Fungi, like bacteria and cyanophytes, occur in most waters. As early as the nineteenth century, botanists (Zopf, Cornu, Dangeard) described numerous fungal forms found in pools, lakes and rivers. It was soon established that some of them were genuine aquatic fungi unable to develop in other environments. There are also, of course, fungi which are viable in a variety of habitats; thus, some soil fungi grow in rivers and lakes, and some aquatic fungi can grow in damp soils.

All fungi are heterotrophic organisms and, therefore, dependent on the presence of organic material. In water, saprophytic forms are found as well as parasites which attack a great variety of aquatic plants and animals. There are fungi which are only capable of either a saprophytic or a parasitic existence; but there are also facultative parasites which, according to circumstances, take their nutrients from dead material or parasitize living organisms. Others again are able, by means of sophisticated mechanisms, to catch protozoa, rotatoria or nematodes and make use of these animals for nutritional purposes. This mode of life is not, strictly speaking, parasitic, and fungi behaving in this way are referred to as 'predacious'.

Most aquatic fungi require free oxygen. Besides protein, sugars, starch and fats, they can break down pectins, hemicellulose, cellulose, lignin and chitin. Some fungi grow in acid as well as in alkaline waters, at pH values of 3.2–9.6 (ZoBell, 1946a); for instance *Achlya racemosa* and *Saprolegnia monoica*. Other fungi are limited to either acid or alkaline waters (Sparrow, 1968). Suzuki (1960b, 1961) found five different phycomycetes in Japanese volcanic lakes growing at pH values of 1.9–2.9. In the same way, fungi differ a great deal with regard to their temperature requirements. The overwhelming majority are mesophilic. Psychrophils occur but do not seem to play an important role. Some aquatic fungi can grow over the temperature range 1–33 °C (ZoBell, 1946a).

Fungi are much more varied morphologically than bacteria and have much larger cells. They are eucaryotes, i.e. they possess proper nuclei and they often produce complicated fruiting bodies. There are unicellular fungi, but also multicellular ones with extensive mycelia.

The following groups occur in waters:

Slime moulds	Lower fungi	Higher fungi
Labyrinthulomycetes (reticulate slime mould) Plasmodiophoromycetes (parasitic slime moulds)	Phycomycetes (algal fungi)	Ascomycetes (ascomycetous fungi) Basidiomycetes (basdiomycetous fungi) Fungi Imperfecti or Deuteromycetes (imperfect fungi)

The lower fungi (see Table 3) are the aquatic fungi proper. As a rule, they are unicellular; more highly developed forms are able to produce mycelia, though without septa. The majority of phycomycetes are confined to an aquatic mode of life. Re-

Vegetative Reproduction		Generative Reproduction
	Mastigomycotina	
	Chytridiomycetes	
Zoospores with 1 flagellum	Chytridiales	
	Blastocladiales	
	Monoblepharidales	
	Hyphochytridiomycetes	
	Oomycetes	
with 2 flagella	Lagenidiales	
	Saprolegniales	
	Leptomitales	
	Peronosporales	
	Zygomycotina	
	Zygomycetes	
Conidia	Mucorales	
	Entomophthorales	
Myxamoebae	_Trichomycetes_	

production is by means of zoospores which are motile with one or two flagella. The fungi colonize living creatures or dead organic material, present in the water or in the sediments; the zoospores, by contrast, seek the free water. This holds also for the forms living in soil and those which parasitize plants on dry land; these need at least dew drops or a little rain to swarm out. In the Peronosporales (downy mildews) we find formation of both zoospores and conidia; they mark the transition to a terrestrial life which is finally accomplished in the Aplanatae. Nonetheless, the Aplanatae contain several forms which lead an aquatic existence, for example various _Ancylistes_ species (Entomorphthorales), which are parasites of planktonic algae (Desmidiaceae) (Sparrow, 1960). These may, however, have changed over to an aquatic life secondarily — like some higher fungi. _Ancylistes_ does not produce any zoospores, but non-motile conidia instead. These are situated on the hyphae which, as conidiophores, grow out of the water and shoot the conidia into the air (Figure 25).

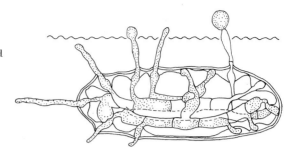

Figure 25   *Ancylistes netrii* in a desmid
(*Netrium* sp.) with coni-
diophores in different
developmental stages. The
conidia rise above the
water surface (wavy line).
(From Sparrow, 1960,
modified)

Of the higher fungi, mainly representatives of the Ascomycetes and the Fungi
Imperfecti are found in water; Basidiomycetes play only a minor role in the aquatic
habitat.

Only the yeasts (Saccharomycetaceae), i.e. the most primitive ascomycetes, and
the yeast-like Fungi Imperfecti grow in free water. All other higher fungi colonize
solid structures in the water, particularly living and dead water-plants, wood, sediments, mussel shells and stones. A small number can parasitize plants or animals.
Fungal spores are found in almost all surface waters; some of these spores are those
of purely terrestrial forms which do not grow in water. Myxomycetes, except for the
Labyrinthulae, which are here of greater importance, hardly ever occur in water
(see Section 10.3).

5.1. Fungi of Inland Waters

In the microflora of subterranean waters fungi do not play a prominent part. In clean
groundwater they are almost entirely absent because of lack of nutrients. They
grow only where there are enough nutrients, i.e. in areas with a high water table
and in the subterranean streams of Karst mountains.

Also, fungi are rarely present in clean spring water. On the other hand, they occur
regularly in streams and rivers. Some colonize streams relatively poor in nutrients,
others prefer the more or less eutrophic flowing waters. Thus, most of the members
of the genus *Sapromyces* which belongs to the order Leptomitales (Table 3) are found
on submerged branches in clean, cool streams, but *Leptomitus lacteus*, which belongs
to the same order, is a distinct sewage fungus and sometimes shows massive development in polluted waters. It forms a dense lawn on river beds and, under appropriate
conditions, may cause considerable mycelial drifting (see Section 15.1).

In rivers, there are usually a number of parasitic phycomycetes which attack
not only planktonic algae and small animals, but also the eggs and larvae of crustaceans and fish. Numerous parasites of fresh water algae belong to the order Chytridiales (Table 3). These are very primitive fungi consisting, as a rule, only of a small
vesicle with several fine rhizoids. This develops into a zoosporangium and, on maturation, it releases a fairly large number of uniflagellate zoospores (Figure 26). These
fungi grow either on or in the host organism. Several *Olpidium* species, for example
O. euglenae and *O. rotiferum*, are widely distributed. Members of the genera *Polyphagus*
and *Chytridium* are also not infrequently found in flowing water, and some of them
parasitize other phycomycetes while others again live as saprophytes. Various Saprolegniales occur frequently in rivers, and some of them already possess a branching
mycelium and barrel- or tube-shaped zoosporangia with numerous biflagellate
zoospores. During sexual reproduction an oogonium with one or more egg cells is
produced; from the antheridium penetration tubes grow into it, and each egg cell
is fertilized by the nucleus of a male gamete (Figure 27). Most of the Saprolegniales

4*

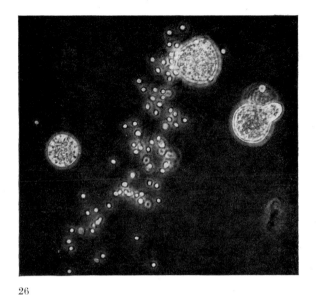

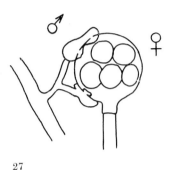

26   27

Figure 26   Sporangia and zoospores of a Phycomycete from the sediment of the Kiel Fjord
             (× 350). (J. Schneider)

Figure 27   Sexual organs of *Saprolegnia monoica*

are saprophytes. Some live parasitically on fish, amphibia and their eggs, on some
lower animals and on the roots of aquatic plants (Sparrow, 1960). Suzuki (1960a)
frequently found members of the genera *Saprolegnia*, *Achlya* and *Aphanomyces* in
rivers; the genus *Pythium* belonging to the Peronosporales also occurs in this habitat.

On foliage in well-aerated streams there is frequently found a species-rich flora of
deuteromycetes with strikingly branched spores (see Figure 32; Ingold, 1976).

Yeasts may also be encountered in many flowing waters and are particularly
common in rivers which carry much sewage (Rheinheimer, 1964). Higher ascomycetes
and Fungi Imperfecti are present in abundance and are found mainly on dead plant
material and on wood.

In the microflora of lakes phycomycetes are very important. Here also one finds
predominantly Chytridiales and Saprolegniales, represented by both saprophytic
and parasitic species. In addition, there are members of almost all other orders of
phycomycetes.

As the flora and fauna of the individual types of lakes differ greatly, their fungal
flora shows corresponding differences. Thus, the Desmidiales are mainly distributed
in dystrophic and oligotrophic lakes and bogs (Fott, 1970), and here also their parasites
like *Olpidium endogenum* and *Micromyces mirabilis* are most frequently found; the
same holds for *Ancylistes closterii* and *A. netrii*, which live in *Closterium* and *Netrium*
species respectively (Sparrow, 1960). Those groups of algae and of animals which
predominate in other types of lakes again have their specific parasites amongst the
phycomycetes. Thus it could be shown that in Lake Windermere, with the rise in
population density of *Asterionella formosa* in April and May, the parasite *Rhizophydium planctonicum* also increased vigorously (Canter and Lund, 1948). *Rhizophydium pollinis-pini* (Figure 28a, b), which lives on submerged pollen grains, is
frequently found in lakes. Widely distributed, particularly in eutrophic lakes, is
Polyphagus euglenae, which has a somewhat more complex structure. From a spindle-
or club-shaped sporangium an elongated zoosporangium develops and the *Euglena*

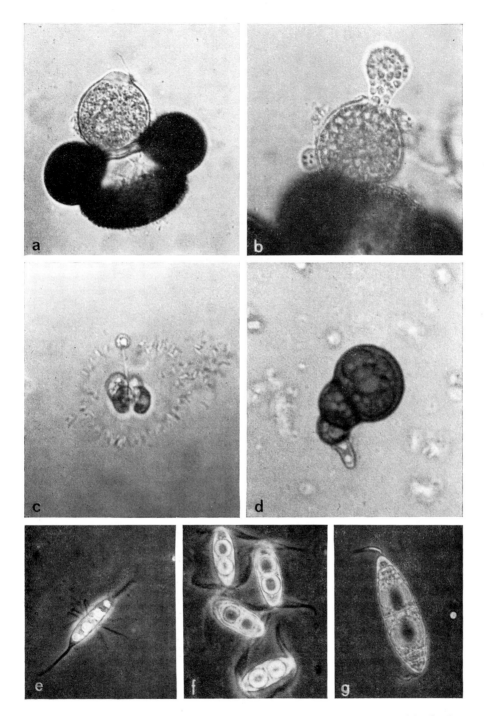

Figure 28   Aquatic fungi from inland waters (a–c) and from the sea (d–g). (a) *Rhizophydium pollinis-pini*, sporangium on a pollen grain with an apical papilla and rhizoids (× 800). (b) As (a), emergence of the spore mass through two pores of the sporangium (× 800). (c) *Rhizophydium fulgens*, young fungal thallus with long rhizoid on the slime coat of a green alga (*Gloeococcus* sp.) (× 800). (d) *Cirrenalia macrocephala*, conidia (× 750). (e) *Corollospora maritima*, ascorpores (× 450). (f) *Halosphaeria quadriremis*, ascospores (× 450). (g) *Haligena spartinae*, ascospores (× 350). (J. Schneider)

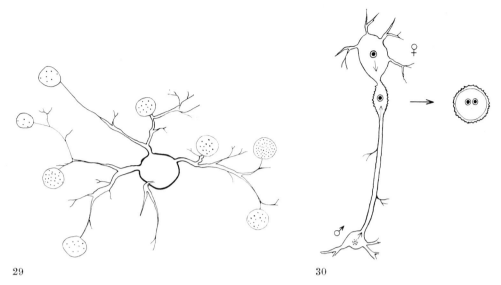

29   30

Figure 29   *Polyphagus euglenae.* Central vesicle and rhizoids with killed *Euglena* cells. (Based on Gäumann, 1964, modified)

Figure 30   Sexual reproduction of *Polyphagus euglenae.* (Based on Sparrow, 1960, modified)

or *Chlamydomonas* cells near it are held together by the rhizoids of the fungus (Figure 29). In addition to the usual egg-shaped zoospores, spherical resting spores with thick spiny walls are also produced, particularly if there is a scarcity of nutrients. In preparation for this process, the fungus produces two gametangia, unequal in size, and a copulation tube grows from the smaller towards the larger gametangium. At the end of the tube a vesicle is formed and the contents of both gametangia flow into it (Figure 30). The zygote, with two nuclei, becomes the resting spore; the nuclei fuse when the spore germinates (Sparrow, 1960).

The fungus *Ectrogella bacillariacearum* belonging to the Saprolegniales, attacks various fresh water diatoms and, consequently, is frequently found in lakes where these grow abundantly. In addition, members of the genera *Saprolegnia, Leptolegnia, Achlya* and *Aphanomyces* occur frequently in lakes (Sparrow, 1968).

Zoophagus insidians (Figure 31), which feeds on rotifers, lives on filamentous green algae in rivers and lakes. From the long hyphae of this fungus numerous short hyphae branch off; these, when touched by the cilia of a rotifer, secrete a sticky substance and thus hold the animal. With great rapidity the hyphae then grow into its mouth and inside form a mycelium which gradually absorbs the contents of the animal's body. The closely related *Zoophagus tentaclum* has on each of its short hyphae four or five tentacles which serve to catch the prey.

Numerous ascomycetes and Fungi Imperfecti also occur in lakes, and yeasts are particularly common in polluted lakes. The other higher fungi live predominantly in sediments and on dead plants or wood. In small woodland lakes, twigs, fruit and much foliage collect on the bottom; these are colonized by phycomycetes, ascomycetes and particularly by numerous Fungi Imperfecti (Figure 32). These can form here a dense fungal lawn which, however, is hardly visible to the naked eye. Some important genera are *Anguillospora, Tricladium, Tetracladium, Dendrospora, Tetrachaetum, Clavariopsis,* and *Lemonniera.*

Some *Arthrobotrys* species which are common in soil are present also in the sediments of inland waters; these, too, belong to the Fungi Imperfecti. They are animal-catching

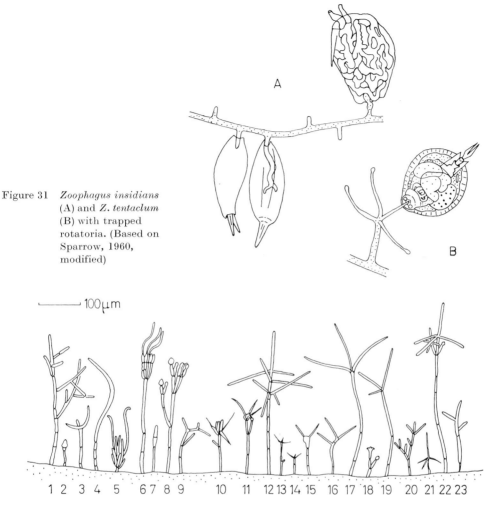

A

B

Figure 31   *Zoophagus insidians*
(A) and *Z. tentaclum*
(B) with trapped
rotatoria. (Based on
Sparrow, 1960,
modified)

100 μm

1 2 3 4 5   6 7 8 9   10   11   12 13 14 15   16 17 18 19   20 21 22 23

Figure 32   Aquatic Fungi Imperfecti on decomposing foliage: 1. *Varicosporium elodeae*; 2. *Piricularia aquatica*; 3. *Lunulospora curvula*; 4. *Anguillospora longissima*; 5. *Flagellospora
curvula*; 6. *F. penicilloides*; 7. *Piricularia submersa*; 8. *Margaritispora aquatica*;
9. *Tricladium angulatum*; 10. *Tetracladium setigerum*; 11. *T. marchalianum*; 12. *Dendrospora erecta*; 13. *Alatospora acumiata*; 14. *Heliscus longibrachiatus*; 15. *Clavariopsis
aquatica*; 16. *Tricladium gracile*; 17. *Tetrachaetum elegans*; 18. *Heliscus aquaticus*;
19. *Articulospora inflata*; 20. *A. tetracladia*; 21. *Triscelophorus monosporus*; 22. *Lemonniera aquatica*; 23. *Tricladium spendens*. (Based on Ingold, 1953; modified)

fungi, specialized mostly to catch nematode eelworms. Particularly interesting here
is *Arthrobotrys oligospora* (Figure 33), a predacious fungus which lives in rotting plant
material in the soil and in waters. The fungus produces a fine net of hyphae shaped
like loops and hooks which serve to trap the nematodes, which are then consumed.

Frequently, various soil fungi which are viable only for a limited time in water are
also found in inland waters, for instance *Penicillium* and *Aspergillus* species. Their
spores are present in almost all waters and can germinate on plant debris. They
often get into rivers and lakes with dead plant material and remain alive there for
some time. Normal reproduction, however, is not possible in water.

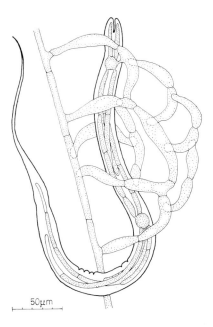

50μm

Figure 33   *Arthrobotrys oligospora* with a trapped
nematode

A number of fungi known to occur in certain areas of the sea are found in salt lakes with low salt concentrations. Anastasiou (1963) found eight ascomycetes and ten Fungi Imperfecti in Salton Sea in California, which offers conditions similar to those in the Caspian Sea. Sparrow (1968) isolated from the same lake three phycomycetes which had been found earlier to occur in the Adriatic. According to Scholz (1958), *Rhizophydium halophilum* grows frequently in salty habitats inland or on the coast. There are no reports of the occurrence of fungi in salt lakes with extremely high salt concentrations.

5.2. Marine Fungi

Until a few decades ago, hardly any notice had been taken of the fungi living in the sea. Since then, however, it has been established that they are widely distributed in the marine habitat. Representatives of all four classes were found and it has become evident that some of them grow in sea water only and require NaCl. Besides these halophilic marine fungi, there are many salt-tolerant forms of limnic or terrestrial origin.

Myxomycetes are represented in the sea only by the peculiar Labyrinthulae, which produce a network of cytoplasmic filaments, the 'net plasmodium'. The spindle-shaped uninucleate cells, singly or in cord-like aggregates, glide along the filaments, which arise on the end of the cells. The *Labyrinthula* species are parasites which grow in algae and higher aquatic plants and in the roots of terrestrial plants. The marine *Labyrinthula coenocystis* is found mainly on *Cladophora*, *Fucus* and *Zostera*, but *in vitro* it can live entirely on bacteria and yeasts (Klie and Schwartz, 1963).

Like the phycomycetes of inland waters, those of the marine habitat include not only saprophytes but also numerous parasites which attack a great variety of marine plants and animals. The most important here are again the orders of Chytridiales and Saprolegniales. Various organisms of the genera *Olpidium*, *Rozella*, *Chytridium*, *Rhizophydium*, *Sirolpidium* and *Ectrogella* play a role as parasites of marine creatures (ZoBell, 1946a), but some of these genera contain saprophytic species also. Widely

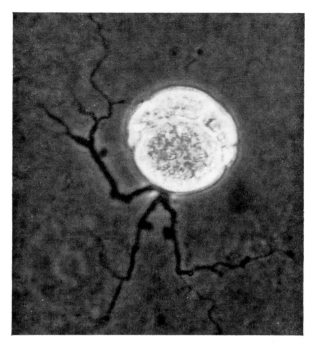

Figure 34   *Thraustochytrium kinnei.* Sporangium with rhizoids on yeast extract peptone agar. Spore formation begins in the apical part of the sporangium ($\times$ 1 200). (J. Schneider)

distributed in the Baltic and the North Sea (and in other marine regions) are members of the genus *Thraustochytrium* (Saprolegniales), which live on marine algae (Figure 34) (Gaertner, 1964; Schneider, 1969; Ulken, 1964, 1965). These are very similar to the chytrids but their zoospores are biflagellate. *Leptolegnia marina* produces a mycelium and, not infrequently, parasitizes mussels and crabs (Johnson and Sparrow, 1961). Marine forms occur also in most other orders of the phycomycetes. Of some importance are several *Pythium* species (Peronosporales) which live in crayfish eggs or in algae.

Yeasts (Figure 35a, b) and yeast-like Fungi Imperfecti (Figure 35c, d) also occur in the sea. Kohlmeyer and Kohlmeyer (1979) named 177 species which were found in the water, sediments, algae, animals or detritus of the sea; of those only 26 species are regarded as obligate marine forms. Of the remaining yeasts, it was simply a matter of salt-tolerant species which had reached the sea and some of which had originated in sewage (or from ships also). (Figures 36 and 37). They are found, therefore, especially in the vicinity of land and apparently remain viable in the sea for only a relatively short time. The most important genera of true marine yeasts are *Metchnikowa*, *Kluyveromyces*, *Rhodosporidium*, *Candida*, *Cryptococcus*, *Rhodotorula*, and *Torulopsis*.

Higher ascomycetes (Figures 28d–g) and Fungi Imperfecti have been isolated from the sea in considerable numbers (Kohlmeyer and Kohlmeyer, 1964–1968). They are mostly saprophytic and live mainly on organic material in the water and in sediments rich in organic substances. Hence they occur mainly near the coasts. The fungal flora on driftwood, which may travel over considerable distances, is very varied. Most of these forms break down cellulose and lignin. But in animal substrates such as exoskeletons, shells, dwelling tubes, higher fungi have also been found (Kohlmeyer, 1969; Kohlmeyer and Kohlmeyer, 1979). Johnson 1968) states that the following fungi which colonize wood have a worldwide distribution:

Ascomycetes:	*Halosphaeria mediosetigera* *Ceriosporopsis halima* *Lulworthia* div. sp.
Fungi Imperfecti:	*Monodictys pelagica* *Humicola alopallonella* *Cirrenalia macrocephala*

According to Meyers (1968), such higher fungi should regarded as marine organisms only if their growth and reproduction occur preferentially or exclusively in the sea, or optimally at the normal salt concentration of the sea (i.e. 2.5–4 per cent). Numerous terrestrial forms are, of course, found in the sea; they may even grow there to a certain extent, but cannot reproduce normally. Hughes (1974), in a biogeographical investigation, showed that, although the salt content is of local importance, the distribution of higher marine fungi over large areas appears to be dependent chiefly on the temperature.

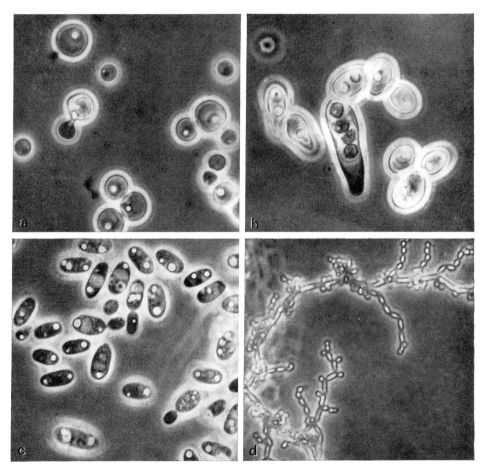

Figure 35   Yeasts (a, b) and yeast-like Fungi Imperfecti (c. d) isolated from the Baltic Sea. (a) *Debaryomyces hansenii* (×2500); (b) *Hansenula anomala*, cells and sporangia (×2500); (c) *Rhodotorula mucilaginosa*, older cells with enlarged central vacuoles and reserve substances (×2000); (d) *Candida parapsilosis*, budding mycelium (×1000). (H. G. Hoppe)

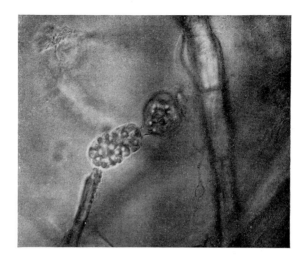

Figure 36 *Nowakowskiella* sp. from the Kiel Fjord. Sporangium with energing zoospores (× 450). (J. Schneider)

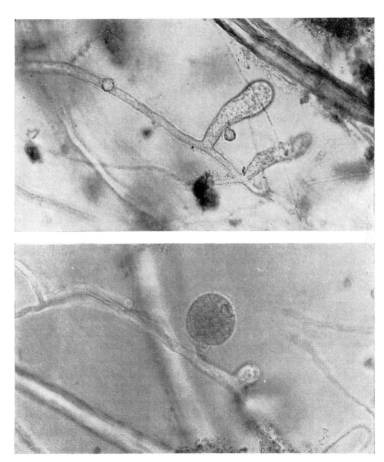

Figure 37 *Pythium torulosum* from the Kiel Fjord. Top: hyphae with club-shaped zoosporangia (early stage). Bottom: hyphae with vesicle which contains the zoospores (× 450). (J. Schneider)

The currently known obligate marine higher fungi include

Ascomycetes	149 species
Deuteromycetes	56 species
(Fungi Imperfecti)	
Yeasts	26 species
Basidiomycetes	4 species

The great majority of the ascomycetes belong to the Pyrenomycetes and Loculo-ascomycetes; namely, 91 and 51 species (in 36 and 23 genera) respectively.

Of the four Basidiomycetes, *Nia vibrissa*, *Digitatispora marina* and *Halocyphina villosa* are wood-inhabiting species, whilst *Melanotaenium ruppiae*, belonging to the Ustilaginales, is parasitic in the leaf bases of *Ruppia maritima* (Kohlmeyer and Kohlmeyer, 1979).

According to Kohlmeyer (1977), in the deep sea there is a characteristic fungal flora, which is distinguishable from that of shallow water regions. Thus he was able to isolate from wood exposed for two and three years at depths of between 1615 and 5315 m various higher fungi and to describe the ascomycetes *Bathyascus vermisporus* and *Oceanitis scuticella* as well as the Fungi Imperfecti *Allescheriella bathygena* and *Periconia abyssa*. Parasitic fungi of characteristic deep sea animals were also observed, but not investigated in any further detail.

In areas of brackish water, halophilic fungi are found which grow optimally at salt concentrations of between 0.5 and 2.5 per cent. A typical fungus of brackish water is *Olpidium maritimum*. Its salt optimum for propagation is 1.3–1.7 per cent; it does not grow in fresh water. Some *Thraustochytrium* and *Pythium* species have similar salt optima (Höhnk, 1969a).

Numerous higher fungi have also been observed in subtropical and tropical brackish water areas on *Spartina townsendii* and *Rhizophora mangle* (Kohlmeyer and Kohlmeyer, 1979).

CHAPTER 6

6. *Viruses*

As we know today, viruses occur in many waters. With viruses we are no longer dealing with independent organisms because they make use of living plant or animal cells for their multiplication and can neither divide themselves nor are they able to grow. Their size is so minute that they can pass through bacterial filters and they possess only one kind of nucleic acid — either DNA or RNA. Accordingly one may distinguish DNA and RNA viruses.

Viruses contain much nucleic acid — that is, genetic material. This is surrounded by a protein coat which is called a capsid.

The structure of some bacteriophages (Figure 38) is more complicated as they are differentiated into head and tail and so in part still possess tentacles (see Section 9.3). Bacteriophages may be regarded as bacteria-specific viruses, which are widely distributed in waters. Bacteriophages have been detected for autochthonous and also for allochthonous bacteria. Thus, Ahrens (1971) found in the western Baltic Sea various agrophages, which attack the halophilic *Agrobacterium* species, as well as coliphages. Very probably there are corresponding bacteriophages for all water bacteria.

Torella and Morita (1979) found more than 10000 bacteriophages per ml occurring free in the water in Yaquina Bay on the coast of Oregon, USA. It is supposed that similar amounts of phages occur also in other more or less eutrophicated coastal waters.

Cyanophages, i.e. the viruses of cyanophytes (see Padan and Shilo, 1973), also play an important role in waters. These are very similar to bacteriophages. Likewise, they are DNA viruses which consist usually of head and tail (see Figure 38).

In recent years viruses have been discovered also for unicellular and multicellular algae (see Brown, 1972; Dodds, 1979) as well as for numerous water animals. These include both DNA and also RNA viruses. More is known, however, about those viruses

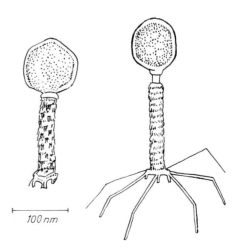

100 nm

Figure 38   Cyanophage N-1 (left) and coliphage
T-2 with tail fibres (right)

which are responsible for diseases of crabs and fish and thereby cause economic losses (see Section 9.3).

Viruses are found both in the different inland waters as well as in nearly all regions of the sea. In addition to viruses of water organisms, human pathogenic viruses have been detected which cause different virus diseases of man and likewise animal pathogenic viruses which cause diseases of wild and domestic animals (see Section 15.2).

Even though different viruses occur in inland waters and in the sea corresponding to their host organisms, no distinction between marine and fresh water forms can be recognized as is the case with real organisms which may have differing salt requirements.

7. Distribution of Micro-organisms and their Biomass

Although bacteria and fungi occur in almost all waters, their distribution with regard to numbers and kinds differs a great deal and there are probably no two rivers, lakes or regions of the sea where it corresponds exactly. Micro-organisms are components of particular biological communities whose composition and size are, in turn, dependent on a variety of physical and chemical conditions. Heterotrophic bacteria and fungi can develop only to the extent to which organic substances, synthesized and re-synthesized by other living creatures, are available as nutrients; they compete for them with one another and with other organisms (non-pigmented algae, protozoa and metazoa). In their turn, they play an important role as food, particularly for protozoa and filter-feeders. While the nutritional factor is of paramount importance, other factors associated with the biological environment exercise considerable in-fluence, for example the production of favourable or inhibitory substances and the changes of pH and oxygen tension. Also, the direct effects of physical and chemical factors — light, temperature, pressure, chemical reactions in the water and in the sediments — are important. The distribution of micro-organisms in a body of water thus always results from the interaction of all biotic and abiotic factors and, like these, is constantly subjected to changes. Frequently, however, a dominating role must be ascribed to one factor or a small group of factors. These may be different according to the type of water, so that the distribution of bacteria and fungi in rivers, lakes and seas shows characteristic differences.

The active spread of bacteria and zoospores by means of flagellar movement is restricted to a relatively small space on account of their small size. For their size, however, flagellated bacterial cells can attain high speeds. Vibrios can achieve tran-siently up to 12 mm per min. This corresponds to approximately 3000 times the body length of these bacteria (Schlegel, 1981). Nevertheless, the self-movement of micro-organisms does not play any notable role in their distribution in waters over large areas. This also holds for the sinking velocity of bacteria occurring free in water, which amounts to only 0.5–3.2 mm daily, according to investigations by Jassby (1975) in a subalpine lake. On the other hand, rapid distribution of micro-organisms can take place due to currents which cause the sedimentation of detritus and phytoplank-ton as well as to animals over distances of a few meters up to many kilometers.

Of course, our knowledge of the distribution of bacteria, cyanophytes and especially of fungi in different waters is still very incomplete and we know much less about it than of most other groups of plants and animals. The reason for this is above all a question of methods. The determination of these micro-organisms is both expensive and difficult (see Schlieper, 1968) and is only possible to a limited extent in the case of water investigations. Consequently, with bacteria, often only total numbers are determined and also, sporadically, the amount of forms belonging to different physiological groups (for example, proteolytic bacteria, cellulose decomposers, nitrifiers, sulphate reducers, etc.).

The total number of bacteria is determined directly by counting bacterial cells under the microscope. The most reliable results are obtained with fluorescence micro-scopic methods employing polycarbonate filters (see Zimmermann, 1977). In this

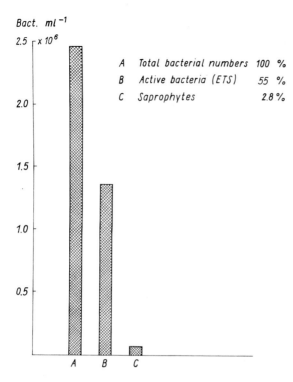

Bact. ml^{-1}

A Total bacterial numbers 100 %
B Active bacteria (ETS)   55 %
C Saprophytes   2.8 %

Figure 39   Total bacterial numbers, number of bacteria with active electron transport systems and saprophyte numbers in a water sample from the Kiel Fjord of 5 May, 1975

way one comes very close to the actual number of bacteria in the water sample. There are certain difficulties, however, with water containing much detritus. Total bacterial numbers include equally actively living and inactive cells as well as dead ones. Therefore the number of active bacteria on occasion is again determined separately (see Figure 39). That can be done by means of a colour reaction (reduction of a tetrazolium salt to a red formazan in cells which have an active electron transport system) or with the aid of microautoradiography (see Iturriaga and Rheinheimer, 1975; Hoppe, 1976). It is simpler to determine the so-called saprophyte number, which is obtained indirectly by the use of culture methods (Koch's plate or MPN methods) (see Schlieper, 1968). In the older literature it was known as the total viable count or in more recent times also appropriately as the colony count. Since the saprophyte numbers represent well reproducible values they suffice to provide a valuable insight into bacterial distribution in waters and are used for their hygienic appraisal. Kusnezow (1959) states that high saprophyte numbers indicate intensive decomposition of organic material. This is confirmed by investigations of Gocke (1974) in the Kiel Fjord, which revealed a significant correlation between the saprophyte numbers and the uptake of ^{14}C-labelled nutrients such as glucose, acetate, aspartic acid, etc. by the bacterial population of the corresponding bodies of water (see Figure 40). Several authors (Rasumov, 1932; Jannasch, 1955; Kusnezow, 1959) have, in various waters, determined the proportions of saprophytes to the total number of microbes and are agreed that it is an indicator for the degree of trophism of the water. Thus, the 'saprophyte' proportion in waters extremely rich in nutrients (carrying sewage) is relatively great with figures of 1 : 5 to 1 : 100, whereas in clean rivers and lakes and also in the open sea it is much smaller and may be 1 : 100 to 1 : 10000. Figure 39 gives an example of the ratios of total bacterial numbers, number of all active bacteria and saprophyte numbers in a coastal water (Kiel Fjord).

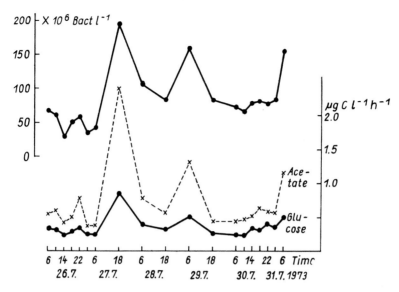

Figure 40   Saprophyte numbers (top) and maximal uptake rate of glucose and acetate (bottom) in water of the inner Kiel Fjord on repeated daily investigations during the period from 26 to 31 July, 1973. (Gocke, 1974)

Even more problematic is the determination of the number of fungi. Gaertner (1968c) counted, by means of the MPN method, the number of so-called infectious units for phycomycetes, as the separation of zoospores and fungi proper was not possible. Yeasts may be counted directly and also indirectly by means of the plate or the MPN method. The numbers of spores of higher fungi are also easily determined. It is more difficult to find the number of mycelia on their respective substrates. Hardly any work has been done so far on the distribution of fungi in waters, so that we know even less about it than about that of bacteria. Up to a point this is due to the fact that, with the exception of yeasts, almost all fungi grow in or on solid substrates (see Chapter 5) and only their spores and broken-off pieces of mycelium are found in free water.

Of great interest is the determination of the bacterial biomass. The cell volume of individual aquatic bacteria, according to recent investigations, varies between less than 0.005 to more than $5.0 \mu m^3$. The bacteria in a water sample, therefore, can differ in their volume in the ratio of about $1:1000$ (Rheinheimer, 1977a; Zimmermann, 1977).

The large differences are apparent in Figure 41. This presents scanning electron micrographs on Nucleopore filters of bacterial populations from four different waters. Tabulated below are the actual amounts of filtered water (AFW), the total bacterial numbers (TBN), bacterial biomass (BBM) and the mean cell volume (MCV) of the corresponding samples:

	AFW ml	TBN $\times 10^9 \, l^{-1}$	MCV μm^3	BBM $\mu g \, C \, l^{-1}$
a. Atlantic	10	1.23	0.131	15.99
b. Forest stream	50	0.70	0.120	8.32
c. River (Elbe)	5	9.54	0.149	141.86
d. Sewage	0.5	253.26	0.357	9038.85

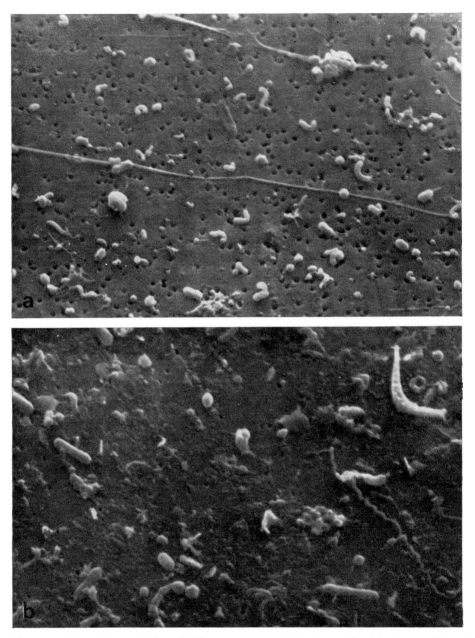

Figure 41   Bacterial population of different waters: a. Atlantic, north of the Azores, 20 m deep,
4. 6. 83; b. Forest stream in the Solling, 27. 7. 82; c. Elbe, at Lauenburg 16. 5. 83;
d. Sewage of the town of Kiel, 20. 12. 82 (× 8000). Scanning electron micrography by
R. Schmaljohann

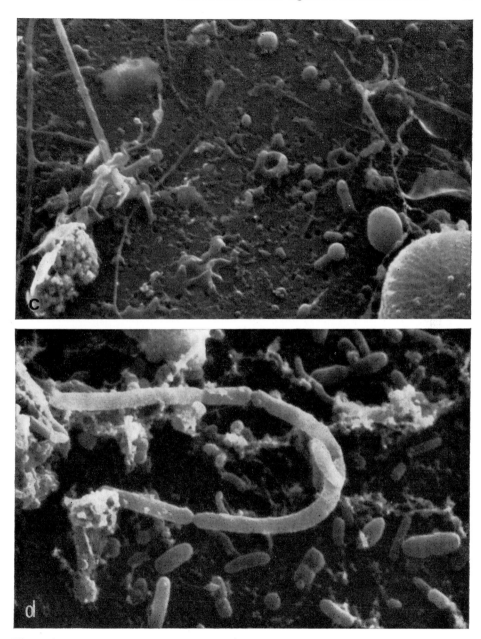

Figure 41

The bacterial population of the sea water sample consists throughout of relatively small cells. These are likewise predominant in the water of the forest stream, in which, however, larger cells may also be seen as well as the very small ones. However, a somewhat lower mean cell volume was found here. In the Elbe river water a considerable proportion of larger cells occurs. In the sewage sample, on the other hand, very large forms predominate. Thus, there is a connection with the prevailing nutrient concentration of the water and this becomes even more apparent in the case of the bacterial biomass, which is lowest in the forest stream water.

ZoBell (1963) assumes an average volume of $0.2\,\mu m^3$. The wet weight would then be 2×10^{-12} mg, the dry weight 4×10^{-11} mg and the carbon content 2×10^{-11} mg. Kusnezow and Romanenko (1966) based their calculations of the bacterial biomass in the Rybinsk reservoir on an average cell volume of $0.5\,\mu m^3$. They assumed the dry weight to be 7.5×10^{-11} mg and the carbon content 3.75×10^{-11} mg. Detailed investigations of water samples from the Baltic Sea, north Atlantic and the Elbe estuary, by means of fluorescence and scanning electron microscopy, however, gave smaller values (Zimmermann, 1977). To obtain the most reliable results, the average cell weight for each body for water investigated must be calculated separately. For this purpose, however, the measurement of the length and width of all bacteria would be necessary, in addition to counting the cells. This time-consuming procedure can be shortened substantially by measuring only the length and then separating the bacteria into groups according to their average size and volume; the average width of rods would be based on estimates. The following classification by Rheinheimer (1977 b) provides an example for biomass determination in north German rivers:

	Rods					Cocci			Others	
Length	0.3	0.6	1.2	2.0	2.8	0.2	0.6	1.2	2.8	μm
Width	0.2	0.3	0.5	0.6	0.7	0.2	0.6	1.2	0.7	μm
Volume	0.009	0.042	0.236	0.565	1.078	0.004	0.113	0.905	1.078	μm^3

Occasionally, however, it is recommended to measure the 'remaining bacteria' (others) separately because these represent a very heterogenous group. The values obtained in this way may approximate very well to the actual bacterial biomass. Recently it has been found that biomass determinations can also be achieved by means of automatic image analysis. Calculation of the mean cell weight (dry weight) of the bacterial population in the lower Elbe at Brunsbüttel yielded 2.9×10^{-11} mg and in the Trave at Nütschau 2.4×10^{-11} mg on the basis of 14 and 15 investigations respectively at about monthly intervals. In oligotrophic inland waters and in the open sea, the average cell weight might be still smaller.

According to Hamilton and Holm-Hansen (1967) the bacterial biomass can also be estimated by determining the ATP (adenosine triphosphate) content. This varies between 0.5 and $6.5 \times 10^{-9}\,\mu g$ ATP per cell, which corresponds to 0.3–1.1 per cent of the cell carbon, and the average for all bacteria tested was 0.4 per cent.

Except for yeasts there are still no biomass estimations for aquatic fungi.

7.1. Distribution in Springs and Rivers

In spring water there are, as a rule, few bacteria, because nutrients are in short supply. The total bacterial counts range from several thousand to some hundred thousands per millilitre and the saprophyte numbers are generally between ten and a few hundred. Because of the low nutrient concentration, small cocci and short rods are often predominant and these can only be detected, for the most part, by the light microscope after fluorescence staining.

In many springs — especially at the margin of the spring basin — cyanophytes are also found. Their occurrence in thermal springs is particularly interesting. Here they can form thick coverings on the bottom and on the margins. The species composition depends on the temperature and on the mineral composition. According to Brock (1967a, b) *Synechococcus lividus* is found in the hot springs of the Yellowstone National Park even at 73–74 °C. The greatest biomass is produced at about 60 °C. In Hunter's Hot Springs in Oregon, USA, various *Synechococcus* species form thick green crusts at temperatures of 53–73 °C. Below that another orange- or flesh-coloured layer of phototrophic bacteria is found (see Castenholz, 1973). At temperatures below 53 °C *Oscillatoria terebriformis* then develops. At 47–48 °C the looser *Oscillatoria* lawns are consumed by the small mussel crab (Ostracoda) *Potamocypris* and the *Oscillatoria* is replaced by *Pleurocapsa* and *Calothrix*, which form firm layers at the base of the spring. In the hot spring sources of the Yellowstone National Park *Mastigocladus laminosus* appears at 57 °C in place of *Synechococcus lividus* and at 46 °C is usually replaced by *Calothrix*. In Iceland and New Zealand *Mastigocladus laminosus* is found even at 63–64 °C. This temperature represents the upper limit for the occurrence of green plants.

In thermal springs with temperatures of more than 50 °C, as a rule only bacteria occur in addition to the cyanophytes. Thus, this is an environment which is colonized by procaryotes alone.

In rivers determinations of total bacterial numbers by direct counting have been done much too infrequently to give a comprehensive picture in view of their differing hydrography. Jannasch (1955) found 352000 to 9.8 millions of bacteria per millilitre of water in the Fulda (Germany). On repeated investigations in the Rio Negro (Brazil) Schmidt (1970) obtained nearly constant values of between 200000 and 300000 bacteria per ml. Daubner gives 455000 as an average value for the Danube at Bratislava and 1194400 for the March at Devin (Slovakia).

The following values for total bacterial numbers (TBN), mean cell volume (MCV) and bacterial biomass (BBM) per litre of water were determined for different flowing waters of the Weser mountains (Germany):

	$TBN \times 10^{-9}$	MCV (μm^3)	BBM (μg C)
1. Spring at Amelith	0.08	0.035	0.28
2. Spring stream at Amelith	0.16	0.096	1.54
3. Forest stream 1, above Schönhagen	0.70	0.120	8.42
4. Forest stream 2, above Schönhagen	1.39	0.099	13.73
5. Schwülme above Lippoldsberg	3.63	0.148	53.81
6. Weser, at Karlshafen	7.66	0.183	140.05

From these values, the large differences in amounts of bacteria in the clean streams of the forest hills (2–4) and in the polluted rivers of populated areas (5 and 6) become evident.

According to Geesey *et al.* (1978), bacteria (epilithic) growing on stones can play a bigger role in mountain streams than those occurring freely in the water. They are frequently found in the slime of epilithic algae, on the photosynthetic products of which they evidently feed extensively.

In a small north Swedish river and various north German rivers, the yearly course of total bacterial numbers and bacterial biomass has been established by employing fluorescence microscopic methods (Müller-Haeckel and Rheinheimer, 1983; Rheinheimer, 1977a, b). The following mean values have been determined from the latter:

	Elbe km				Stör at Wewels fleth	Trave at Nütschau
	569	580	668	699		
Total bacterial numbers	4.650	4.850	3.900	6.870	5.030	5.970×10^{-9} l^{-1}
Bacterial biomass	0.668	0.646	0.549	0.896	0.646	0.709 mg l^{-1}
Number of samples	14	11	10	12	10	15

Figure 42 shows the yearly course of total bacterial numbers and bacterial biomass in the scarcely polluted north Swedish Ängeran. The values show very definite summer maxima. On the other hand, in the relatively heavily polluted Elbe estuary (Figure 43), the curves reveal a distinct winter maximum and relatively low values in summer. The yearly rhythm can be disturbed by extreme hydrographic conditions especially by high water or very low water supply. Taking into account the amount of sewage, the seasonal fluctuations in bacterial content may usually still be detected even in such cases. Particularly with smaller rivers, disturbances are fairly frequently caused by irregular loading, for example on the occasion of heavy rains which wash in soil and forest litter or by temporary introductions of sewage.

The load of bacterial biomass was estimated for the Elbe at Lauenburg for the year 1975. In this section of the river on the days of the investigation it varied between 173 and 973 g s^{-1} or 14947 and 84067 kg per day. The total amount of bacterial mass carried came to about 11000 tons in the year 1975. This corresponds to about 1100 tons of carbon. The bacterial biomass, therefore, acquires a considerable significance for the biology of rivers. In the temperature climatic zone this is especially true in the

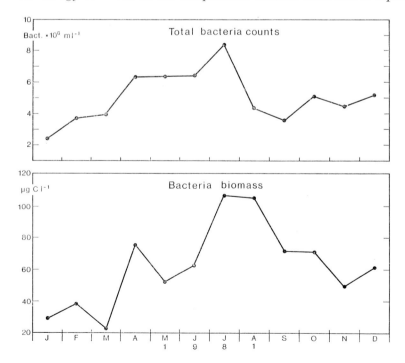

Figure 42   Yearly course of total bacterial numbers (top) and bacterial biomass (bottom) in the Ängeran (north Sweden)

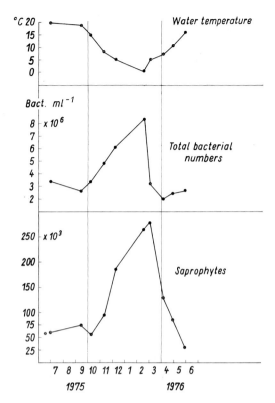

Figure 43   Water temperature, total
bacterial count and saprophyte
count in the Elbe at Stein-
deich (km 668)

winter months, if the amount of phytoplankton declines sharply — bacteria and also
the protozoa (ciliates and flagellates), which use these as food, increase (Schulz, 1971).

Much more often saprophyte numbers have been ascertained by means of the Koch
plate method. Particularly abundant is information for a stretch of the lower Elbe
about 200 km length from Schnackenburg (km 474) to the mouth of the Stör (km 678)
(Rheinheimer, 1965a). Examinations were carried out here once or twice a month,
at 8–16 locations, during the years 1956–1964, and saprophyte numbers varied
between a few hundred and several hundred thousand per ml of water. The longitudinal
profile of rivers always shows great fluctuations in bacterial numbers, due mainly to
the effect of tributaries, sewage, harbour installations, etc. Thus the Elbe below
Schnackenburg showed almost always a decrease in bacterial counts because of the
natural self-purification of the river (see Section 15.3), followed by a marked rise in
the vicinity of the city of Hamburg and a considerable drop again further downstream (Figure 44). In areas of dense housing with more pollution, as in the middle
course of the Elbe and along the middle and lower Rhine, the fluctuations are markedly greater; in non-polluted areas of the river they are presumably less. In contrast
to the longitudinal profile, the transverse and the vertical profiles show relatively
small differences in bacterial numbers (Rheinheimer, 1965b), as the currents, the wind
and the shipping cause a continuous mixing of the water which leads to a comparatively
even distribution of bacteria and their nutrients. However, along the banks and immediately above the bottom, deviations of some magnitude may be found.

Investigations carried out over eight years in the Elbe permitted the recognition
of a marked yearly rhythm of saprophyte numbers, with a clear-cut winter maximum
and a summer minimum. The summer and winter averages depicted in Figure 44
show this very clearly. The higher bacterial numbers obtained during the cold season

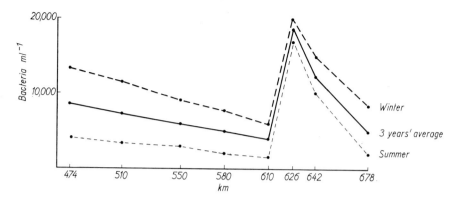

Figure 44   Three-year average of saprophyte numbers in the lower Elbe for the period from
October, 1956 to September, 1959, together with the corresponding summer and
winter curves

are due to the favourable conditions at lower water temperatures for the nutrition
and life of the saprophytes which come mainly from sewage (see Section 15.3). Similar
figures were obtained for the putrefying bacteria, coliforms and yeasts, most of which
get into the river by way of sewage, whereas the nitrifying bacteria have their maximum in summer (see Section 11.5). Particularly high saprophyte numbers were
counted during the cold winter 1962/63 when the Elbe, upstream from Hamburg,
carried a solid ice cover for several weeks (Rheinheimer, 1964). In February 1963,
at almost all locations, values of several hundred thousands per ml were found.

During summer floods the saprophyte numbers may also rise markedly, though
only temporarily, as the flooded lands cause much additional contamination and are
responsible for supplying nutrients; but even under these conditions, the values found
in the lower Elbe always remained considerably below those obtained during the
winter 1962/63 under the ice.

On the other hand, in rivers whose loading with sewage is negligible, a seasonal
rhythm with a winter maximum and a summer minimum can hardly be expected.
Here, total bacterial counts and saprophytes will depend much more on the nutrients
produced in the river itself, particularly by the phytoplankton, and hence their maximum — as in clean lakes (see Section 7.2) — will not occur in winter but in spring
and autumn or late summer, i.e. at the time of greatest production. In tropical rivers,
where conditions of light and temperature are fairly even throughout the year, no
marked fluctuations in production occur. Consequently, no seasonal variations in
bacterial numbers will be found.

Only part of the bacterial population is found in free water; the rest live growing
on solid substrates, usually on floating particles. Usually older rather than young
populations of phytoplankton are colonized by bacteria, while organisms of the
zooplankton seem almost always to carry numerous bacteria both internally and on
the surface (see Section 10.1). This holds also for higher animals that can transport
bacteria upstream against the current. In this way, infection with pathogens is also
possible upstream from the mouth of a sewage inlet.

Little is known, so far, about the numbers of fungi in river water. In the lower
Elbe during 1956/57 monthly counts of yeasts were made, and for some locations
the yearly average, and also summer and winter averages, were ascertained (Rheinheimer, 1965a):

Elbe (km)	474	510	550	586	610	626	666	
Summer average	42	35	20	16	14			Yeast cells/ml
Winter average	70	65	58	60	46			
Yearly average	56	50	39	38	30	111	42	

The number of yeast cells in river areas which are loaded with sewage is usually relatively great. Thus, within the section of the Elbe examined the highest values were regularly found in the area of Greater Hamburg (km 626). In clean streams, on the other hand, yeasts were hardly ever found.

Spores of higher fungi are abundantly present in river water and are particularly numerous in areas loaded with sewage. To a greater or lesser extent, however, they are those of terrestrial fungi.

The numbers of hyphomycete conidia in an unpolluted north Swedish river in the course of a year ranged from about 100 to 1100 per litre of water. In this case, the highest values were found in March and October and the lowest in July and December, whilst total bacterial numbers showed a distinct summer maximum (Müller-Haeckel and Rheinheimer, 1983).

The lower fungi are of greatest importance; the composition of their populations and their numbers differ depending on the amount of organic matter present. In sewage-loaded parts of a river some so-called sewage fungi may multiply copiously, developing into large mycelial drifts (see Section 15.1). However, detailed investigations of fungi in rivers are still non-existent.

7.2. Distribution in Lakes

More results are available on the distribution of bacteria in lakes than in flowing waters. Determinations of total bacterial numbers were carried out regularly for several years in some lakes of the USSR, and various central European countries (Rasumov, 1932; Kusnezow, 1959, 1970; Overbeck, 1968; Olah, 1968a, b, Niewolak, 1974a). In natural lakes they ranged from 50000 to several millions. Particularly high values were found in eutrophic reservoirs as is shown by the following results from Kusnezow (1970):

	Total bacterial numbers	
Oligotrophic lakes	50–340	$\times 10^{-3}$ ml^{-1}
Mesotrophic lakes	450–1400	$\times 10^{-3}$ ml^{-1}
Eutrophic lakes	2200–12300	$\times 10^{-3}$ ml^{-1}
Eutrophic reservoirs	1000–57900	$\times 10^{-3}$ ml^{-1}
Dystrophic lakes	430–2300	$\times 10^{-3}$ ml^{-1}

The saprophyte numbers, on the other hand, which were determined at the same time, were always much lower: in eutrophic lakes between several hundred and several hundred thousand, in oligotrophic ones not infrequently below one hundred. In clean lakes the highest bacterial counts are found at the time of the greatest production of nutrients by the phytoplankton, in spring and early autumn or late summer. Thus Overbeck and Babenzien (1964a) observed in a small body of water in the Mark Brandenburg (German Democratic Republic) two bacterial maxima parallel to the phytoplankton development, one from April to May, the other from September to October (Figure 45). Similarly, maxima were found in May to June and in September by Niewolak (1974b) in the Ilawa lakes in the Mazury region of Poland. Investigations by Aliversieva-Gamidova (1969) in the Mektheb lake in Dagestan (USSR) on the

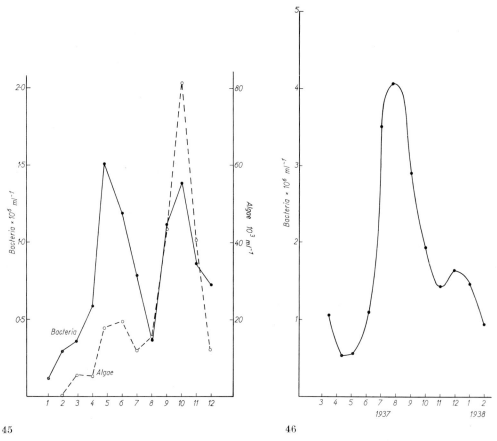

45   46

Figure 45   Total bacterial counts and numbers of algae (*Scenedesmus quadricauda* and centric
             diatoms) in a small body of water in the Mark Brandenburg (German Democratic
             Republic) in 1960. (Based on Overbeck and Babenzien, 1964a, modified)

Figure 46   Total bacterial numbers in the water (at a depth of 2 m) of Tschornoje Lake in Kossino
             (USSR). (Based on Kusnezow, 1959, modified)

yearly cycle of total bacterial numbers, saprophyte numbers and generation time
gave the following results:

Time (months)	Water temperature (°C)	Total bacterial numbers ($\times 10^{-6}$ ml^{-1})	Saprophyte numbers ($\times 10^{-3}$ ml^{-1})	Generation time (h)
XII–I	0–2	0.55	5–12	19–27
II–III	3–7	1.00	22–103	17
IV–V	17–23	1.40	14–21	10–12
VI	26	3.00	31	7.1
VII	28	2.50	28	6.5
VIII	21	3.10	81	6.3
X	11	1.10	17	23
XI	3	0.27	0.4	18

In various lakes of the central USSR only one bacterial maximum was found (Figure 46).
The highest total bacterial numbers were obtained for Tshornoje Lake in August
and for Glubokoje Lake in November (Kusnezow, 1959). In the latter the maximum

occurs at the time when much of the phytoplankton dies off and there is a large supply of nutrients. In one of the Mazurian lakes, Niewolak (1974a) determined the biomass of the bacterial plankton as 1.42–83.31 g C per m³, with the maxima being found in September and the minima in February. In lakes polluted by sewage the saprophyte numbers — as in the case of rivers — usually increase markedly in the winter. Overbeck (1974) shows the relationships between primary production, saprophyte numbers and total bacterial numbers based on investigations in four Holstein lakes (average for 1962–1964 in 1 m depth):

	mg C m⁻³ 24 h⁻¹	Saprophyte numbers ml⁻¹	Total bacterial numbers × 10⁻⁶ ml⁻¹
Plußsee	57	1000	1.1
Gr. Plöner See	33	650	1.0
Schluensee	17	350	0.5
Schöhsee	16	300	0.5

In newly constructed retention dam reservoirs, after flooding, a marked rise in the amount of bacteria could at first be observed throughout, and in subsequent years a clear decline. Total bacterial numbers increased, for example, in the Krementshug reservoir of the Dnieper (Ukraine) from 5.4 to 10.4 million per ml in the years 1961 to 1964, and by 1978 had fallen to 2.1 million per ml. The bacterial biomass increased accordingly from 4.6 to 9.1 mg l⁻¹ and then fell to 1.1 mg l⁻¹ (Michajlenko, 1981).

The strong influence of unusual natural events on the microbiological conditions of lakes was observed by Staley et al. (1982) in the area of the volcanic eruption of Mount St. Helens in the north-west of USA of May 13, 1980. Total bacterial numbers increased by an order of magnitude to more than 10 million per ml of water and saprophyte numbers to more than 1 million per ml, as against fewer than 10000 per ml. The originally oligotrophic lakes had been transformed in a short time into eutrophic lakes by the addition of organic matter from the destroyed forests.

The vertical distribution of bacteria in lakes of the temperate climatic zone also exhibits considerable seasonal variation. During the time of the summer stagnation characteristic thermal and chemical stratification takes place in the water with the consequent development of stratification of the populations of algae and bacteria. Not only are the total numbers of bacteria in these zones very different but also their composition with regard to species. Particularly striking are the zonal differences in eutrophic lakes, where the oxygen completely disappears in the hypolimnion and hydrogen sulphide is produced. As an example, the bacterial distribution in Lake Pluss in east Holstein is shown in Figure 47 as it was determined by Anagnostidis and Overbeck (1966) on October 7th, 1964. There is one maximum for the number of heterotrophic bacteria in the region of the thermocline and a second immediately above the bottom. These maxima are, however, due to different bacterial populations; in the first, proteolytic organisms predominate, in the second, methane producers and sulphate reducers. In the layer which is still sufficiently lit but already contains H_2S, immediately under the thermocline, numerous photoautotrophic sulphur bacteria (mainly Chlorobacteriaceae) are found while in the depths without light non-pigmented sulphur bacteria predominate. As long as oxygen is still present in the deeper layers of the lake there are also very many methane oxidizers, which, according to Anagnostidis and Overbeck (1966), consume a large proportion of the oxygen. As the anaerobic zone extends upwards so also the methane oxidizers rise higher and higher.

In the September of 1970, Gorlenko and Kusnezow (1972) observed in a mesotrophic Karst lake, the Kononjer Lake in the Mariyskaya ASSR, a copious development of *Metallogenium personatum* in the 10 m layer which was already very low in

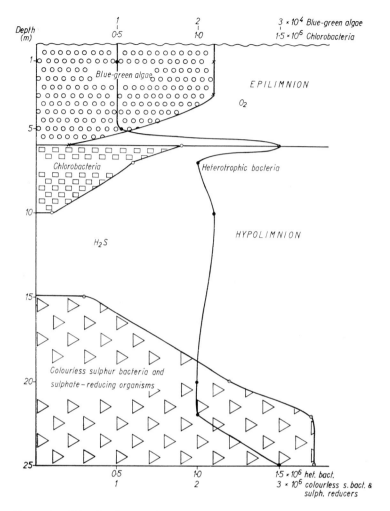

Figure 47   Distribution of blue-green algae (*Oscillatoria*), heterotrophic bacteria, chlorobacteria
and colourless sulphur bacteria in Lake Pluss in east Holstein on October 7, 1964.
(Based on Anagnostidis and Overbeck, 1966, modified)

oxygen. At the depth of 10.75 m, where H_2S was already present, purple sulphur
bacteria such as *Rhodothece conspicua* and *Thiocapsa* sp. had their maximum. Below
that — at a depth of 11–11.5 m — green sulphur bacteria (*Pelodictyon luteolum*)
predominated. In the 11–12 m zone, numerous aggregations of the brown *Pelochroma-
tium roseum* and of *P. roseoviride* were also found. Here also the symbiotic *Chlorochromatium consortium* Skuja, which contains gas vacuoles, develops together with
some examples of *Chlorobium aggregatum* (see Figure 78). At a depth of 13–22.5 m,
Peloploca pulchra Skuja was also found with the maximum immediately above the
bottom, which again is in good agreement with the observations of Anagnostidis
and Overbeck (1966) in Lake Pluss.

According to Gorlenko (1977) the highest cell numbers of photoautotrophic bacteria
were obtained in eutrophic salt lakes, with 48 million; in oligotrophic fresh water
lakes, on the other hand, there were 3.5 million per ml.

In the Dead Sea, Oren (1983) observed a massive development of extremely halophilic Archaebacteria of the genus *Halobacterium* in the topmost 10–25 m. Their numbers reached 19 million per ml in the summer of 1980 and gave rise to a red coloration of the water. In autumn, the cell numbers declined and stabilized at 5 million per ml. However, the population showed only low activity.

In lakes without a pronounced thermocline, the highest total bacterial counts are usually obtained in the zone with the most profuse development of algae and also above the bottom. At the times of autumn and spring circulations the water is turbulent, which results in a much more even distribution of bacteria. At the same time there is an increase in the available oxygen with a strong decline of the anaerobic sulphur bacteria. In larger lakes, the longitudinal and transverse profiles also show great variations in the bacterial counts. Streams and rivers entering a lake affect it to a large degree, but the differences decrease, as a rule, with increasing distance from the banks. After heavy rains, bacterial counts and also those of fungal spores rise steeply, though only temporarily, particularly in small lakes (Collins and Willoughby, 1962).

During calm weather, a characteristic neuston flora develops in the thin surface film of the water. It consists, apart from some algae, predominantly of bacteria (Babenzien, 1966).

Cyanophytes are widely distributed in inland lakes. In oligotrophic lakes, of course, their portion of the phytoplankton is often very small. However, it increases, as a rule, with increasing eutrophication. In eutrophic lakes cyanophyte blooms occur in the summer months fairly frequently (see Section 4.1) and give rise to an intense greenish colour of the water. They concentrate first usually at the uppermost 1–2 m of the body of water and later can spread to the whole epilimnion (see Figure 47). Thus, in the moderately eutrophicated Lake Erken in Sweden in June 1970, Granhall and Lundgren (1971) counted some hundreds of cyanophyte filaments at a depth of 0–2 m and in August at a depth of 0–4 m, 8000–13000 trichomes of *Anabaena* spp. and 13000–21000 of *Aphanizomenom flos-aquae* in 1 litre of water. Other species played only a very subsidiary role at that time. The maximum was found in September with about 200000. Anagnostidis and Overbeck (1966) found between 5000 and 22000 trichomes of *Oscillatoria redekii* in a litre of water in the epilimnion of Lake Pluss in October 1964 (see Figure 47). In more heavily loaded lakes and ponds still greater concentrations often occur. Increasing eutrophication also leads to changes in the cyanophyte population. For example, this is the reason for the spread of *Oscillatoria rubescens* (Carr and Whitton, 1973). All the remaining phytoplankton species can be nearly completely displaced in the process. Due to the intensified eutrophication of many waters during the last few decades, cyanophyte blooms have become more and more frequent in lakes in all parts of the world. Not infrequently there are marked fluctuations in the cyanophyte populations, since the blooms often break down and lyse just as quickly as they have been formed. In this, cyanophages might often play an important role (see Chapter 6 and Section 7.3).

The distribution of lower fungi in English lakes has been studied by Canter and Lund (1948, 1951, 1953) and by Willoughby (1962). The first two authors studied the development of plankton parasites in relation to that of their host cells (see Section 10.3). Willoughby determined the number of spores of Saprolegniales in several lakes of the English Lake District; the surface water at the edge of Lake Windermere contained between 25 and 5200 spores per litre; the middle of the lake, on the other hand, only 11–100. The highest figures were obtained close to the banks after heavy rain when the water level had risen. Most frequent was the genus *Saprolegnia*, follwed by *Achlya* and *Aphanomyces*. *Dictyuchus* and *Leptolegnia* were observed only occasionally. Yeasts are particularly widespread in eutrophic lakes. Thus, Niewolak (1977) found up to 82 yeast cells per ml in slightly polluted North Poland lakes but up to 2310 ml^{-1} in heavily polluted waters.

7.3. Distribution in the Sea

Due to lack of nutrients in large parts of the open sea, very small forms often dominate in the bacterial flora there. The determination of total bacterial numbers is thereby rendered more difficult. It has been possible to obtain reliable values only by using fluorescence microscopic methods.

As a rule, the highest total bacterial numbers are found in surface regions and the numbers decline more or less markedly with increasing depth of water. An example of this is shown by a vertical profile from the western Biskaya. Here total bacterial numbers decreased from 804300 at the surface to 100900 at a depth of 1000 m. The bacterial biomass behaves similarly. At that station, it fell from 14.21 to 0.73 μg per litre. Occasionally, however, there is also an increase again, even if relatively small, in total bacterial numbers and bacterial biomass at depths of between 500 and 1000 m. A stronger increase, not infrequently, may also be observed above the sediment.

In investigations of vertical profiles, differing amounts of bacteria can be encountered in different bodies of water. This is so, for example, to the west of the Straits of Gibraltar, where the inflowing Mediterranean water may be recognized not only by its higher temperature and salinity but also by changes in the bacterial population (Rheinheimer and Schmaljohann, 1983). In antarctic waters of the eastern South Pacific, values of between 100000 and 1000000 per ml were found. The proportion of cells in division occurring there lay between 3 and 16 per cent (Hanson et al., 1983).

In the deep sea the concentration of bacteria is generally low and, as a rule, total bacterial numbers may be less than 100000 per ml. However, some years ago in the region of thermal vents near the Galapagos Islands in the Pacific, at a depth of about 2500 m, high bacterial numbers from 0.5 to 1000 million per ml of water were determined. For the most part, this was a matter of chemoautotrophic sulphur bacteria which oxidize the H_2S originating from volcanic springs in the mixing zones. But, in addition, iron and manganese oxidizers as well as heterotrophic bacteria are found. These amounts of bacteria are a source of nutrition for the extensive mussel banks, large worms and other animals on the margins of the crater (Karl et al., 1980; Ruby et al., 1981).

In coastal waters of the North and Baltic Seas total bacterial numbers between some hundred thousands and several millions, as a rule, were found. The total bacterial numbers determined by Zimmermann (1977) in Kiel Bay in 1974, when water samples were examined monthly, ranged from 450000 to 5240000 per ml. The following mean values were obtained for five locations in Kiel Fjord and Kiel Bay:

	Inner Kiel Fjord	Outer Kiel Fjord	Southern Kiel Bay	Middle Kiel Bay I	Middle Kiel Bay II
2 m	3100000	2580000	1430000	1320000	1460000 ml^{-1}
above bottom	2470000	2310000	1320000	1370000	1580000 ml^{-1}

Seasonal fluctuations are relatively small in the case of locations further away from the land — nevertheless in spring, summer and autumn somewhat higher values may be found than in the winter months. These seasonal differences are still more pronounced in the more intensely eutrophicated Kiel Fjord.

Zimmermann (1977) calculated the bacterial biomass in 1974 in the inner Kiel Fjord at between 5.1 and 24.2 μg C per litre and in the middle Kiel Bay at between 1.5 and 11.3 μg C per litre.

The number of active bacteria in the Kiel Fjord varied between 170000 and 2100000 and in Kiel Bay between 55000 and 1200000 (Hoppe, 1977). Its proportion of the

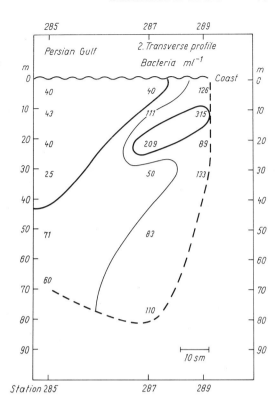

285   287   289

Persian Gulf   2. Transverse profile
                    Bacteria ml⁻¹

m   Coast   m
0   0
   40   40   126
10 43   111   315
20 40   209   89
30 25   50   133
40
50 71   83
60
70 60
80   110
90   10 sm
Station 285   287   289

Figure 48   Distribution of bacteria
            (saprophyte numbers) from
            the coast of Iran to the
            middle of the Persian Gulf

total bacterial count fluctuated in the course of the year between 10 and 65 per cent with minima in winter and maxima in summer.

In contrast to the as yet not very numerous but reasonably reliable total bacterial counts, saprophyte numbers are available from more extensive investigations of various marine regions. They show very much stronger regional and seasonal fluctuations than the total bacterial numbers (see Rheinheimer, 1977c). In coastal waters saprophyte numbers are fundamentally higher than in the open sea and, as a rule, diminish rapidly with increasing distance from the coast (Figure 48) (Gunkel, 1966; Rheinheimer, 1966). Apart from heavily polluted harbours and creeks, the highest figures are usually found in the breaker zone, where a great deal of organic material is ground up. In the North Sea and the Baltic, this zone often yields saprophyte numbers of between ten thousand and several hundred thousand; in the adjoining area, there are between a thousand and some tens of thousands, and at a somewhat greater distance from the coast (several sea miles) less than a hundred to several thousand. In the oceans the saprophyte numbers are lower still, between less than 1 and 100 per ml of water (ZoBell, 1946a; Gunkel, 1966; Rheinheimer, 1966).

With regard to vertical distribution, the highest bacterial numbers are found almost always in the productive euphotic zone; the maximum, however, occurs not in the top zone — if we disregard the neuston — but at a depth of 10–50 m. ZoBell (1946a) states that it is often a little below that of the phytoplankton (Figure 49). Here, Fuhrman and Azam (1982) detected also the quickest bacterial growth.

Below 200 m only very small numbers are found, and below 1000 m depth the saprophyte numbers no longer vary; but, not infrequently, they rise again immediately above the bottom. Sorokin (1964) determined in parallel the numbers of saprophytes and the total number of microbes in the tropical Pacific, and found good correlation

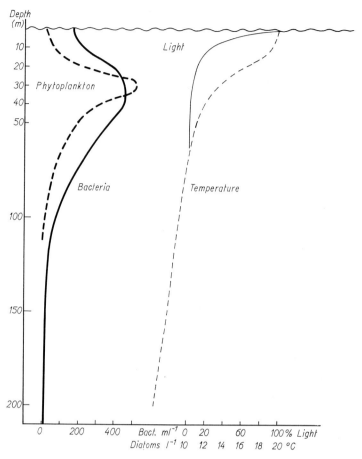

Figure 49   Vertical distribution of bacteria (saprophyte numbers per ml), phytoplankton (diatoms per litre), light and temperature in the Pacific, off the coast of southern California (average values). (Based on Zobell, 1946a, modified)

in the vertical profile, although the proportion of the saprophytes was very small, 1 : 1000.

In sea regions with marked differences in density, the highest bacterial counts are found usually in the zones of the thermocline and of the sudden change in salinity (Figure 50). As such zones between water masses of different density present an obstacle for the sedimentation of bacteria and debris, they just accumulate there. This, in turn, makes the nutritional conditions more favourable, so that bacteria not only accumulate there but can also multiply more vigorously. In shallow coastal waters, the stratification is often destroyed by strong winds, and the high turbulence then causes a relatively uniform vertical distribution of bacteria. When the weather becomes calmer, the original stratification is gradually restored. Occasionally, heavy gales may stir up the sediment and this may cause a temporary but considerable rise in the bacterial content of the water. Also, the tide often affects the distribution of bacteria to a great extent. Examination of the water at the mouth of the Elbe near Cuxhaven (Kühl and Rheinheimer, 1968) and at Brunsbüttel (Rheinheimer, 1968b) showed that with strong tidal currents the distribution of bacteria, plankton and debris is relatively uniform from the surface to the bottom. When the current

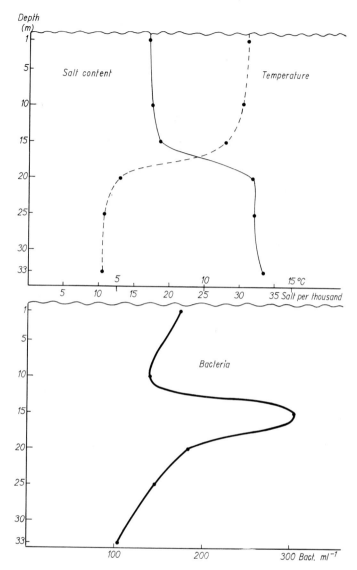

Figure 50   Salt content, temperature and bacterial content (saprophyte numbers) on June 7, 1966, in the southern Kattegat (location 56° 24′ N 11° 22′ E). Salinity and temperature curves show a marked discontinuity layer between 15 and 20 m depth. This area contains the maximum number of bacteria

slackens, particularly at low tide, numbers decrease in the surface water due to sedimentation, and increase markedly in the deeper parts; as the current increases again, bacteria are stirred up, and the turbulence equalizes the bacterial numbers over the whole water column (Figure 51).

Wright and Coffin (1983), in the tidal areas of river estuaries on the east coast of north America, found a concentration of nutrients caused by mixing processes, which, in the warmer seasons, led to a greater rise in total bacterial numbers than in the adjacent limnic and marine areas.

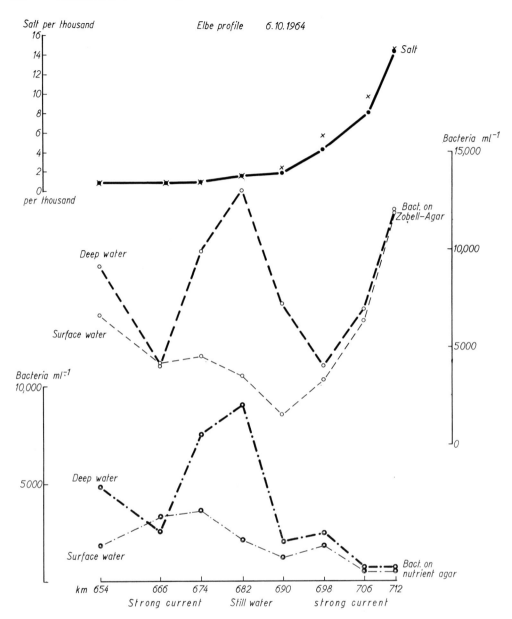

Figure 51   Salt content and saprophyte numbers of two different media in water from the surface and from the deeper parts of the Elbe estuary. Whilst in a turbulent current the vertical distribution of bacteria shows hardly any differences; in still water the bacterial content in the deeper water increases markedly

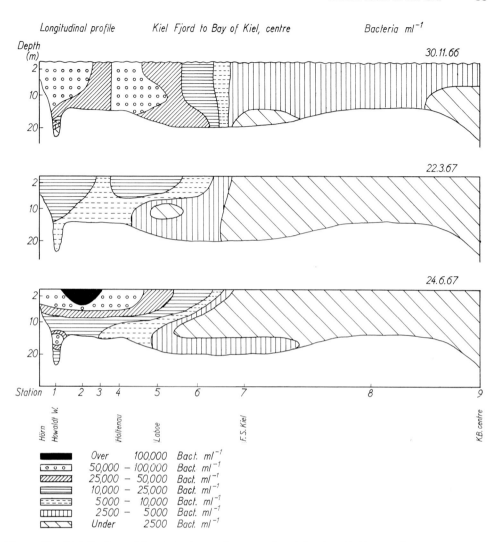

Figure 52   Distribution of bacteria (saprophyte numbers) in the Kiel Fjord (locations 1–6) at various times of the year. The great differences are due to the lively exchange of water with the open Baltic Sea (location 7–9)

In bays and fjords also the bacterial content of the water is much influenced by currents due to the wind. This operates particularly in landlocked seas where the role of the tides is insignificant.

Investigations in the Kiel and Flensburg Fjords (Rheinheimer, 1968a, 1970b) showed that the bacterial content of the water may change within a short time by a factor of 10 or even 100, dependent on the meteorological conditions. In calm weather the saprophyte numbers rise rapidly, particularly in the inner parts of the fjord, due to increasing pollution; on stormy days, water from the Baltic which is poor in microbes enters and the saprophyte numbers are much reduced (Figure 52).

In the temperate climatic zones, the sea — like inland waters — shows seasonal fluctuations in bacterial numbers. Thus, the saprophyte numbers in the western Baltic have two maxima, one in spring (April/May), the other in autumn (October/

6*

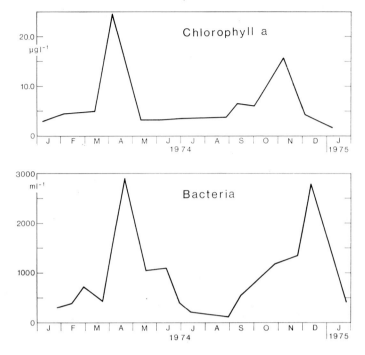

Figure 53   Yearly course of chlorophyll a concentration as a measure of the amount of phyto-plankton (top) and the saprophyte numbers (bottom) in the Kiel Bay.

November). The highest saprophyte numbers, then, occur only after the production maxima of the phytoplankton, when much of it is dying off (Figure 53). The saprophyte minimum was found always at the height of summer. Similar observations were made by Gunkel (personal communication) in the German Bight. In badly polluted bays and harbours, with little exchange of water, an unequivocal winter maximum — as in corresponding rivers and lakes — was observed.

The study of longitudinal and transverse profiles occasionally yields strikingly high saprophyte numbers in some samples, independent of the time of the year. These, as a rule, are probably due to a local supply of food, for example dead plants or animals (ships also in the surface zone). In those parts of the oceans poorest in nutrients, such a sudden supply of food is probably the only chance of multiplication for many bacteria.

Recently, parallel determinations carried out in the Baltic Sea of total microbial and saprophyte numbers showed clearly that the total numbers show very much smaller fluctuations both spatially as well as in time than the saprophyte numbers. Thus, for example, on monthly crossings from the relatively heavily loaded Kiel Fjord into the comparatively clean middle of the Kiel Bay, a decrease of the total microbial count to half, as a yearly average, was found — in contrast, the saprophyte numbers decreased to a fiftieth. Similarly, in self-purification experiments, after the addition even of small amounts of nutrients the proportion of saprophytes rose distinctly more markedly than the total microbial count and also fell again more quickly. The number of active bacteria determined by means of microautoradiography, as a rule, lies between these two values (Hoppe, personal communication).

Sieburth (1971) and Tsyban (1971) state that a clear-cut bacterioneuston can also develop in the sea, so that occasionally very high saprophyte numbers may be encountered in the top 150 μm. With that might also be connected the relatively large bacterial content of drops of water flung into the air by bubbles, the bacterial concentration in which often amounts to many times that in the body of water from which they arose (Blanchard and Syzdek, 1970).

Kjelleberg *et al.* (1979) determined saprophyte numbers of between 740 000 and 1 300 000 per ml in lipid films on the water surface off the Swedish west coast. These values were some 2–4 orders of magnitude greater than those of deeper waters. Kim (1983) also found, in the Kiel Fjord, greater differences in numbers of bacteria of different physiological groups in the surface film and in water from 1 m depth. Thus, for example the annual mean values for these two sites ran to 6440 and 2390 for starch degraders, 3380 and 590 for lipolytic bacteria and 1430 and 430 ml^{-1} for oil degraders.

Cyanophytes play an important role especially in the phytoplankton of the sea. Small unicellular forms of the *Synechococcus* type are widely distributed. They are rod-like cells, 0.6–0.8 × 0.9–1.5 μm in size. They can be clearly distinguished from bacteria by their own characteristic orange-yellow fluorescence and can be separately counted.

In the north Atlantic between England and the Azores, up to 60 000 cells per ml of water were found in June, 1983. In the Skagerrak and the Baltic, in the period between September and November, 1982, from 1600 to 23 000 per ml were counted, which corresponded to 0.5–2.6 per cent of the total bacterial numbers but to about 1.0–5.2 per cent of bacterial biomass. In winter and spring, in this sea area, the values were very much lower than in summer and autumn (Schmaljohann, 1984).

Members of the genus *Trichodesmium* are strongly represented, particularly in tropical regions. They can form plankton blooms there where a salt content of about 3.5 per cent and water temperatures of more than 25 °C prevail (Barth, 1967). These occur both on the surface and also in deeper regions of the photic zone. *Trichodesmium* blooms have been observed in the relatively nutritious planktonic regions (e.g. near the Indian coast) as well as in regions of the Indian Ocean poor in nutrients and of the Sargasso Sea. Not infrequently, they are particularly extensive in coastal waters; thus Wood (1967) describes a cyanophyte bloom which extended between the Australian coast and the Great Barrier Reef to a distance of 1600 km and an area of 52 000 km². It had a thickness of 0.5 m — the greatest cell density, however, was found at the surface. According to Hart (1966) 140–560 trichomes were found in 1 ml of water — this corresponds to 5.9–50.0 μm³ ml^{-1}.

In coastal waters containing brackish water with salt contents below 2 per cent, blooms of fresh water species develop. In the middle Baltic Sea the copious occurrence of *Aphanizomenon flos-aquae* and *Nodularia spumigena* was frequently observed and recorded by satellite photographs (Figure 54). They have their origin usually in creeks or bays and spread out from there far into the open sea. Often they occur on the water surface in long streaks or bands.

Rinne *et al.* (1977) determined the biomass of the heterocyst-forming cyanophytes in the surface water of the northern Baltic Sea and in the late summer of 1974 and 1975 found average values of between 21 and 275 mg m^{-3}.

However, cyanophytes were encountered not only in the photic zones but also at greater depths of the sea. For example, representatives of the genus *Nostoc* and a very small *Dactyliococcopsis* species were found in the Atlantic and Indian Oceans as well as in the Mediterranean Sea. A form similar to *Nostoc planktonicum* was found at a depth of 1000 m. These cyanophytes evidently are here capable of a heterotrophic existence. The population density runs to 400–12 600 trichomes per ml (Bernard, 1963).

On rocky coasts various cyanophytes often form black bands, which can attain a thickness of 5 mm. The rock pools occasionally contain populations with many species and high cell numbers.

Figure 54   Satellite (Landsat) photo of the Baltic Sea between the islands of Bornholm and
              Rügen on 7 August, 1975, which reveals a massive accumulation of cyanophytes in
              the water surface. (Spectral regions MSS 4 — processed on DIBTAS by K. A. Ul-
              bricht, DFVLR)

In shallow water areas near to coasts as, for example, the Baltic Sea bays, benthic
cyanophytes temporarily cover large areas of the bottom and here have a considerable significance as nitrogen fixers and primary producers (Hübel and Hübel,
1974 b).

The distribution of marine phycomycetes was investigated by Gaertner (1967,
1968 a, b) in the North Sea and the north-eastern Atlantic. Up to 2000 fungi per
litre of water were found, the highest numbers occurring near land; in the open sea,
the figures were rather low, 1–12. At five locations in the German Bay the phycomycetes were counted once or twice monthly during 1965 and 1966, and great fluctuations
in numbers could be demonstrated (Gaertner, 1968 b). In the Arcachon Basin on the
French south-west coast, Ulken (1969) found great seasonal differences. Thus, in
April there were numerous Chytridineae, which were already disappearing towards
the end of May and were completely absent in October. Thraustochytriae, on the
other hand, seem to be present all the year round.

In degradation experiments with *Spartina alternifolia*, Marinucci *et al.* (1983)
observed a marked rise in fungal biomass, which increased to about 3 per cent of the
ash-free dry weight. The bacterial biomass amounted only to about 10 per cent of
that of the fungi. Furthermore, in the decomposition of the benthic algae, especially
the dead material, fungal numbers and biomass can also increase greatly.

In various regions of the sea the distribution of yeasts was also studied. For the
most part, these are facultatively marine forms of terrestrial origin (see Section 5.2).
Their numbers are, therefore, relatively high in sewage-loaded coastal areas and
decrease rapidly further seawards. Nevertheless, some salt-tolerant yeasts have still
been demonstrated in the open ocean (van Uden and Fell, 1968). Fell (1967) found
living yeasts, for instance, in the Indian Ocean from the surface down to a depth

of 2000 m. Of 179 samples examined, 65 were positive, containing between 1 and 513 per litre of water. The relative frequency of the various species was as follows:

Candida atmosphaerica	21%	Candida polymorpha	10%
Rhodotorula rubra	20%	Rhodotorula glutinis	6%
Candida spp.	12%	Other kinds	19%
Sporobolomyces odorus	12%		

In coastal waters, up to several thousands of yeast cells per litre of water were found (Roth et al., 1962; Meyers et al., 1967). In heavily polluted waters there could be considerably more. Regular sampling of the Schlei (western Baltic) sometimes gave yeast counts of several tens of thousands.

Schaumann (1975) investigated the higher fungal flora in the German Bight near Helgoland and found ascomycetes and Fungi Imperfecti especially on wood substrates. He was able to identify 46 species in this part of the sea. The number of species declined to 13 with the salt content in the brackish water of the Weser estuary.

Schneider (1976) also found a larger number of wood-attacking fungi in the western Baltic Sea. There, more species developed on indigenous woods (beech and pine) than on tropical varieties. The formation of fruiting bodies is dependent on the time spent in the water and also on its temperature.

7.4. Distribution in Sediments of Inland Waters

The top layers of the mud in lakes have a high content of organic nutrients and are, therefore, colonized by a very large number of micro-organisms. Kusnezow (1959) determined by direct microscopic counts the total number of bacteria in several lakes of the USSR, and found several hundred millions to several thousand millions per g of damp mud.

	Millions bacteria per g		Percentage ash content
	Damp mud	Dry mud	
Byeloye Lake	2326	54219	54.1
Tshornoye Lake	1285	35109	49.3
Swyatoye Lake	922	29790	18.6
Lake Gab	1883	15680	81.9
Krugloye Lake	1110	7991	79.2

As a rule, the microbiological conditions of sediments in the flat litoral near the shore differ from those in the deep regions of the profundal. Thus, Jones (1980) found higher total bacterial numbers and greater microbial activity in the surface sediment of the profundal in the northern English lakes, Blelham Tarn and Lake Windermere, than in the shore zone. This was attributed, in particular, to the higher concentrations of carbohydrates, proteins and amino acids in the interstitial water.

The saprophyte numbers are generally several tens of thousands to several hundred thousand per g of damp mud. In polluted waters, however, they can quite often amount to many millions. In the dry mud deposited on the stones on the banks of the upper Weser even more than a thousand million saprophytic bacteria per cm^3 have been found.

At a depth of even a few centimetres, the bacterial content of the mud — as also its content of organic substance — is reduced, the saprophyte numbers falling more rapidly than the total numbers. At a depth of 1 m the bacterial counts are only a fraction of those on the surface of the sediments and then slowly decrease further. Exerzew (1948) states that this holds equally for both aerobic and anaerobic forms.

In most lake sediments, besides eubacteria, actinomycetes can also be detected. As a rule, their number likewise declines with increasing depth — although Cross (personal communication) found in Lake Windermere still living actinomycete spores in the sediment zone the age of which amounted to between 1000 and 2000 years.

Also, the number of fungi in the mud of lakes decreases with increasing depth of the sediment. The fungal flora consists here predominantly of phycomycetes and Fungi Imperfecti, and is particularly abundant in the mud of shallow areas near the banks. The fungi grow in the mud as well as on its surface. Submerged foliage and twigs often carry a characteristic flora of hyphomycetes (Sparrow, 1968).

Aliversieva-Gamidova (1969) studied the annual cycle of different bacterial groups in the sediment of Lake Mektheb in Dagestan (SSR) and in the course of the year found considerable changes in the microflora:

Month	Sapro-phytes $\times 10^{-3}\,g^{-1}$	Actino-mycetes $\times 10^{-3}\,g^{-1}$	*Clostridium pasteurianum* $\times 10^{-3}\,g^{-1}$	Hydrogen bacteria $\times 10^{-3}\,g^{-1}$	Methane formers $\times 10^{-3}\,g^{-1}$	Sulphate reducers $\times 10^{-3}\,g^{-1}$
XII–I	790	80	10	10	1	2
II	2700	0	10	10	40	1
III	4800	300	10	10	240	10
IV	3100	0	100	1	140	1300
V	1800	100	10	10	15000	100
VI	4800	300	10	10	15000	1200
VII	3800	0	10	100	2000	5000
VIII	5400	0	100	10	3000	1500
IX	2700	0	10	100	13	1
XI	80	0	1	1	10	0

Nearly all the micro-organisms studied show a clear winter minimum. The methane producers have a clear maximum in May/June, the sulphate reducers, on the other hand, in July.

Ward and Frea (1980) investigated the occurrence of different methane bacteria with the aid of immunofluorescence techniques, in sediments of the North American Lake Erie. The highest numbers, 10^6–10^9 cells per g dry sediment, were found for *Methanobacterium ruminantium* in the nutrient-rich mud. The values for *Methanosarcina barkeri* varied from 10^6 to 10^7. This micro-organism occurred in sand, sandy mud and clay sediments.

According to observations by Suzuki (1961) in the Japanese Lake Nakanuma, the numbers of fungi in the mud at the bottom display seasonal fluctuations, with a summer minimum and a winter maximum; there is a correlation with the oxygen content. *Pythium* was found all the year round, but with a winter maximum. *Achlya flagellata, A. racemosa, Dictyuchus* sp., *Aphanomyces* sp. and *Saprolegnia* spp. were found only during the circulation periods and the winter stagnation, when the water at the bottom still contained oxygen.

No comparative investigations on the distribution of bacteria and fungi in lake sediments of differing composition are available so far. It may be assumed, though, that they will vary greatly depending on the oxygen content and the mineral composition of the sediments.

7.5. Distribution in Marine Sediments

Marine sediments are inhabited by bacteria and fungi right into the deep sea. Most micro-organisms are adsorbed on to sediment particles, which makes the determination

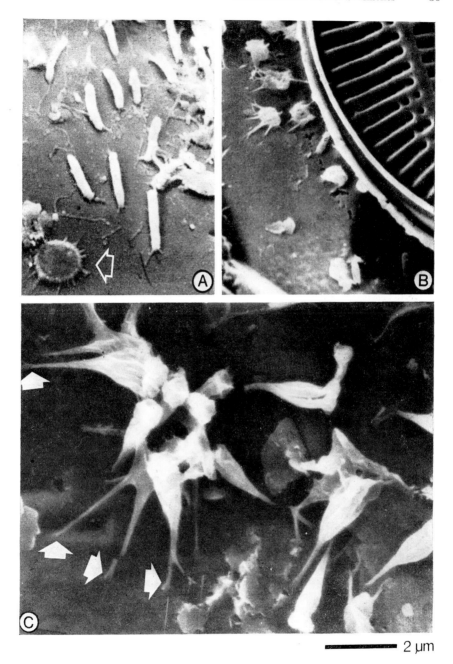

2 μm

Figure 55   Bacterial aufwuchs on sand grains. (A) Small colony of rods with slime filaments. Below left (arrow): discus-shaped bacterium with fimbriae. (B) Pleomorphic bacteria with star-shaped slime secretions on a diatom shell. (C) Stalked bacteria with branched cell appendages (arrow). Scanning electron micrograph by W. Weise

of their numbers rather difficult. Here, as in the open water, they play a very important role in the remineralization of organic substances and probably also as food for the fauna of the sea bed. The directly determined total bacterial counts in the top layers vary depending on the kind of sediment and the depth of the water; they are between a few hundred thousand and many thousands of millions per cm³.

In the sand sediment of the middle Kiel Bay, for example, total bacterial numbers of between 682 million and 2300 million per cm³ of fresh sediment in 12–14 m depth of water were obtained by means of fluorescence microscopy. Of these, 49–64 per cent were found as surface growths on sand grains and 36–51 per cent free-living in the interstitial water (Weise and Rheinheimer, 1978a). A scanning electron microscopic investigation showed that the micro-organisms of the surface growth on the sand grains are found preferentially in grooves and holes where they are largely protected from the mechanical effects caused by movement of the sand. Exposed areas and edges of the sand grains, on the other hand, are avoided. The surface bacteria possess to some extent very effective attachment mechanisms, with which they are firmly fixed to their substrate (see also Weise and Rheinheimer, 1978). In addition to morphological adaptation through flattened cell forms, slime threads and fimbriae also are frequently present (see Figure 55).

In the uppermost zone (0–1 cm) of mud sediments of the middle Kiel Bay, total bacterial numbers of between 46.7×10^9 and 77.7×10^9, and bacterial biomass values

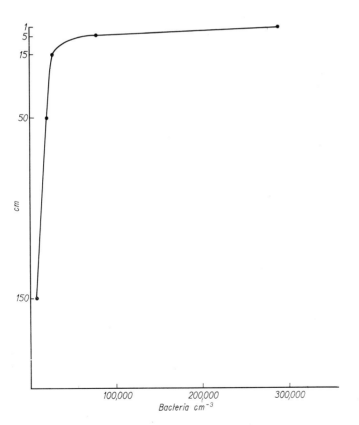

Figure 56   Distribution of bacteria (saprophyte numbers) in a sediment core from the southern Persian Gulf. (Based on Rheinheimer and Gunkel, 1975, modified)

of between 451 and 816 μg C per cm^3 of moist sediment were found in August 1980. Saprophyte numbers, as a rule, lie between a thousand and several millions, with the highest numbers being found in shallow coastal waters. Towards the sea, numbers rapidly decrease. Sediments of coarse sand contain, as a rule, considerably fewer bacteria than do fine silts. The following table from ZoBell (1938) provides an example of the connections between grain size and bacterial content (saprophyte numbers):

	Grain size μm	Water content %	Bacteria $\times 10^{-3}$ g^{-1}
Sand	50–1000	33	22
Silt	5–50	56	78
Clay	1–5	82	390
Colloidal sediment	<1	>98	1510

Particularly low saprophyte numbers have been found in constantly disturbed sands where organic material cannot accumulate. As the depth of the water increases, the numbers of saprophytes (Bansemir, 1969) and of lower fungi (Gaertner, 1968a) in the sediments decrease because the longer the descent of the organic material towards the bottom the further it will be broken down, resulting in less favourable nutritional conditions at the bottom. The highest numbers of bacteria and fungi are almost always found in the top few centimetres of the sediments, and mostly on their surface. Even 10 cm below the surface, the numbers of bacteria are, not infrequently, already reduced to a few per cent; below 100 cm from the sediment surface, the total bacterial and saprophyte numbers hardly change over the next several metres (Figure 56). Where, however, the sediments are layered unevenly, zones of low bacterial content may be followed by others with a higher one, though only within the top 100 cm.

In spite of the often quite high bacterial numbers in sediments rich in nutrients, their proportion of the total biomass is usually relatively small. This is shown in the following table of average numbers and biomass (wet weight) for 1 m^2 of a marine silt sediment in the top 5 mm (from ZoBell, 1963):

Organism group	Numbers	Biomass g
Large macrobenthos	2.8	3.75
Small macrobenthos	230	3.30
Meiobenthos	146×10^3	1.15
Protozoa	283×10^6	0.02
Diatoms	590×10^6	0.06
Bacteria	355×10^9	0.07

Weyland (1969, 1981) detected actinomycetes in numerous marine sediment samples. In the North Sea and Baltic Sea, their numbers ranged from 23 to 2909 cm^3. At 34 sites an average of 480 actinomycetes per cm^3 of sediment was estimated, which corresponded to 0.29 per cent of the saprophyte numbers. For four sites at more than 200 m deep the corresponding values were 780 actinomycetes per cm^3 and 11.88 per cent. Positive findings were also obtained from 9 out of 12 samples examined from the Atlantic, west of Africa. For example, 138 actinomycetes were counted in 1 cm^3 of sediment from a depth of 3362 m at about 175 km from the West African coast.

Gaertner (1967, 1968a) found in the uppermost sediment zone (0–10 mm) 10000–18000 lower fungi per litre of sediment in the southern part of the North Sea, and

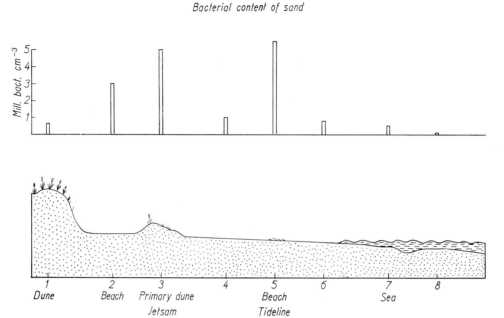

Figure 57   Distribution of bacteria (average saprophyte numbers) in the sandy beach of Bott
Sands, western Baltic Sea. The most bacteria are found below the tide line, the
fewest where the waves buffet the beach

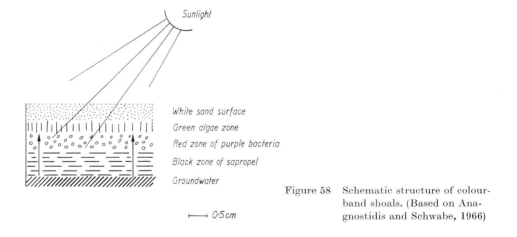

Figure 58   Schematic structure of colour-
band shoals. (Based on Anagnostidis and Schwabe, 1966)

1000–10000 in the northern part of the North Sea and the northern Atlantic. At
the mouth of the River Weser the figures were higher still, between 10000 and about
60000 during most of the year (Gaertner, 1968b).

Höhnk (1969b) states that calcareous hard parts of marine animals (mussel shells,
snail shells, cuttle-bones) found in sediments are colonized by a characteristic fungal
flora. Of 139 appropriate samples from 51 locations in the North Sea and the bordering
shoals, 133 samples from 48 locations contained fungi. Nine kinds of fungi were
isolated from that sort of substrate and, of these, five seem to be obligatory inhabitants of such calcareous hard parts of lower animals.

Yeasts also have been found in marine sediments. They occur particularly in the topmost few centimeters and, according to Suehiro (1963), they are more frequent in black ooze than in sandy sediments. Several hundred living yeast cells per cm^3 were found in the damp mud from the Kiel Fjord (Hoppe, 1970).

Of interest also is the distribution of micro-organisms in sandy beaches. Figure 57 shows, as an example, the average saprophyte numbers of a beach profile on the coast of the western Baltic. In damp sediments of pure sand there are several hundred thousand to several million. Underneath the tideline and under jetsam the sapro-phyte numbers may rise locally to more than 20 million per cm^3. Where the waves beat against the beach, the bacterial content of the sand is, as a rule, lowest and may be reduced to a few tens of thousands at times of rough seas. On the other hand, the bacterial numbers in the relatively nutritious sand shoals of the North Sea lying in front of the northern coast of Sylt are rather high, with several million per cm^3 (Westheide, 1968). The investigation of vertical profiles shows that bacterial maxima are frequently found in the topmost 20 cm. Below that the bacterial content often falls sharply and reaches the lowest values in the inflow region of the groundwater. The reason for this lies both in a decrease in the nutrient concentration as well as in the activities of the fauna living in the interstices in the sand and which is largely absent from the uppermost zone of the beach sand.

Not infrequently, so-called colour-band sand shoals develop above the *Arenicola* zone (Schulz, 1937; Hoffmann, 1942). Typically these show four colour bands (Figure 58); white sand on the surface, the green cyanophyte and algal band, the zone of red purple bacteria and, below, the black zone of sulphate reduction where H_2S is produced, particularly by *Desulphovibrio* (see Section 11.6). By abrasion (erosion by breakers or the wind) or by human or animal influences respectively, these layers can also lie side by side as coloured bands.

8. The Influence of Physical and Chemical Factors on Aquatic Micro-organisms

The growth of aquatic micro-organisms is affected by a great variety of physical and chemical factors which, in a multitude of ways, may also act with or against one another. They influence not only the size and composition of the microbial populations, but also the morphology and physiology of the individual bacteria and fungi. Thus, in some species, temperatures above or below the optimum and salt concentrations or pH values above or below the optimum may lead to considerable changes in metabolism, cell morphology and reproduction. Normally rod-shaped bacteria, for example, turn into cocci or long filaments; occasionally, irregular cell division or branching may occur. The synthesis of enzymes and, in consequence, the ability to break down certain substances may be either promoted or inhibited. In fungi, the formation of fruiting bodies and sporangia is affected.

Although in any natural habitat there are always numerous factors of different kinds which act on the living creatures present, some of these factors are particularly important, namely, those which limit the possibility of life to within a certain area.

Thus, the concentration of certain nutrients and active substances may be too low or too high for some or even for all micro-organisms; and this holds also for the pH values and the temperature. Many living creatures can grow only within a narrow range of pH and of temperature with a very limited optimal range.

The life of organisms is passed between the cardinal points of minimum, optimum and maximum, which differ for the individual species and strains. These points however, are not rigid but may vary due to the effect of other factors. In nature, many species do not always develop best within their optimal range, as they may be inhibited by competition with other organisms. This is also the reason why some inhabitants of extreme habitats show, in pure culture, an entirely different optimum than one might have expected from their natural environment.

In the following, the effect of the most important physical and chemical factors on the growth of aquatic bacteria and fungi will be considered. Our knowledge regarding these matters has been widened considerably within the last few years, but it is often difficult to interpret results obtained *in vitro* in the laboratory in relation to organisms within their natural habitat. Thus we know, for example, that many non-pigmented bacteria are inhibited by light. In some cases we even know what intensity of light is necessary to kill certain bacteria within a certain time. We also can measure the intensity of light in water. Nevertheless, we are unable to assess the effect of the light factor on the bacterial flora of, say, a river, as the turbulence of the water continually changes the location of the bacteria. Even if some day this question should be cleared up, we still do not know how light affects competition within the whole bacterial population. Some very searching investigations will, therefore, be required both in the natural habitats and in the laboratories to aid our understanding. Here, investigations of suitable microcosms can be of further help (see Chapter 12).

8.1. Light

Light is an important ecological factor in water, as it is on land. Its intensity, however, quickly diminishes as it penetrates into the water. Thus, in the North Sea, the light intensity at a depth of 20 m is only about 1 per cent of that on the surface, whereas in the Mediterranean, which is much clearer, it is still about 7.5 per cent (Gessner, 1957). Depending on the degree of turbidity of the water, biologically effective light intensities may be found in the top 10–100 m, in some places down to 200 m. An indication of this is often obtained by the lower edge of algal growth. It is measured accurately by the so-called compensation level, which is the light quantity where the assimilation and respiration of green plants cancel each other out, so that both O_2 and CO_2 contents remain steady. The compensation level can be determined by exposing cultures of algae in glass bottles for 24 h at various depths in the water. These should be pure cultures free of bacteria, as heterotrophic contaminants would consume additional oxygen. Petterson *et al.* (1934) found that in the Gullmarfjord, on the west coast of Sweden, the compensation level is at a depth of 19 m. with a light intensity of 400 lux. This is approximately 1 per cent of the light on the surface. The green plants pass their lives in the zone above the compensation level and practically all the organic substances are produced here — if one disregards the proportion produced by the chemoautotrophic bacteria which is, so far, unknown.

For the relatively small number of photoautotrophic bacteria — i.e. the chloro- and purple bacteria — light provides energy as it does for the green plants which reduce CO_2. However, as they are anaerobic organisms they are not capable of dissociating water; instead of H_2O they have to use H_2S or various organic compounds as hydrogen donors. It follows that such bacteria can grow in all anaerobic waters where they still have sufficient light for a positive assimilation balance. Kusnezow (1959) and Anagnostidis and Overbeck (1966) report that there is massive growth of purple sulphur and chlorobacteria in eutrophic lakes during the summer stagnation; it takes place in the upper region of the hypolimnion, immediately below the thermocline (Figure 59). Not infrequently, these bacteria turn up also in H_2S-containing pools of brackish water, in the water as well as on the surface of the sediment. If the radiation is strong, the chloro- and purple bacteria occupy a region not near the surface but a little deeper, as their life processes in general, and photosynthesis in particular, are inhibited by strong radiation. This is true also for most other phytoplankton species, which have their optimum depth in summer not at the surface but several metres below.

Photoautotrophic bacteria can thrive on very small amounts of light. Thus, purple bacteria can still develop at 50 lux and chlorobacteria even at only 5 lux (Biebl, personal communication). Gorlenko (1977) determined the most favourable light conditions in stratified lakes as between 500 and 3000 lux.

Some of the cyanophytes can also utilize quite small amounts of light for assimilation. With increasing depth of water, however, the light quality also changes so that chromatic adaptation is induced (see Chapter 4), which permits a better utilization of the available light for photosynthesis.

The damaging effect of light operates even more strongly on non-pigmented bacteria. It is due not only to the ultraviolet part of the spectrum — as was assumed for a long time — but also to the light of visible wavelengths. Blue light particularly of the wavelength in the range of 366–436 nm inhibits nitrite oxidation by *Nitrobacter winogradskyi* (Müller-Neuglück and Engel, 1961). The lower limit of inhibition lies at 200–300 lux. Red light, in contrast, has no inhibitory effect. In addition, heterotrophic bacteria like *Micrococcus denitrificans* are inactivated and finally killed by light (Mütze, 1963). Bock (1965) found that in *Nitrosomonas europaea* and *Nitrobacter*

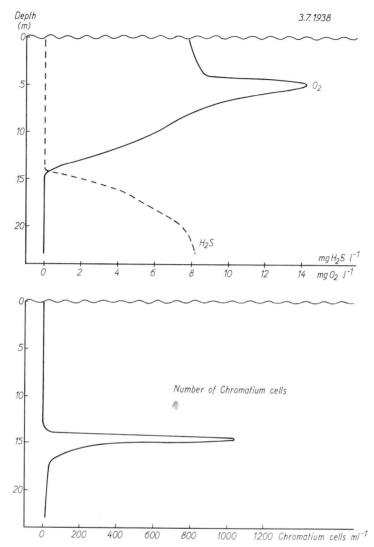

Figure 59   Vertical distribution of oxygen and hydrogen sulphide (top) and of *Chromatium* cells
(bottom) in Lake Belowod on July 30, 1938. (Based on Kusnezow, 1959, modified)

winogradskyi the vital cytochromes are destroyed, so that the cells finally perish.
Nitrobacter was killed by 54000 lux after only 4 h, but *Nitrosomonas*, on the other
hand, only after 24 h. Thus, the sensitivity of the individual species varies widely.
Harms and Engel (1965) found that denitrification by resting cells of *Micrococcus
denitrificans* decreased markedly with an illumination intensity of 58000 lux and
after 24 h had fallen to barely a third of that in non-illuminated parallel cultures.
In one experiment 99 per cent of the illuminated bacteria were killed after 16 h.

Bailey *et al.*(1983) studied the inhibiting influence of sunlight on amino acid uptake
by bacteria from Chesapeake Bay (USA) using microautoradiography in combination
with epifluorescence microscopy. An increase in solar radiation resulted in a decrease
in amino acid uptake and in the survival rate of the bacteria. The assimilatory-active
proportion of the population varied thereby between 93 and 20 per cent.

7   Rheinheimer, Microbiology, 3. A.

Pigmented bacteria which contain carotenoids are light-tolerant to a considerable degree, and are not inhibited by light of normal intensity. For this reason, air contains mainly organisms with strong red, yellow or orange pigmentation, for example *Sarcina aurantiaca*, *S. lutea* and various pigmented *Micrococcus* species. These organisms are also more resistant to ultraviolet rays than non-pigmented bacteria. An inhibitory effect on bacterial life in the surface zone of water must be assumed to operate where radiation from strong sunshine reaches the water; the neuston particularly is affected. In turbid waters, inhibition due to the radiation flux will occur only in the uppermost few centimetres, but in clear lakes and ocean regions the inhibitory effect of light may extend several metres down. The relatively low bacterial numbers found immediately under the surface in some regions of the sea may be due, amongst other things, to the effect of the strong radiation flux.

Bacteriological investigations carried out on the research ship 'Meteor' during the International Indian Ocean Expedition showed that the proportion of pigmented bacteria in the saprophyte flora of the relatively shallow Persian Gulf is considerably greater than of the much deeper Gulf of Oman (Rheinheimer and Gunkel, 1974). This might also be due to effect of radiation. Sieburth (1968a) found in the Narragansett Bay of the American atlantic coast a clear-cut correlation between radiation flux and the number of pigmented flavobacteria. Organisms with orange pigment had their maximum at the time of highest solar radiation, those with yellow pigment at the time of the lowest. Examining the effect of sunlight on pure cultures of non-pigmented, yellow or orange flavobacteria, it could be shown that the death rate due to sunlight decreased with increasing pigmentation.

Some aquatic fungi also are inhibited by light (Sparrow, 1968). Again, the effect of the blue and green part is greater than that of the red part of the spectrum. Thus, blue and green light suppress the formation of the oogonal rudiments in *Saprolegnia ferax* completely, red light only partially (Krause, 1960). A similar effect on growth and the formation of resting spores was observed in *Blastocladiella emersonii* and in *B. britannica* (Sparrow, 1968). Cultures of *Rhizophlyctis rosea* exposed to light produce considerably more pigment than if they are kept in the dark; and the breakdown of glucose by this fungus is impaired by light (Haskins and Weston, 1950). By contrast, no injurious effect of light on growth and reproduction could be observed in various other lower fungi (Sparrow, 1968).

Taken altogether, what we know of the influence of the light factor on the life of micro-organisms in water is still very little. All we know with some certainty is that sunlight may be inhibitory to non-pigmented bacteria and some fungi in lakes and in the sea within a shallow surface zone. The depth of this zone depends on the intensity of the radiation and the turbidity of the water. The effect is, therefore, greater in clear waters of arid regions than in very turbid waters exposed to less radiation, for example the North Sea. In addition, the light effect is modified by other factors, such as temperature. One should not, however, overestimate the importance of light, even in clear waters. The ultraviolet part of the spectrum can penetrate only a small distance, at most 1 m; blue light can penetrate considerably further, but its intensity falls so quickly within a few metres of water that an inhibitory effect on micro-organisms can hardly be possible at greater depths.

8.2. Temperature

The life processes of all micro-organisms are affected by temperature. Thus, bacteria, cyanophytes and fungi can grow only within a limited range of temperatures lying between the extremes —10 and +90 °C. Within this range, the temperature affects the growth rate, the nutritional requirements and, to a lesser extent, the enzymatic and chemical composition of the cells (Ingraham, 1962; Precht *et al.*, 1973). The cardinal points of the various life processes are not identical. Biosynthesis, for example, has sometimes a lower optimum than bioenergetics. In both cases, however, the optimum is nearer the maximum than the minimum. Three groups of bacteria may be distinguished by their temperature requirements:

	Minimum (°C)	Optimum (°C)	Maximum (°C)
Psychrophils or cryophils	—10 to +5	+10 to +20	+13 to +25
Mesophils	+10 to +15	+30 to +40	+30 to +50
Thermophils	+25 to +45	+50 to +75	+75 to +100

These groups, however, are not sharply distinguished, and the boundaries between them are hazy. Moreover, many micro-organisms may adapt to higher or lower temperatures. Under increased pressure, some bacteria may even develop at temperatures above 100 °C. The cardinal points are also affected by other factors, such as supply of nutrients, salinity, pH, products of metabolism, etc.

Psychrophilic bacteria are very conspicuous in by far the most extensive part of the sea and inland waters of the cold climatic zones where temperatures of below +5 °C predominate. Their optimum is frequently between 10 and 15 °C. As a rule, they have a maximum at 20 °C or below. Morita (1975) even found individual bacterial strains with a maximum of 13 °C. Of course, these may be exceptions, as psychrophilic bacteria can develop also in waters of temperate climatic zones. Temperatures above 25 °C though are, as a rule, lethal. Some psychrophils die after only a few hours at temperatures of 18–20 °C. For this reason, the isolation of obligate psychrophilic bacteria is difficult and in early experiments was frequently unsuccessful. This may also be the reason why these organisms have been often overlooked and described as rare in the older literature. More recent investigations by Rüger (1982) showed that, with increasing depth and correspondingly lower temperatures, the proportion of psychrophilic bacteria clearly increased in the sediment from areas of upwelling water movements off north-west Africa. At depths of about 1500 m and at 7–12 °C, their number was 4.5 to 9.6 times greater than that of the mesophilic bacteria.

Also, the widespread idea that all psychrophils have a very long generation time is incorrect, as Morita and Albright (1965) point out for *Vibrio marinus*; this organism has a generation time of 80.7 min at 15 °C and of 226 min at 3 °C. Morita and Haight (1964) drew up growth curves which showed that the strain of *Vibrio marinus* examined had its temperature optimum at 15–16 °C, and failed to grow at 20 °C; at temperatures above 30 °C the cells perished after a short time (1.5 h).

In inland waters of the warmer zones and in sea water near the surface, psychrophils are scarce, and mesophilic bacteria and fungi predominate; but Sieburth (1968a) found, during investigations in the Narragansett Bay, that in temperate climatic zones when the water becomes cold in winter the proportion of mesophils declines in favour of psychrophils.

Thermophilic organisms have no great importance in waters. They can accumulate only in thermal springs. Here, at temperatures above 50 °C only some few species of bacteria and cyanophytes live (see Section 7.1). Thus, the coccoid sulphur bacterium *Sulfolobus acidocaldarius* develops in hot sulphur springs at temperatures of 65–93 °C. The optimum is 75 °C (Weiss, 1973). This facultative chemoautotrophic micro-organism, therefore, might well possess the highest temperature requirements of all known living creatures.

Temperatures exceeding the maximum cause quick death, as the cytoplasm suffers irreversible damage. By contrast, temperatures lower than the minimum are rarely lethal; often metabolism just comes to a halt and many micro-organisms may remain in this state of suspended life for a very long time. Nevertheless, freezing may cause death in a more or less large proportion of the affected bacteria, cyanophytes and fungi. The cause of death by freezing is not entirely clear. Often it is assumed that it is due to the formation of fine ice crystals in the cells. This may be true for fungi but, as Ingraham (1962) points out, there is no certain evidence that this occurs in bacteria. Microscopic examination revealed neither ice crystals nor any mechanical damage in the cells. The absence of vacuoles in most bacteria may be relevant in this context. The speed of freezing may affect the survival rate less than was originally assumed; the attempt was made, therefore, to explain death through cold by a rise in the osmotic pressure of the freezing suspension. This would be particularly important when the temperature drops only a little below freezing, and might explain why the bacterial content of the ice on frozen waters changes rather rapidly; below —40 °C hardly any further death occurs. Rapid warming up does not cause as big a decrease in microbe numbers as a slow rise in temperature.

The size of the temperature range within which individual kinds of micro-organisms can live differs a great deal. It may amount to 15 °C or to more than 50 °C. Within the temperature range for viability of an organism, increasing temperature promotes its biological reactions. As in chemical reactions, van't Hoff's law of a doubling or trebling of the rate of reaction for every 10 °C rise in temperature holds true. Thus the generation times of a heterotrophic bacterium — not specified — are as follows, at five different temperatures (Christophersen, 1955):

0 °C	18.4 h	25 °C	0.773 h
6 °C	7.0 h	30 °C	0.695 h
12 °C	2.71 h		

Temperatures near the minimum or the maximum may cause morphological changes in various micro-organisms. Hoffmann (1966) states that *Escherichia coli* produces filamentous cells at 7 °C. *Agrobacterium luteum* grows optimally at 25 °C and shows filamentous growth at 30 °C (see Figure 60 and Ahrens, 1968).

Sieburth (1968a) established a temperature-dependent life cycle in *Arthrobacter*, with a gram-negative myceloid stage below 20 °C, gram-positive arthrospores between 20 and 26 °C, and a gram-positive coryneform stage above 26 °C.

In pure cultures under optimal conditions, the effect of temperature on the biological reactions of bacteria *in vitro* is entirely unambiguous; in nature, however, it can be observed only with difficulty or not at all. A great number and variety of living creatures are present whose life processes interact, and the factor of competition thus introduced does not always allow the temperature effect characteristic of the particular species to assert itself. Furthermore, different temperature effects may be superimposed on one another. This may be particularly pronounced in waters with a heavy sewage load; a rise in temperature results in increased activity and a reduction of the generation time, but toxic effects are also increased and autolysis is accelerated. Micro-organisms for which the living conditions in a particular water are favourable will quickly multiply at summer temperatures; but others for which they are unfavourable will quickly perish, as can be observed with many bacteria from sewage, fresh water or soil when they get into brackish or sea water. At low winter temperatures, on the other hand, all reactions are slowed down, so that the time of survival of these bacteria in an alien environment is prolonged. This explains why the proportion of bacteria which grow in meat extract peptone agar from sewage-loaded lakes, rivers and coastal waters is always higher in winter than in summer (Rheinheimer, 1965a; 1970b). Numerous other micro-organisms, though, multiply vigorously at summer temperatures. The seasonal temperature fluctuations, therefore, cause a change of population in any case. The total bacterial count often rises in winter as the water

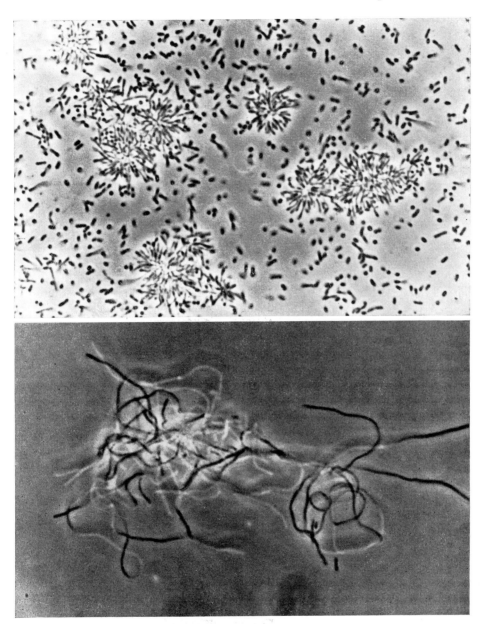

Figure 60   *Agrobacterium luteum* from yeast extract peptone medium. Top: normal cell morphology at 20 °C; Below: formation of long filamentous cells at 30 °C. (R. Jeske)

temperature drops in strongly polluted waters but decreases in clean waters (see Section 15.3). The effect of temperature on the microflora of lakes, rivers and seas can, therefore, be assessed only when all other factors are precisely known.

8.3. Pressure

The hydrostatic pressure rises by about 1 atm ($= 101,325$ kPa) for every 10 m depth. Accordingly, extraordinarily high pressures are obtained in the depths of the sea and of some lakes; in the deep Pacific trenches they may reach more than 1100 atm. ZoBell (1946a) states that around 90 per cent of the seas of the world are deeper than 1000 m and therefore have hydrostatic pressures of more than 100 atm, and, in 24.5 per cent of the seas even more than 500 atm. The pressure-depth gradient is between 0.099 and 0.105 atm per metre, depending on the density of the water, and is about 5 per cent greater in the deepest parts of the sea than in shallow waters.

The high pressure in the deep sea is an important ecological factor which considerably influences the life of micro-organisms.

ZoBell and Morita (1959), during the 'Galathea' deep sea expedition, found numerous viable bacteria at depths of more than 10000 m. In sediment from the bottom of the Philippine trenches, 10000 to 1 million organisms per g wet weight were counted. Some of these deep sea bacteria, which have an optimum above 500 atm, grow either not at all or very badly at normal atmospheric pressure; ZoBell and Johnson (1949) suggest the term barophils for these strains. In addition, there are organisms which also grow well at atmospheric pressure; these are called barotolerant. Most soil and fresh water bacteria do not grow at pressures of more than 200 atm, and this is true also for marine bacteria and lower fungi isolated from shallow areas of the sea; they are, therefore, called barophobic.

Barophilic bacteria of the deep sea are, as a rule, also psychrophilic (see Section 8.2) and grow best at high pressures and low temperatures (3–5 °C), i.e. they are in large measure adapted to the conditions of their habitat. Morita (1967) states that most barophilic bacteria grow very slowly. Thus, with sulphate reducers from the Sunda deep which were kept at 700 atm and 5 °C, sulphate reduction could be demonstrated only after 10 months. In controls at 1 atm and 5 °C it could not be shown even after several years. It is astonishing that many barophilic bacteria survive decompression to 1 atm (ZoBell, 1964a). Nevertheless, one must accept as a fact that some of the more delicate forms cannot be brought to the surface alive with the usual sampling equipment. Apparently contradicting this were laboratory experiments which showed that cells of *Escherichia coli* survived unharmed decompression following immediately on compression to 1000 atm within a few minutes, nor was there any decrease in the number of viable cells. Various other soil and marine bacteria behaved in the same way. Most of them, however, died if kept at 1000 atm for several hours.

Some bacteria sensitive to pressure showed morphological changes at very high pressures (ZoBell and Oppenheimer, 1950). *Serratia marcescens* grows optimally at atmospheric pressure as a short motile rod. At 600 atm it grows only very slowly, loses its motility and produces filaments up to 100 μm long, though with a normal diameter. Similar observations were made with *Bacillus* and *Vibrio* species and with *E. coli*. If culturing was then continued at normal atmospheric pressure, the original size and shape of the cells were restored after a few hours or days. From this, it seems that high pressure interferes with the normal mechanism of reproduction, as the bacteria go on growing but do not divide. Heden (1964) suggests that this phenomenon is associated with changes in the volume of the DNA molecules which, in turn, interferes with DNA replication so that cell division is stopped. The DNA content of the biomass under greatly increased pressure is smaller than under atmospheric pressure, while the RNA content is increased, and the relative proportion of protein and nucleic acids remains unchanged (ZoBell and Cobet, 1964).

Barophilic deep sea bacteria, by contrast, are able to synthesize DNA even at very high pressures, and can thus reproduce normally at great depths. Motile forms are flagellated only at high pressures and lose their flagella at atmospheric pressure.

The effects of hydrostatic pressure on the physiology of bacteria and fungi are many and varied. The luminescence of luminous bacteria is an example. Cultures of *Photobacterium fischeri, P. harveyi* and *P. phosphoreum* are killed within a few days at pressures of more than 500 atm. Pressures of 50–500 atm, at temperatures a few degrees above the optimum, promote luminescence, while at temperatures below the optimum luminescence is inhibited. It can also be shown that pressures increasing up to 330 atm prevent at 34 °C the thermal destruction of the luminescence system of *P. fischeri* which has a temperature optimum of 21 °C. This fact is important in nature only in the regions of deep sea thermal springs, as elsewhere the water temperatures throughout are considerably lower (see Section 2.2). High pressure at optimal temperature also may cancel out inhibition due to various chemicals if the inhibitory agent is not too overwhelming. The effect of pressure on the luminescence of luminous bacteria is an effect on the appropriate enzyme system (Brown *et al.*, 1942; Johnson and Eyring, 1948; Johnson *et al.*, 1954).

Other enzyme systems are similarly affected, to a greater or lesser degree, by pressure. Thus, succinic acid dehydrogenase is completely and irreversibly inactivated at 1000 atm in *E. coli* (Morita and ZoBell, 1956) but not in barophilic deep sea bacteria. Ureases are not inactivated at 1000 atm, but their activity is slowed down (ZoBell, 1964a). Also, bacterial nitrate reduction is affected by pressure, except in some obligatory barophilic bacteria from the greatest depths of the sea; it is slowed down with increasing pressure due to inactivation of the enzyme, nitrate reductase, and to reduction of the velocity of reaction (ZoBell and Budge, 1965). The rate of inactivation is greater at 10 °C than at 40 °C. Other metabolic processes, for example methane and hydrogen formation, sulphate reduction and phosphatase activity are also slowed down by high pressures (ZoBell, 1964a).

In barophobic procaryotic micro-organisms, protein synthesis is particularly sensitive to the increase of hydrostatic pressure (Pope and Berger, 1973). This results in a corresponding inhibition of growth.

Jannasch and Wirsen (1977) studied the uptake and respiration of an amino acid mixture by bacterial populations which had been removed, without decompression, from a depth of 2600 m in the sea by means of a suitable apparatus. They found that both processes proceeded more rapidly at atmospheric pressure than at the site pressure of 260 atm. Clearly, they were dealing with a population of barotolerant bacteria.

The microflora of deep sea animals, on the other hand, shows high metabolic rates. In this case, the barophilic forms play a greater role (Morita, 1979b; Colwell *et al.*, 1983) (see Section 11.3).

The effect of hydrostatic pressure on the biological processes of micro-organisms determines, then, in what regions of the sea they will find their proper niche, i.e. whether they can live only in the upper or only in the deeper zones, or in either.

8.4. Turbidity

The turbidity of the water also affects the life of aquatic micro-organisms. It is caused by the seston, which is defined as the total of living and dead material suspended in the water, which finally leads to the formation of sediment. The seston consists of the following three components (Dietrich *et al.*, 1975):

1. The fine particles of mineral material which originate on land and are transported from land to the waters;

2. Detritus consisting of inorganic and — predominantly — organic material finely ground up;

3. Plankton — plant and animal creatures of small size, floating in the water.

As the distinction between the fine particles of mineral material and detritus is often hardly possible, these two components are sometimes collectively called tripton.

The turbidity can be determined optically by means of any apparatus which measures transparency. A good indication is provided in the first instance by measuring the depth of visibility with a so-called Secchi disc. Often the quantity of seston is also determined gravimetrically.

Turbidity differs very markedly in individual waters. In clean streams, oligotrophic lakes and in the open ocean it is generally very low, whereas it can be extraordinarily high in some river mouths. Thus, in the north Atlantic, for example, visibility depths of 20–30 m have been measured; in the lower Elbe and lower Weser, on the other hand, less than 0.5 m. Many waters in the temperate zone show considerable seasonal variation, and this is also true for the relative proportion of tripton and plankton. Also, the proportion of organic matter in the seston may vary greatly. In springs and some rivers, it can amount to a few per cent, or to almost 100 per cent in certain lakes and sea areas.

The seston plays an important role as substrate for many micro-organisms. Detritus particles in particular frequently carry a surface flora of numerous bacteria and fungi (see Figure 61 and Melchiorri-Santolini and Hopton, 1972). Not only are the organic components, which may be directly used as food, colonized by micro-organisms, but also inorganic particles. Suspended particles, whether of organic or mineral origin, adsorb to their surfaces nutrients which are present in the water in very high dilution only; in consequence, the microbes find a more favourable nutritional environment here than in free water. This is the more pronounced, the poorer in nutrients the surrounding water.

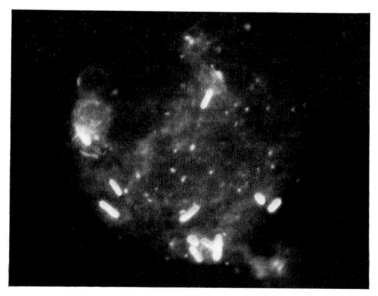

Figure 61   Detritus particles from the Kiel Fjord with bacteria (× about 4000). (R. Zimmermann)

Occasionally, coccoid starvation forms of saprophytic bacteria, which have higher nutrient requirements than the oligocarbophilic forms occurring in the water, first invade. Under favourable conditions they quickly grow into larger rods (see Figure 61).

In recent years, much greater attention has been paid to the physicochemical reasons for the attachment of micro-organisms to solid substrates. Accordingly, there are hydrophilic and hydrophobic bacteria which show correspondingly different behaviours towards attachment. Surface active molecules evidently play a role in attachment processes. The cells become surrounded by a polymer coating (e.g. lipopolysaccharides) which, at the same time, acts as a protection against unfavourable, especially chemical, influences. Consequently, the activities of attached bacteria differ from those of free-living micro-organisms in the water (Fletcher and Marshall, 1983; Kjelleberg et al., 1982; Marshall, 1976).

Inhibitors and poisons can also be rendered harmless by adsorption on detritus particles. In this way, suspended matter has a promoting action, especially on bacterial development. In addition there is also a certain protective effect against the harmful influence of light (see Section 8.1). This is especially the case in the detritus-rich turbidity zones occurring in the estuaries of different rivers.

In an area of maximum turbidity, for example in the Elbe estuary between Glückstadt and Brunsbüttel, the depth of visibility is reduced to just a few centimetres, so that only the organisms immediately below the water surface are exposed to light to any extent. Here one also encounters considerable tide-dependent fluctuations in the quantity of suspended matter; sedimentation takes place during the slack of the tide, and the sediment is stirred up by the currents of high and low tide. These fluctuations also affect the vertical distribution of bacteria (see Section 7.3) and cause a remarkable parallelism between turbidity and bacterial content of the river water (Koske et al., 1966). Such a parallelism was also found in the western Baltic (Figure 62), and it is clear-cut in the zones of the thermocline and of sudden discontinuity in salinity of lakes and coastal waters (Rheinheimer, 1970c). However, increase or decrease of detritus is not always accompanied by a similar fluctuation of the bacterial numbers. In less dynamic waters differences in turbidity do not always affect the bacterial counts as long as the nutritional conditions remain steady. It may be concluded, therefore, that an increase in turbidity accompanied by a vigorous rise in the bacterial counts is due, at least in part, to an increase in the amount of suspended organic matter. If, on the other hand, bacterial numbers change little with changes in turbidity, the latter is predominantly caused by inorganic suspended particles. Comparison of turbidity measurements and total microbial counts thus permits one to draw some conclusions with respect to the kind of substances responsible for the turbidity.

The proportion of micro-organisms growing on detritus of the total microflora can be very variable. Thus on the average only a few per cent of attached bacteria were found in detailed investigations in the relatively clear water of the Kiel Fjord — in contrast, more than 50 per cent of the total bacterial numbers were found on detritus in the strongly turbid waters of the mouth of the Elbe (Zimmermann, 1977).

In the upper Bay of Fundy, in south-eastern Canada, the proportion of attached bacteria increased, even to 94 per cent of the total bacterial numbers. There was also a strong correlation with the turbidity (Cammen and Walker, 1982). Bell an Albright (1981) found large differences in bacterial biomasses and heterotrophic activity in the estuarine area of the River Fraser in British Colombia (Canada). There, the proportion of bacteria, about 60 per cent, occurring on detritus in the very turbid river was distinctly higher than in the clearer water of the Georgia Strait.

Turbidity also influences the composition of the microflora of waters. Thus, according to observations by Kang and Seki (1983) in a Japanese sewage pond, the proportion of gram-positive forms of 37.6 per cent in the yearly average of attached bacteria

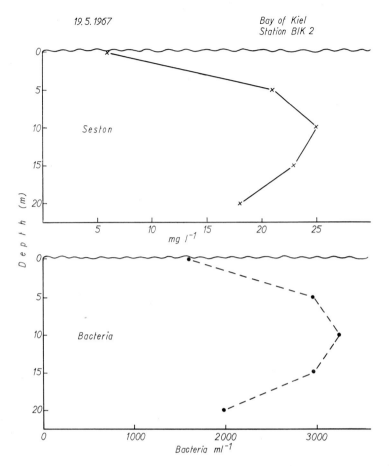

Figure 62   Vertical distribution of seston and of bacteria (saprophyte numbers) in the Kiel Bay
often shows good correlation

was distinctly greater than the 19.8 per cent of the free-living forms. Seki and Taga
(1963) found more chitin-degrading bacteria on detritus than on plankton or in free
sea water in the Aburatsubo Bay in Japan. The same is probably true for cellulose-
decomposing micro-organisms. Other bacteria, by contrast, prefer free water or
grow in the slime of plankton algae. On the whole, the effect of the turbidity factor
is exercised mainly indirectly by way of its influence on nutritional and light conditions.

8.5. Hydrogen Ion Concentration and Redox Potential

The growth and reproduction of micro-organisms is much affected by the hydrogen
ion concentration (i.e. the pH) of the medium. Most bacteria can grow only within
the range pH 4–9. Only very few can still grow at pH 3 or below, as for example the
acidophilic *Thiobacillus* species *thiooxydans* and *ferrooxydans*, which tolerate a reaction
of pH 1. There is also the thermophilic *Sulpholobus acidocaldarius*, which is found
in hot sulphur springs with pH values of between 1.6 and 3. The pH range lies between
0.9 and 5.8, with the optimum at pH 2–3 (Weiss, 1973). The optimum for most other

aquatic bacteria is between pH 6.5 and 8.5. This corresponds to the pH range of most of the larger bodies of water. Thus, the average reaction of many lakes is pH 7, that of many large rivers, for example the Elbe, is pH 7.5 and the surface zone of the sea 8.2. There may, of course, be deviations of varying magnitude. Thus, the Schlei Fjord (Baltic) at a time of vigorous plankton bloom had a pH of 9.5 (Nellen, 1967). Relatively large fluctuations of the pH are found also in eutrophic lakes, where the pH may vary between 7 and 10. This naturally affects the composition of the bacterial and fungal populations.

In the water of the Pacific, Maeda and Taga (1980), in addition to alkali-tolerant bacteria showing very good growth at pH 7.3–10.6, also found alkaliphilic bacteria which developed only at pH 10.0–10.6. Studies of the oligotrophic Adirondack lakes in the mountains of the North-American State of New York showed that, at pH 5 and pH 7, quite different bacterial populations occur. Of 1200 isolated strains, less than 10 per cent grew at pH values below 5. Bacteria isolated originally at pH 5 also developed at pH 6 and 7. Bacteria isolated at pH 7, on the contrary, preferred this reaction. Some 98 per cent also grew at pH 6, but only 44 per cent at pH 5 (Boylen et al., 1983).

There are more acidophilic fungi than bacteria; consequently, in acid waters and sediments — as also in soil — the proportion of fungi in the microflora is greater than in those with neutral or slightly alkaline reaction. The majority of aquatic fungi can grow in a relatively wide pH range. Nevertheless fresh water fungi are more often encountered at pH 7.0 or below, whereas marine fungi prefer a weak alkaline medium (optima at pH 7.5–8.0).

Gunkel et al. (1983), however, on decomposing brown alga, Desmerestia viridis, in a model aquatic system, observed a rapid increase in yeasts to more than 1 million per ml of water, with bacterial numbers simultaneously declining. This could be the result of a low pH value (about 1), due to the high sulphuric acid concentration of the cell sap of this alga. Also on natural sites of different algae, when they decompose, a rise in yeast numbers was observed concomitant with a lowering of the pH.

More pronounced deviations from the optimal hydrogen ion concentration cause not only physiological but, not infrequently, also morphological changes. Some micro-organisms tend to produce involution forms whereby cells that are normally rod-shaped become enlarged and show irregular swelling and branching.

The oxidation-reduction (= redox) potential of waters and sediments is also of considerable ecological importance. It is a measure of the readiness to part with electrons. The redox potential E_h represents the difference of potential between the medium and a hydrogen electrode and is usually given in mV (millivolts). The rH value represents the negative logarithm of the hydrogen ion concentration producing the existing oxidation-reduction conditions. It thus involves both the E_h and the pH values:

$$rH = \frac{E_h}{0.03} + (2 \times pH)$$

Kusnezow (1959) determined the redox potential of various lakes in the USSR. In the epilimnion at a pH of 5.7–7.2 it varied between +410 and 510 mV, in the hypolimnion at a pH of 5.6–6.5 between 120 and 459 mV. In the surface water of the Atlantic, Cooper (1937) found E_h values of 402–435 mV, at pH 7.62–8.15.

For marine sediments ZoBell (1946b) reports E_h values of between +350 and —500 mV, at pH 6.4–9.5.

As a rule, definite physiological groups of micro-organisms can grow only within a particular range of E_h (Figure 63). Thus aerobic bacteria require higher E_h values than anaerobes. The activity of micro-organisms alters the redox potential to a varying extent. Bacterial consumption of oxygen, for instance, always leads to a

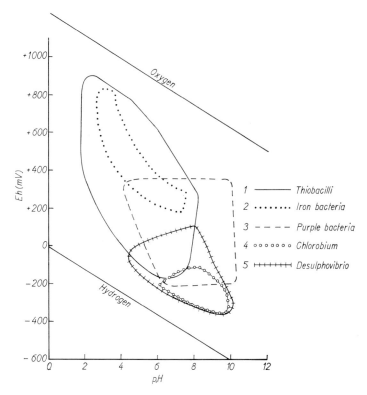

Figure 63   The pH and E_h range of some groups of aquatic bacteria. (Based on Baas Becking and Wood, 1955, modified)

lowering of the E_h value. This is particularly pronounced if H_2S is produced (Figure 64). On the other hand, a supply of oxygen causes a rise in the E_h value (Jakob, 1970).

8.6. Salinity

The degree of salinity determines to a particularly large extent the living communities in waters. The relatively high NaCl concentration of sea water (Section 2.2) led to the development of physiologically different fresh water and sea organisms. Only few living creatures can thrive both in fresh water and in the sea. This holds for bacteria and fungi just as for green plants and animals. Accordingly, most micro-organisms in clean lakes and rivers are more or less halophobic and cannot, under natural conditions, grow in waters with more than 1 per cent salt. Only relatively few are salt tolerant and thrive at higher salinity. On the other hand, the over-whelming majority of marine bacteria and fungi are halophilic, i.e. they require a definite amount of salt for their growth and cannot, therefore, develop normally in fresh water habitats. The decisive factor is not, as was formerly assumed, the higher osmotic value of sea water; it is the Na^+ ions which are the vital necessity for most marine creatures and for some, in addition, Cl^- ions (MacLeod, 1965, 1968). As halo-philic marine bacteria contribute 95 per cent to the saprophyte numbers in the open sea, the proportion of salt tolerants then is only 5 per cent, or even less. On media prepared with fresh water the bacterial growth is often less than 1 per cent of that

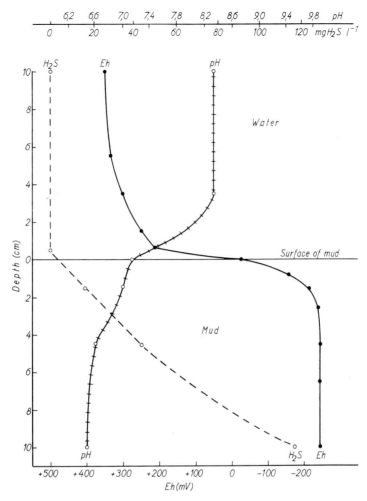

Figure 64   E$_h$, pH and H$_2$S content in a beach sulphuretum on the Baltic islands of Hiddensee. (Based on Suckow, 1966, modified)

obtained with the corresponding samples on sea water media. Only very near the coast is the proportion of salt tolerants greater; but even in the heavily polluted Kiel Fjord it rarely rises above 20 per cent (Rheinheimer, 1968a). Relatively high proportions were found, by contrast, in an area of upwelling water movement on the West African coast during the Rossbreiten expedition of the research ship 'Meteor' in 1970. Rüger (1975) established that, in sediments of the north Atlantic at depths of more than 1000 m, salt-tolerant *Bacillus* species occur abundantly; these species develop well in fresh water media.

Salt-tolerant bacteria in larger or smaller numbers are found in almost all inland waters, but they are particularly numerous in urban sewage and in badly polluted rivers and lakes. Here are also found osmophilic bacteria, i.e. those which develop optimally at increased osmotic values, but require neither Na$^+$ nor Cl$^-$ ions. Osmophilic bacteria also occur in smaller numbers in brackish water and in sea water. Meyer-Reil (1973), for example, was able to isolate osmophilic strains from the Kattegat and the Baltic Sea.

Turbidity

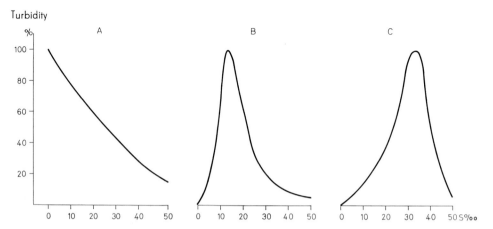

Figure 65   Growth curves in salt solutions for (A) a halotolerant fresh water bacterium, (B) an obligate halophilic brackish water bacterium and (C) an obligate halophilic marine bacterium, all isolated from the Baltic Sea. (Salinity, S, given in parts per thousand)

The salt optimum for most halophilic bacteria (ZoBell and Upham, 1944) and fungi lies between 2.5 and 4 per cent (Figure 65). This range includes the natural salt content of the sea which is, on average, in the open sea, 3.5 per cent. In areas of brackish water like the Baltic, there are in addition, halophilic organisms with salt optima between 0.5 and 2 per cent (Rheinheimer, 1970a).

Salinity which deviates to some degree from the optimum prolongs the generation time in all bacteria and fungi. Often morphological and physiological changes are also observed. Thus, some marine bacteria which are rod-shaped or comma-shaped at optimal salinity become much longer at salt concentrations of more than 5 per cent and turn finally into filaments. Luminous bacteria isolated by the author from the Arabian Sea grew optimally at about 3 per cent salinity as slightly curved rods, 1–2 μm long, at 1 per cent as cocci and at 7.5 per cent as filaments up to more than 100 μm long. The rise in salinity — like a rise in pressure (see Section 8.3) — interferes here with the normal reproductive mechanism. The cells can still grow but are unable to divide. In phycomycetes, too, reproduction is impaired sooner than the vegetative growth (Johnson and Sparrow, 1961).

The physiological effects of the salt concentration are also very diverse. Luminous bacteria in more than 50 per cent dilution of sea water lose their luminescence. At further dilution this loss becomes irreversible, and finally the cells lyse. McElroy (1961) regards this effect of NaCl on luminescence as due to inhibition of luciferase synthesis; luminous bacteria in media with only 1 per cent NaCl contain only little of this enzyme, but after transfer to a medium with 3 per cent NaCl vigorous luciferase synthesis sets in. The NaCl may be replaced to a considerable extent by other salts or even by organic substances, but a certain quantity of sodium chloride is indispensable. Also other metabolic activities, for example oxidation of organic acids and sugars or production of indole, are encouraged by certain salt concentrations (MacLeod, 1965), but this effect cannot be observed in cell-free suspensions; indeed, some enzymes involved in these reactions are inhibited at salinities optimal for the intact cells. Some free enzymes, such as aconitase and isocitrate dehydrogenase, require a raised osmotic pressure, but this can be supplied by K^+ or the ashes of the bacterial extract. As in the intact cells Na^+ improves oxidation of the intermediate products of the tricarboxylic acid cycle but has no such effect on the isolated enzymes, it may be concluded that the Na^+ ion plays a role in the transport of substances through the cell membrane, possibly by forming complexes, for example with intermediates of the tricarboxylic acid cycle (MacLeod, 1965).

Many (but by no means all) marine bacteria lyse when transferred to fresh water. MacLeod (1968) found that the lowering of the salt concentration destroyed the mucopeptide layer in the cell wall of a marine *Pseudomonas*; this weakens the wall sufficiently for the osmotic pressure in the cell to burst the remaining wall layers.

Beside the weakly halophilic bacteria of brackish water and the sea, a small number of organisms exist with much higher salt requirements. Larsen (1962) subdivides them into moderately halophilic organisms, with optima of 5–20 per cent, and extreme halophilic organisms, with optima of 20–30 per cent (see Section 3.1). Members of both these groups occur in certain coastal areas, for instance in splash ponds and salt pans, and also in inland salt lakes.

Whereas most micro-organisms grow only within a rather narrow range of salt concentration (1–5 per cent), a small group of bacteria and fungi can grow over a very wide salinity range (15–25 per cent). These are, on the one hand, organisms which grow optimally in fresh water, and, on the other hand, organisms with an optimum between 2 and 4 per cent. The author's own observations show that bacteria belonging to the latter group often occur abundantly in sandy beaches and shallows and may there represent the largest group of bacteria. Because of the strong fluctuations in salt concentration caused by alternating drying up, flooding and washing out by rain, they have an advantage over most micro-organisms.

The salinity range of bacteria is widest at their particular optimal growth temperature, but is narrowed down in varying degree by higher or lower temperatures. Temperatures above the optimum cause an increase of the NaCl requirements, temperatures below the optimum a reduction (Meyer-Reil, 1972).

Under culture conditions fresh water bacteria may adapt to a higher salt concentration and originally halophilic marine bacteria to a lower one (ZoBell, 1946a). Marine *Desulphovibrio* strains, for instance, after appropriate adaptation, can grow in fresh water and fresh water strains in sea water (Baars, 1930). How far this may be possible under natural conditions cannot, at this stage, be judged. Besides microorganisms with adaptable salt optima, there are many others whose optima remain constant after years of culture in higher or lower salt concentrations (MacLeod, 1965; Rheinheimer, 1970a). According to Meyer-Reil (1972), adaptation is largely restricted to within certain salinity ranges characteristic for the particular ecological groups, i.e. the bacteria with the greatest salt tolerance are, as a rule, also the most adaptable.

Of the lower fungi (Phycomycetes), except for some parasitic forms such as *Dermocystidium marinum* and *Haliphthoros milfordensis*, the Thraustochytriaceae, which are closely related to the Saprolegniales, should be regarded as obligately marine organisms. They can grow only in salt water (optimally at about 2.0–3.5 per cent salinity) and they require at least 0.8 per cent salinity for reproduction (Jones and Harrison, 1976).

On the other hand, of the Saprolegniaceae isolated by Harrison (1972) from the sea, only two species formed sexual organs at a salinity of more than 1 per cent, whilst zoospores were produced only at below 1 per cent salinity. According to Höhnk (1953), several *Pythium* species were able to develop fully in a wide range of salinities. Testrake (1959) supposed that, besides hydrographic factors, the nutrient supply also influenced the reaction of lower fungi to particular salt concentrations.

Higher fungi (Ascomycetes and Deuteromycetes) of terrestrial or marine origin are capable of reproduction, apparently, in a considerably wider salinity range than the lower fungi mentioned. Naturally, the optima for terrestrial fungi are likely to be found in fresh or brackish water (up to 1 per cent salinity) whereas those from marine regions generally developed better at higher salt contents (Jones and Byrne, 1976; Henningson, 1978).

8.7. Inorganic Substances

The life of micro-organisms in water is affected by other inorganic substances besides NaCl. Of particular importance are the inorganic nitrogen and phosphorus compounds which, in the productive zone of many waters, represent the limiting factor for plant life. In oligotrophic lakes and in regions of the sea poor in nutrients, ammonia, nitrite, nitrate and phosphates can hardly be detected because, as soon as they are released, they are immediately bound again by the phytoplankton. Under these conditions competition may arise for these inorganic nutrients between bacteria and planktonic algae. In the depths of large lakes and of the sea, however, the activity of hetero-trophic micro-organisms causes enrichment of nitrate and phosphate; consequently, regions where water from the deeper parts which are rich in nutrients wells up to the surface show high productivity and a more abundant growth of bacteria and fungi.

While in the photic zone of tropical and subtropical seas there is a shortage of nitrogen- and phosphorus-containing compounds all the year round, in the temperate climatic zones marked seasonal fluctuations are observed. Thus, in the western Baltic, the nitrate content of the water rises markedly in late autumn and winter and declines dramtically in March or April because of the development of phytoplankton (Rhein-heimer, 1967). Ammonia and nitrite play an important role in the supply of energy for nitrifying bacteria, and the oxygen bound in nitrate can be used by the numerous bacteria capable of denitrification under anaerobic conditions for the oxidation of organic material (see Section 11.5). The other inorganic nutrients necessary for micro-bial life are present in most waters in sufficient quantities. Trace elements like iron and cobalt are needed — although in very small amounts — as constituents of impor-tant enzymes; during the bloom of algae a shortage of these substances may arise in some waters (Wood, 1965).

Heavy metals, on the other hand, may impair life in waters, as some of them are toxic for many micro-organisms even in relatively small concentrations (see Section 15.3). Ions of heavy metals or organic heavy metal complexes seem to contribute to the bactericidal effect of salty sea water for non-marine creatures (MacLeod, 1965), and likewise to the inactivation of viruses (Mitchell, 1972). Not infrequently, copper and mercury get into rivers, lakes and coastal waters by way of industrial sewage and waste, and can completely destroy their natural living communities. The toxic effect of these metals is due to the binding of the sulphydryl (—SH) groups of enzymes. According to Cobet et al. (1971) nickel causes a distinct alteration in the cell size and fine structure of *Arthrobacter marinus*. At a nickel concentration of 4×10^{-4} M, the rods which are normally $2 \times 4 \mu$m in size become spherical and enlarged to a diameter of $10–15 \mu$m on account of severe plasmolysis. In addition the innermost mucopeptide layer of the cell wall disappears.

Cyanides which also occasionally get into water destroy plants and animals. The CN group binds iron, thereby blocking the cytochrome oxidase, and is, therefore, a powerful respiratory poison (Schlegel, 1981).

8.8. Organic Substances

Organic substances, dissolved or suspended in water, are particularly important as food for heterotrophic micro-organisms (see Seki, 1982). The size and composition of the bacterial and fungal populations of a water depend to a large degree on the concentration and composition of those substances. Organic compounds have, how-ever, other important roles, as activating and inhibiting factors.

The water of oligotrophic lakes, clean rivers and wide regions of the sea contains only very little organic material: in the open ocean, for example, 0.3–2.0 mg l^{-1}, of which about 10 per cent is present in particulate form. This concentration of nutrients is for most bacteria and fungi at the lower limit of what can be taken up and utilized. Near the coast the content of organic substances in the water rises. In the Baltic, average values of 2.7–6.5 mg organic substance per litre (2–5 mg C per litre) were found (Ehrhardt, 1969). In polluted bays and harbours these values may be higher still, but they are probably always below 100 mg organic substance per litre. This is true also for most inland waters. Birge and Juday (1934) found in American lakes between 3 and 50 mg organic matter per litre of water. The investigation of more than 500 lakes in the state of Wisconsin (USA) gave average values of 15.2 mg l^{-1} for dissolved organic matter and 1.4 mg l^{-1} for particulate matter. That corresponds to a ratio of 11: 1 (Wetzel, 1975). The dissolved organic matter is composed of

  83.7 % carbohydrates
  15.6 % protein ('crude protein')
   0.7 % lipids

Steinberg (1977) in a comparative investigation of four lakes in east Holstein (Germany), established that the peptide fraction (< 10000 mw) constituted the largest portion of the amino nitrogen (43–93 per cent). Larger amounts of free amino acids were found in only two of these lakes. The composition of the organic substance in the individual lakes can be very variable and a more or less large part of it consists of substances that are difficult to degrade. The easily assimilated substances, like protein, carbohydrates and organic acids, are normally decomposed in the water or on the surface of the sediment (see Section 11.2).

According to Overbeck (1982), the concentration of bound carbohydrate in Lake Pluss is distinctly higher than that of free carbohydrates, as the following example of a gas chromatographic determination of carbohydrates in a water sample from 1 m deep on May 23, 1976, shows:

	Free	Bound
D-Ribose	1.05	12.46 µg C per litre
D-Mannose	3.48	62.21
D-Fructose	4.98	11.61
D-Galactose	1.93	16.48
D-Glucose	2,10	14.19

It is assumed that the low concentrations of free carbohydrates can be attributed to the vigorous microbiological turnover.

Investigations by Geller (1983) in the small eutrophic Lake Mindel (south Germany) showed that 50–70 per cent of the dissolved organic material consisted of macromolecules (molecular weight 1500) and 30–40 per cent of oligomers (molecular weight 400–800) which, independent of the season, were only attacked to a small extent by the bacterial population present in the lake.

According to Liebezeit and Dawson (1982) the following concentrations of different groups of substances were found in sea water:

	µg C · l^{-1}		µg C · l^{-1}
Free amino acids	10	Amino sugars	2
Bound amino acids	50	Phenols	2
Free carbohydrates	10	Indole	1
Bound carbohydrates	200	Urea	10
Fatty acids	5	Vitamins	0.007
Hydrocarbons	5	others	10

These constitute at least 34 per cent of the dissolved organic carbon. But there are also great differences in the marine region in the different bodies of water. Thus, in May 1976, in 20 bodies of water from the Baltic Sea between the Kiel Fjord and the Gotland basin, Dawson and Pritchard (1978) found 4.5–84.0 μg l^{-1} of free amino acids and in five samples 438–805 μg l^{-1} of bound amino acids.

In many waters the organic material is the limiting factor for the growth of sapro-phytic bacteria and fungi. It follows that there is usually a positive correlation between the numbers of micro-organisms and the concentration of organic substances; and wherever organic material accumulates, as, for instance, in the zones of discontinuity of physical properties or in the water at the bottom, high total counts and sapro-phyte numbers are found (see Chapter 7). It is not, however, so much the total quan-tity of organic substance that is decisive but rather the fraction of organic compounds which are easily assimilated, like proteins and their constituents, sugars, starch, organic acids, fats, etc. This is shown by the following table (Kusnezow, 1959), which illustrates the relations between total quantity, assimilable organic substance and total microbial counts for several lakes of the USSR:

Lake	Type of lake	Organic substance (mg O_2 in 1 litre)		Total counts in 1 ml
		total	assimilable	
Lake Kontch	Oligotrophic	15.0	0.3	170 000
Lake Pert	Oligotrophic	15.3	0.5	130 000
Lake Yolowoye	Mesotrophic	—	1.43	342 000
Lake Ilmen	Mesotrophic	—	3.03	1 640 000
Lake Beloye	Eutrophic	32.1	3.06	2 230 000
Lake B. Medweshye	Eutrophic	33.7	5.16	3 420 000
Peat cutting	Dystrophic	226.6	3.96	2 320 000

The table clearly shows that the large total quantity of organic matter in dystrophic waters has no effect on the microbial numbers, but only the assimilable part of it. Where a shortage of organic nutrients exists, many bacteria do not reach their normal cell size but degenerate into coccoid involution forms. These are produced when the cells continue to divide but can no longer grow. Nutritional conditions more favourable than in open water are found on the surface of suspended matter where nutrients become concentrated due to adsorption (ZoBell, 1946a).

Conditions in the sea are similar to those in lakes. Newell et al. (1981) found that about 30 per cent of the detritus originating from phytoplankton was remineralized at 10 °C within 3 days. According to Ogura (1970), in the surface areas of the western North Pacific Ocean about 50 per cent of the total organic material dissolved in the water (i.e. about 0.5 mg C l^{-1}) are quickly utilized by micro-organisms — the residue, likewise of about 0.5 mg C per litre, can be degraded only slowly by bacteria.

A large proportion of the organic matter in the deep sea may be completely resistant to microbial decomposition. Consequently, in this case the assimilable portion must be very small. The concentration of free amino acids, therefore, varies within the nanomole range per litre. The dissolved organic matter in the deep sea can be very old. Investigations of water samples from about 1900 m deep in the north-east Pacific, using radiocarbon dating procedures, showed that the organic matter present therein was well nigh 3500 years' old (see Morita, 1980).

With the aid of growth experiments in the chemostat using a series of marine bacterial strains, Jannasch (1967) found the threshold values of a few limiting carbon sources such as lactate, glucose and glycerol. Below these critical concentrations, growth of the micro-organisms in question is no longer possible. The threshold values vary considerably — according to the species of bacteria, nutrient, growth rate and other factors. Thus, with regard to the growth parameter there are two kinds of micro-organisms, one of which possesses the ability to grow with a very low nutrient concentration and therefore is adapted to the conditions, e.g. in oligotrophic inland waters or in the open sea, while the other shows an adaptation to the high nutrient concentrations as they exist, e.g. in eutrophic waters and sediments. The bacteria belonging to the latter type remain largely inactive in nutrient-deficient water, but are nevertheless able to survive there. The following table by Jannasch (1970) lists for a member of each of the two groups the limiting concentrations of the above-mentioned nutrients for different growth rates:

	Growth rate h^{-1}	Lactate mg l^{-1}	Glycerol mg l^{-1}	Glucose mg l^{-1}
Achromobacter	0.5	0.5	1.0	0.5
aquamarinus	0.1	0.5	1.0	0.4
	0.005	1.0	5.0	1.0
Pseudomonas spp.	0.5	20	50	20
	0.3	50	50	50
	0.2	100	100	>100
	0.1	>100	—	—

Horowitz *et al.* (1983) found that bacterial populations isolated from subarctic waters off Alaska with poor nutrient media, could utilize widely different organic substrates as nutrients, whereas populations isolated on rich nutrient media were more specialized.

According to Amy and Morita (1983), after keeping various marine bacteria under starvation conditions for 4 or 8 months and then adding nutrients, the bacteria quickly consumed the latter and showed normal respiration and assimilation. From this, it may be concluded that survival in a state of starvation is by no means an abnormal condition for oceanic bacteria. They can withstand lack of nutrients for long periods and still retain their capacity for active metabolism.

The composition of the organic material strongly influences the particular microbial population. For instance, in waters loaded with sewage rich in protein, proteolytic bacteria predominate, while in waters containing much cellulose, cellulose-decomposing bacteria and fungi flourish. Consequently, in all waters, rivers, lakes and the sea, the microflora reacts very rapidly to changes in the composition of the organic matter (see Section 15.3).

Micro-organisms in waters may also be affected by the metabolic products of plants and animals present. These may either favour or inhibit individual species or whole groups of organisms.

8.9. Dissolved Gases

Besides salts and organic compounds, small quantities of dissolved gases are also found in water and these may considerably influence the life of micro-organisms. Oxygen, carbon dioxide and nitrogen are those mainly present. In addition, under special conditions, molecular hydrogen, carbon monoxide, hydrogen sulphide and hydrocarbons may also occur. Their solubility decreases with increasing temperature and is somewhat greater in fresh water than in sea water.

Oxygen, carbon dioxide and nitrogen reach the water continously from the air — until the surface of the water is saturated. In addition, dissolved gases may have been produced by biochemical processes in the water or in the sediments. Thus, oxygen is produced during assimilation by green plants, carbon dioxide by respiration, nitrogen by denitrification, hydrogen sulphide by desulphurication and hydrocarbons by fermentation (see Chapter 11). The gas produced in the mud of Lake Plön contains, for example, according to Ohle (1958):

O_2	N_2	CO_2	CO	H_2	CH_4	Other hydrocarbons
0–1.4	9.7–16.8	0.2–0.5	0–1.0	0–0.9	80.6–87.4	0–0.2 vol. per cent

Oxygen is needed by most plants and animals for their respiration. By consuming the O_2 in the water they cause an oxygen deficit of varying degree. In the photic zone, on the other hand, green plants carry out assimilation and frequently release more oxygen than they consume. The water here, therefore, may be temporarily oversaturated. Waters with a pronounced thermocline usually have, in consequence, an accompanying layer of sudden discontinuity of oxygen content (see Section 2.2).

Four groups of micro-organisms may be distinguished with regard to their oxygen requirements:

1. *Obligate aerobes*, which grow only in the presence of oxygen;
2. *Microaerophilic organisms*, which grow optimally at low oxygen concentrations;
3. *Facultative aerobes* (or *facultative anaerobes*), which grow in the presence or absence of oxygen;
4. *Obligate anaerobes*, which grow only in the absence of oxygen; oxygen is toxic for them as, due to the lack of cytochromes and catalase, H_2O_2 may accumulate.

The overwhelming majority of aquatic micro-organisms are facultative anaerobes. This is particularly true for the marine environment. In the oxygen-free hypolimnion and in the sapropel of some eutrophic lakes, obligate anaerobes also play an important role, for example *Chlorobium, Desulphovibrio* and *Clostridium* species.

For obligate aerobes molecular oxygen is vital; it is needed as terminal hydrogen acceptor in respiration. Fluctuations of the oxygen content, however, within a wide range do not affect the bacterial life of the water to any considerable extent. Thus, at 30 °C, the aerobic nitrite bacterium *Nitrosomonas europaea* nitrifies quite normally from saturation with oxygen down to 1 mg O_2 l^{-1} (Schöberl and Engel, 1964) and the nitrate bacterium *Nitrobacter winogradskyi* down to 2 mg O_2 l^{-1}. Only a further decrease of oxygen makes the oxidation rate drop. In *Nitrosocystis* (*Nitrosococcus* sensu Watson) *oceanus* the lowest limit of oxygen concentration for oxidation of ammonia to nitrite is 0.08 mg $O_2 l^{-1}$ (Gundersen, 1966). It follows that obligate aerobic bacteria are impaired in their development only at very low oxygen tension — although there are differences regarding the utilization of small oxygen concentrations. Some micro-aerophilic organisms, on the other hand, are inhibited by higher oxygen concentrations. Therefore, in waters poor in oxygen — in contrast to those with an abundant oxygen supply — relatively small fluctuations may lead to important population changes.

Molecular *nitrogen*, dissolved in water, is utilized by nitrogen-fixing bacteria and cyanophytes. On the other hand, molecular nitrogen is liberated into waters during bacterial nitrate reduction (see Section 11.5); but otherwise this gas would have no effect on the microflora of waters.

Carbon dioxide is needed not only by the photo- and chemoautotrophic organisms but also in small quantities by heterotrophic bacteria and fungi (Wood and Stjernholm, 1962). Thus, some bacteria produce malate or di- and tricarbocylic acids from pyruvate and carbon dioxide. Aquatic plant life as a whole is determined by the carbon dioxide-carbonate system, i.e. the equilibrium of dissolved CO_2, HCO_3^- and CO_3^{2-} ions. This is dependent on the pH, the excess bases, the partial tension of atmospheric carbon dioxide and the temperature (Round, 1968). The carbon dioxide-carbonate system has its origin in the activity of procaryotes and existed long before the development of green plants. Accordingly, bacteria are thereby still of great importance in the oceans. About 90 per cent of the carbon dioxide produced is derived from bacterial activity (Morita, 1975).

Hydrogen sulphide is stable only in an anaerobic environment and may accumulate there through the activity of micro-organisms (see Section 11.6). By binding the iron of cytochrome oxidase it acts as a respiratory poison for all organisms where this enzyme is the terminal link in the respiratory chain. Consequently, the appearance of H_2S is followed always by a complete change of populations. All higher organisms and most micro-organisms die off gradually and, apart from a few H_2S-tolerant organisms, their place is taken by bacteria which use H_2S as a source of energy (chemoautotrophic sulphur bacteria) or as a hydrogen donor (purple sulphur and chlorobacteria) (see Section 11.6). An extreme biological environment comes into being with a living community of only a few species, called a sulphuretum (Baas Becking, 1925).

During anaerobic decomposition of cellulose large quantities of *methane* are produced which leads to an enrichment of methane-oxidizing bacteria like *Pseudomonas methanica* (Anagnostidis and Overbeck, 1966). In marine regions, *Thioploca* can play an important role as a methane oxidizer (Morita *et al.*, 1981). Other gaseous hydrocarbons are broken down particularly by various pseudomonads and nocardias, which may multiply vigorously. Also molecular hydrogen is present in the gas from lakes and bogs and serves as energy source for 'Knallgas' bacteria (*Hydrogenomonas*). Carbon monoxide inhibits respiration by competing with free oxygen at the cytochrome oxidase level (Schlegel, 1981), but CO-tolerant organisms also exist and others, such as *Carboxydomonas* which can oxidize CO to CO_2 and therefore accumulate in an environment which contains CO (see Section 11.1).

9. The Influence of Biological Factors on Aquatic Micro-organisms

Besides physical and chemical factors, biological ones also affect aquatic micro-organisms. Multifarious interrelations exist between the members of a living community, and the organisms may help or inhibit one another. The first case is referred to as synergism, the second as antagonism. Of great importance is the competition for nutrients, between the micro-organisms themselves as well as between micro-organisms and other living creatures. On the other hand, bacteria and fungi serve as food for many lower animals. This means that in some waters frequent and rather far-reaching fluctuations of bacterial numbers occur. Some parasitic micro-organisms may attack bacteria and fungi and destroy them; and bacteriophages can also affect the microflora of a water considerably.

Even less is known about the biological factors than about the non-biological ones, as they are more difficult to explore. Salinity, temperature and pressure, for instance, may be measured very accurately, whereas we can, at best, only estimate the number of bacteria, say, which a mussel in its natural location consumes in a day — not to mention the competition for nutrients between different organisms. While, then, there are numerous observations showing the effect of biological factors on the aquatic microflora, precise figures are available only exceptionally.

9.1. Competition for Nutrients

In every habitat the nutritional competition between organisms plays an important role and influences decisively the composition of the microflora. The organisms which succeed best are those which under the particular conditions are quickest in reaching the available nutrients and taking them up. As a rule, several or even many kinds can live on the food which is on offer, particularly if the nutrients can be broken down readily, for example protein and carbohydrates, and if the conditions are fairly normal. As the various micro-organisms differ in their growth rates, however, some species multiply more rapidly than others; more and more metabolic products are released by them which may, in turn, inhibit competitors and even completely eliminate some of them. This happens, for example, when there are significant changes in the pH, or when antibiotically active substances are produced. If the accumulation of such products of metabolism goes too far, even the producer may be injured and finally succumb to the competition of organisms which show a greater tolerance towards these products or even feed on them.

Not all organisms which feed on the same nutrients must necessarily be competitors. There are examples where the nutrients can be utilized only by the coordinated activity of several kinds of organisms. Thus, *Escherichia coli* and *Proteus vulgaris* utilize a lactose-urea medium jointly: *E. coli* breaks down the lactose, *P. vulgaris* the urea, and the breakdown products of each add to the nutrients for the other, so that they complement each other (Schwartz and Schwartz, 1961).

Under extreme environmental conditions, competition for nutrients plays a subordinate role; in waters with extreme temperature, salinity or pH only a few kinds of organisms — sometimes only one — will be capable of utilizing the nutrients present. Under such circumstances, a nearly pure culture of many individuals may develop.

Similar effects are observed also with substances that are difficult to attack, for example, lignin, cellulose, chitin, hydrocarbons, phenols and others which are only utilized as nutrients by a few specialists. Competition for food does not play any significant role here.

9.2. Interactions of Different Micro-organisms

Frequently cooperation between different micro-organisms in nutrition and growth can be observed. Thus it, has been repeatedly confirmed that mixed cultures develop much better than pure cultures.

According to Meffert and Overbeck (1981), bacteria-free pure cultures of *Oscillatoria redekei* have only a low rate of multiplication which, in contrast, was distinctly greater in cultures containing bacteria. The growth of cyanophytes and heterotrophic bacteria proceeded simultaneously. When the growth of *Oscillatoria* was limited, as in dark cultures, no multiplication of bacteria was detected. Under normal growth conditions, substances are liberated which serve as nutrients for the heterotrophic bacteria. Essentially these nutrients are nitrogenous compounds such as proteins and ammonium compounds. Their concentration was 60 per cent higher in the bacteria-free *Oscillatoria* cultures, which corresponded to the nitrogen content of heterotrophic bacteria that were found in the mixed cultures. These higher concentrations of nitrogen compounds have an inhibitory effect on the growth of *Oscillatoria*. Such two-(or more)-component systems simulate natural ecosystems and, in contrast to unnatural one-component systems, can remain functional for several years without any nutrient addition.

In the degradation of recalcitrant substances such as lignin or cellulose, frequently there is likewise an interaction by various micro-organisms. After the development of specialists, which possess enzyme systems for the decomposition of such substances, other microbes which can utilize the intermediate products follow. In that way the specialists first create the pre-conditions for the development of this latter group of organisms, by whose activities an accumulation of harmful metabolic products is often avoided. A cooperation of this kind by different organisms is known as metabiosis and is widespread in Nature.

Furthermore, changes in pH and E_h values, oxygen concentration, the formation of CO_2, CO, H_2S, H_2 and CH_4 as well as inorganic nitrogen and sulphur compounds by certain micro-organisms, produce conditions for the development of other, mostly specialized, forms. Thus, for example, the formation of H_2S by *Desulphovibrio* makes possible the growth of sulphur-oxidizing bacteria, or production of CH_4 by methane bacteria enables methane-oxidizing organisms to develop (see Chapters 8 and 11).

9.3. Grazers of Bacteria and Fungi

Numerous creatures, in water or in sediments, feed on micro-organisms. Some may live almost entirely on bacteria and fungi, which are high-value protein food. This is, however, possible only in waters with an appropriately high content of micro-organisms, particularly eutrophic lakes, sewage-loaded rivers and coastal waters. In these, they frequently become the basis for a rich meiofauna of bacteria feeders. In oligotrophic inland waters and in large parts of the open sea, however, micro-organisms are so scarce that they can hardly have any importance as food. In the sediments, conditions are more favourable than in open water, and bacteria and fungi may provide a considerable part of the food for the fauna at the bottom. Zhukova (1963), on the basis of investigations in the northern part of the Caspian Sea, estimated that 1–10 per cent of the nitrogen requirements of invertebrates living in the sediment were covered by the uptake of micro-organisms. According to Morita (1979), in the deep sea bacteria form the main food source of the benthos organisms.

Most protozoa feed, at least partly, on bacteria (see Seki, 1966). Research by Luck *et al.* (1931) showed that *Euplotes taylori* can be kept on a purely bacterial diet, but it grows better if fed mixed cultures rather than pure ones. Heat-killed or autolysed bacteria will not do. It could also be shown repeatedly that the supply of certain kinds of bacteria resulted in the death of the protozoa, possibly due to toxic products of metabolism. It is important that the size and shape of the micro-organism should permit them to be taken up by the animals; this process may be impaired by formation of aggregates or by slime production.

Gast (1983) found, in the Schlei (western Baltic Sea), that with increasing eutrophication total bacterial numbers as well as the numbers and biomass of protozoa increased. By using radioactively labelled bacteria, filtration rates of 5.3–57.6 per cent of water per day were measured in March, 1982. In August, 1982, even 70.1 per cent of water per day was filtered in the Baltic Sea, north-east of Bornholm.

Feeding experiments with *Uronema marinum* showed that there was multiplication of the ciliates only after the concentration of bacteria exceeded 1 million per ml. The feeding rate was dependent on the concentration of bacteria and also on the water temperature. With a diet of bacteria, *Euplotes vannus* converted 20 per cent of the bacterial biomass into ciliate biomass.

According to Sherr *et al.*, (1983), in the case of the heterotrophic microflagellate, *Monas* sp., from Lake Genezareth, growth and feeding rates increased with the concentration of bacteria but certain differences were shown in the four species of bacteria used as food. At temperatures of 18–24 °C, the doubling times of the flagellate population and also the rates of ammonia excretion were at their lowest and growth effectiveness at its highest. Under optimal conditions the doubling times were between 5 and 7.8 h, and the mean efficiency amounted to 23.7–48.7 per cent. The mean rates of ammonia excretion varied between 0,76 and 1.23 μmol NH_4^+ per mg dry weight per h.

Many metazoa also feed on bacteria and fungi, particularly animals which are filter-feeders. Sponges thus take up bacteria and digest them. Studies by Sorokin (1973) on the nutrition of reef-forming corals using ^{14}C-labelled food showed that the polyps could consume daily the equivalent of 10–20 per cent of their carbon content in the form of bacteria. Californian mussels (*Mytilus californicus*) and oysters have been shown to grow on a pure bacterial diet *in vitro* (ZoBell and Landon, 1937; ZoBell and Feltham, 1938). The animals were fed for several months on bacteria and during that time increased in size and weight.

Investigations by Prieur (1981, 1982) showed that various bacteria are not all attacked in the same way by mussel enzymes. For example, vibrios can remain in the digestive tract of *Mytilus* for several hours and are then excreted again. They are even capable of reproducing themselves in this time. In their natural habitat bacteria and fungi are taken up along with debris; it is, therefore, rather difficult to estimate the proportion of bacteria in their diet under natural conditions, but it may be considerable, particularly in river estuaries.

Rieper (1978) states that the marine harpacticoid copepods *Tisbe holothuriae* and *Paramphiascella vararensis* could be fed exclusively on bacteria and in their development showed scarcely any difference in comparison with other animals which received a diet of mussels or fish. They required 2.06–7.07 μg bacteria (dry weight) per animal per day; this corresponds to a daily amount of carbon of 1.0—3.5 μg. Hanaoka (1973) found a similar nutrient requirement with *Tigriopus japonicus* and obtained an increase from 20 to 200 copepods in 35 days with a purely bacterial diet. Furthermore, micro-organisms can be important in the feeding of Annelida and also of insect larvae living in rivers and lakes. According to Baker and Bradham (1976), about half of the food intake of the larvae of simulid flies (*Simulium*) and of midges (*Chironomus*) is from bacterial cells. Preference for certain forms in this connection was not established. For *Simulium*, the number of bacteria consumed per day was estimated as 11 or 28 million depending on the method used. This corresponds to a dry weight of 2.2 or 5.6 µg. The zooplankton also uses micro-organisms for food; it has been shown that bacteria play an important role in the feeding of copepods and lobster larvae (Esterly, 1916; Cviic, 1960; Zhukova, 1963).

Above all, fungi can be important in sediments as food for different animals. Cowley and Chrzanowski (1980), in the tidal zone of the American east coast, found in the sediment and in the gut of the fiddler crab (*Uca pugilator*) yeasts (*Rhodotorula*, *Torulopsis*) which promote the development of the crab where the food is deficient in vitamin B and also undoubtedly contribute directly to its nutrition.

If follows that animals that feed on bacteria may affect considerably the aquatic microflora. When such animals first turn up, the immediate effect may be a decrease in bacterial numbers of some magnitude. After some fluctuations, an equilibrium may be established. Copious bacterial growth is, as a rule, followed by a corresponding increase in the numbers of bacterial feeders. Then the bacteria decline until the bacterial feeders also decrease from lack of food; this is then followed by a renewed increase in bacterial numbers. This up and down of both populations may be repeated several times in succession in closed aquatic systems. It can easily be observed in vessels filled with river, lake or sea water which are left undisturbed for a time, and also in mixed cultures of bacteria and protozoa (or of fungi and fungi-consuming animals) see Straskrabova-Prokesova and Legner, 1966).

Cyanophytes likewise represent an important source of nutrients for a variety of animals. Benthic populations are often grazed by rotatoria, nematodes, crustaceans and insect larvae. Planktonic cyanophytes serve as food for the zooplankton (Horstmann, 1975).

However, some bloom-forming species are only consumed reluctantly. For example, this is so for *Nodularia spumigena* and *Aphanizomenon flos-aquae* in the Baltic Sea. Frequently toxic strains of *Microcystis aeruginosa* occur in lakes and, according to observations by Lampert (1982), these cause a reduction of the filtration rates in water fleas (*Cladocera*). The inhibition occurs immediately if the animals come into contact with *Microcystis*, probably due to the rejection of the already filtered particles. The toxicity, therefore, can be regarded as a defence mechanism against being eaten.

The importance of faecal pellets, which often contain numerous bacteria, to the nutrition of various animals in water and sediments should not be underestimated. These occur in large amounts during the mass appearance, for example, of copepods and can find their way to the bottom of waters and there contribute to the feeding of benthos organisms (Sieburth, 1979).

Even amongst plants there are some which consume bacteria and fungi, like the curious net slime moulds of the genus *Labyrinthula*, which feed on living bacteria and yeast cells (Klie and Schwartz, 1963).

9.4. Attack of Micro-organisms by Viruses, Bacteria and Fungi

Like other living creatures, aquatic micro-organisms are attacked by viruses, bacteria and fungi which may penetrate into the host cells, entirely or partially, and destroy them.

Bacteriophages have been detected both in inland waters and also in the sea (see Chapter 6). They attach themselves to bacterial cells by their tails and inject the genetic material present in the head. After the penetration of the phage DNA the metabolism of the bacterial cell is completely transposed — to the production of phage material. Finally, after destruction of the cell wall, the 100–200 newly formed phage particles are released, leaving behind an empty ghost cell which undergoes autolysis. The whole process may take only around 20 min so that, within a short space of time, a great number of bacterial cells may be infected. Bacteriophages are particularly numerous in sewage and probably help in the rapid decrease of bacterial numbers in sewage-loaded rivers, lakes and coastal waters (see Section 15.3). Coliphages have very frequently been isolated from such environments, and also phages which infect pathogens like salmonellae and shigellae. In addition, phages specific for autochthonous aquatic bacteria have been found. In the Kiel Bay, for example, certain *Agrobacterium* strains whose focal point of distribution is that area are infected with bacteriophage (Ahrens, personal communication). Spencer (1963) found bacteriophages in various marine bacteria of the genera *Pseudomonas*, *Photobacterium*, *Cytophaga* and *Flavobacterium*. In the open sea, very few phages are present, corresponding to the paucity of bacterial life, so that it may occasionally be difficult to demonstrate them (ZoBell, 1946a). This is probably true also for oligotrophic inland waters. In rivers and lakes of north England, Willoughby *et al.* (1972) found various actinophages which attack actinomycetes (*Actinoplanes* and *Streptomyces*) and seem to possess a high host specificity. Cyanophages also, whose host organisms are the bluegreen algae, have been isolated from different waters (Padan and Shilo, 1973). They might often play an important role in the sudden dying off of cyanophyte blooms in eutrophic lakes and ponds. In certain fish ponds in Israel, Padan and Shilo (1969) demonstrated an increase in cyanophages of some hundredfold during the summer, if large amounts of cyanophytes were present. It is possible that there is also a connection between the toxin formation of some cyanophytes as, for example, *Microcystis aeruginosa*, and the lysis of the cells caused by attack with cyanophages (see Saffermann, 1973).

Some 20 years ago, an obligate bacterial parasite of bacteria was discovered, *Bdellovibrio bacteriovorus* (Stolp and Starr, 1963), which occurs also in water (Mitchell, 1968) and is a rather small, comma-shaped organism which is motile by means of a powerful flagellum. With its non-flagellated pole it attaches itself to the wall of the host cell and invades it. The infected cell becomes spherical, and the parasite grows in the host and digests the cell contents. After the host envelope has been destroyed, the *Bdellovibrio* cells are released and ready to infect new hosts. Like phage, *Bdellovibrio* is found particularly in sewage-loaded waters with great numbers of bacteria. Mitchell (1968) states that apparently there are halophilic forms of this organism in the sea which are, amongst others, responsible for the rapid decrease of coliform organisms in polluted coastal waters.

Hentzschel (1977) found obligate halophile strains in the North Sea and the Baltic which attack marine bacteria. Their concentration ranged from 20 pfu (plaque forming units) per litre in areas distant from the coast, to 25000 pfu per litre in coastal areas. Experiments by Hentzschel (1980) showed that the composition of the bacterial population of the Fladengrund (North Sea) can be influenced by the activity of *Bdellovibrio* and other lytic bacteria.

Fungi too are infected by bacteria, particularly when the optimal conditions for their growth have ceased to exist.

In addition, there are a whole number of parasitic phycomycetes which attack other aquatic fungi (Figures 66 and 67). Some are highly host-specific and parasitize only a single species or a few closely related species and genera; others infect members of different families. Fungal parasites of both sorts are found, for example, in the

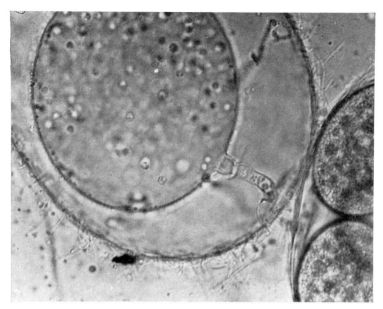

Figure 66   *Olpidiopsis saprolegniae* from the mud on the banks of an inland lake. Three sporangia
in two swollen ends of hyphae of *Saprolegnia* sp. Large sporangium with a discharge
tube (×380). (J. Schneider)

genus *Rozella* (Chytridiales). Some phycomycetes even attack parasitic fungi within
their host cells; this is called hyperparasitism. Thus *Rozella marina* infects the sporangia of *Chytridium polysiphoniae*, which lives as a parasite in red algae (Sparrow,
1960). Although there are a number of fungal parasites amongst the phycomycetes,
they seem to occur only rarely in large numbers.

9.5. Growth Substances and Inhibitors

Growth and inhibiting substances produced by living organisms can also be counted
among the biological factors.

Of particular importance are the *vitamins* which can not be synthesized by many
(especially marine) algae, nor even by different bacteria and fungi themselves. They
must therefore be supplied by other organisms. The chief suppliers are again algae,
bacteria and fungi (especially yeasts). The most important of these vitamins are B_1
(thiamin), B_{12} (cobalamin) and H (biotin) but riboflavin, pantothenic acid, nicotinic
acid and folic acid also have a role to play. Whilst only 4 per cent of the fresh water
cyanophytes so far tested require vitamins, with the marine forms there were 50 per
cent. The same is also true of the diatoms. Vitamins were necessary in the case of
46 per cent of marine diatoms examined and of these, 14 per cent needed B_1, 25 per
cent B_{12} and 7 per cent both vitamins (Wood, 1967). A series of bacterial strains have
been isolated from waters, which required one or more vitamins for their development.
Nevertheless the large majority of aquatic bacteria can synthesize for themselves
all the necessary vitamins. Niewolak and Sobierajska (1971) in investigations of
the water of the Ilawa lakes (Poland) found that species particularly of the genera
Pseudomonas, Aeromonas, Vibrio, Bacillus and *Micrococcus* are capable of synthesiz-
ing vitamin B_{12}. Two strains of *Aztobacter agile* were especially active in this respect —

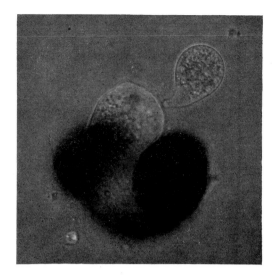

Figure 67   *Septosperma rhizophidii* on the sporangium of another phycomycete which is attached to *Pinus* pollen ($\times$ 600). (J. Schneider)

they synthesized up to 0.16 and 0.17 μg vitamin B_{12} per ml, while the other species mostly produced less than 0.05 μg vitamin B_{12} per ml. According to Hagedorn (1971) in lakes the thiamin concentration increases with the amount of bacteria and reaches a minimum at the maximal development of phytoplankton. This suggests that the vitamin is produced predominantly by bacteria and excreted. Between 50 and 95 per cent of the heterotrophic bacterial species are capable of active thiamin excretion; up to 7 μg l^{-1} have been measured, with the average being 0.5 μg l^{-1}. Thus, taken as a whole, bacteria play an important role as vitamin producers. This is also true for yeasts.

Among the lower aquatic fungi likewise, some forms are found which have an essential requirement for vitamins. For example, the marine phycomycete *Thraustochytrium globosum* develops only in the presence of vitamin B_{12} and shows increasing growth with increasing doses of B_{12} up to a concentration of 100 ng ml^{-1} (Adair and Vishniac, 1958). Also some higher fungi thrive only in the presence of vitamins B_1 and B_{12} (Hawker, 1966).

Because numerous organisms release vitamins into the substrate, these can be detected in varying amounts in the water of many lakes, rivers and regions of the sea. For this purpose one makes use of, as a rule, so-called auxotrophic strains of different micro-organisms, which can not synthesize the vitamin in question and tests their growth, under suitable conditions, in the water samples. Thus, for example, using a biotin mutant of *Serratia marinorubra* (*marcescens*) between 0 and 4.6 ng of biotin per ml were found at different stations in the Gulf of Mexico (Lichtfield and Hood, 1965). Vitamin B_{12} seems to be present in most waters in adequate amounts (see Wood, 1965). This vitamin is particularly abundant in activated sludge. It plays a role as coenzyme in methane formation from carbon dioxide. By uptake of the carbon already reduced to the methyl group, methylcobalamin is formed from which methane is then split off by an ATP-dependent reduction (see Schlegel, 1981).

Different *enyzmes* have also been found in waters and these are derived likewise from algae and micro-organisms. Overbeck and Babenzien (1964b) detected phosphatases, saccharase and amylase in the water of a pond rich in plankton. These enzymes are produced by the plankton and are liberated on their death.

Free cellulases have been identified in the water of lakes. According to Wunderlich (1973) these might play a role in the breakdown of cellulose in waters. There are numerous indications in the literature that free enzymes of very varying kind and function are present both in water and in sediments. Their concentration, however, in most cases is very low but can nevertheless suffice for certain material transformations on suitable suspended matter and sediment particles. In the photic zone enzymes are secreted above all by the phytoplankton, but in the aphotic zone and in sediments by bacteria, fungi, protozoa and metazoa.

Hoppe (1983) determined extracellular enzyme activities in the water of the western Baltic using fluorogenic methylumbelliferyl substrates together with bacteriological parameters. After enzymatic fission of the particular organic substrate component, the fluorescence of the methylumbelliferone was measured. All the enzyme activities examined were greater in the strongly eutrophic Schlei and in Kiel Fjord than in the less polluted open sea. In all the samples, protease activity predominated. The activities of α-D-glucosidase and glucosaminidase indicated greater differences between the Fjord and the open Baltic sea. The following list gives examples of the hydrolysis rates per litre of water per hour by five different enzymes, together with total bacterial numbers, bacterial biomass and bacterial production for three stations in the Kiel Fjord and two stations in the Kiel Bay, as measured on 29–30 March, 1982.

	Kiel Harbour	Fried-richsort	Laboe	Kiel Light-house	Kiel Bay (middle)
α-D-Glucosidase	0.78	0.49	1.30	0.07	$0.09\,\mu\mathrm{g}\,\mathrm{C}\,\mathrm{l}^{-1}\,\mathrm{h}^{-1}$
β-D-Glucosidase	0.52	0.58	0.28	0.08	0.18
Glucosaminidase	0.83	0.64	0.49	0.14	0.08
Phosphatase	2.30	1.60	1.80	1.10	0.90
Protease	5.80	4.60	3.60	1.60	2.50
Total bacterial numbers	2.60	2.48	1.11	1.05	$1.03 \times 10^6\,\mathrm{ml}^{-1}$
Bacterial biomass	20.40	20.80	9.40	9.80	$8.90\,\mu\mathrm{g}\,\mathrm{C}\,\mathrm{l}^{-1}$
Bacterial production	0.32	0.24	0.22	0.12	$0.10 \times 10^6\,\mathrm{ml}^{-1}\,\mathrm{day}^{-1}$

These data show a clear relationship between extracellular enzyme activities and the amounts of bacteria and bacterial production. Similar results were obtained by Meyer-Reil (1983) in sediments of the Kiel Bay. The sedimentation of organic matter — particularly in spring and autumn — led to a corresponding increase in the exoenzyme decomposition of carbohydrates by α-amylase and of proteins by proteases. The maxima of the enzyme decomposition rates corresponded with those of the bacterial numbers.

Antibiotics of various kinds have also been detected in inland waters as well as in the sea. These are produced especially by algae and reach active concentrations particularly in phytoplankton blooms (see Section 7.3). Probably they then play a role, if at the time of the phytoplankton maximum the saprophytic bacteria have reached their annual minimum, as repeatedly established, for example, in the North Sea and the Baltic. Chrost (1975a, b) made similar observations in a eutrophic lake in Masuria. Here a phytoplankton bloom occurred in spring, in summer and in autumn, and at the same time the saprophyte numbers declined markedly. At these times gram-positive bacteria disappeared altogether. It could be shown that the algae produced substances which caused inhibition of growth and respiration in gram-positive bacteria. On the other hand, gram-negative rods proved to be resistant to these substances. In the euphotic zone of the lake, the breakdown of the organic material was also inhibited. Such antibiotically active substances are acrylic acid and polyphenols (Sieburth, 1968b). Higher algae also produce antibiotics. According to investigations by Glombitza (1969), this is particularly true of the Chlorophyceae — but to a lesser extent also of a series of Rhodophyceae and Phaeophyceae. In microsites — especially in sediments and on dead plants and animals as well as on floating particles — antibiotics produced by actinomycetes and fungi might also occasionally influence the microflora. Concentration is the decisive factor in every case. In waters this will only rarely be sufficient to cause inhibition of bacteria and fungi. Small amounts can even be degraded by certain micro-organisms and used as food material. This is true also, for example, of phenols that are frequently used as disinfectants.

Bacteria can also release substances which have an antibiotic effect on other organisms. Thus, for example, Granhall and Berg (1972) found in the culture solutions of two *Cellvibrio* strains a heat-stable agent that induced morphological changes in and lysis of blue-green algae. Distinct changes in the cell wall were also observed.

Various cyanophytes can also excrete substances that inhibit other species. There are numerous although contradictory data in the literature regarding the production of bacteriostatic and antibiotic exudates, particularly with reference to the inhibition of bacteria (Carr and Whitton, 1973). Again and again there are reports of poisonous cyanophyte blooms which cause the death of fish. Gentile and Maloney (1969) found that some strains of *Aphanizomenon flos-aquae* can produce an endotoxin that is capable of killing fish on injection and also when present in the water, after first lysing the cells. *Trichodesmium* blooms are claimed to have caused illness in bathers. Cyclic polypeptides and alkaloids have been detected, to date, as poisons in cyanophytes (Olrik, 1976).

10. Micro-organisms Inhabiting Plants and Animals

Many of the bacteria and fungi occurring in waters live temporarily or permanently within or on the surface of plants or animals, but show great diversity of relationships with their hosts.

Thus they may use plants or animals just as a physical support; or they may be commensals and derive all or some of their food from the food or the metabolic products of their hosts. Generally this does not involve any advantage or disadvantage for the hosts, and is true of many inhabitants of cavities in animals, for example, stomach and guts of fish. There are, however, other microbes which, as parasites, feed on the cells and body substances of their hosts and thereby cause illness and often death. Sometimes the relationships between micro-organisms and host plants or animals may be to their mutual advantage. Such a symbiosis exists, for example, between luminescent bacteria and some molluscs and fish; the animals supply the food for the bacteria and these, in the organs which they inhabit, produce light which serves a variety of uses for the host, for example as a recognition device.

10.1. Aufwuchs

On the surface of numerous plants and animals, bacteria and fungi may settle at least temporarily or, on a more lasting basis, produce a superficial layer of growth

Figure 68   *Caulobacter* on a rod-shaped bacterium from groundwater ($\times 27\,000$). Electron micrograph by P. Hirsch

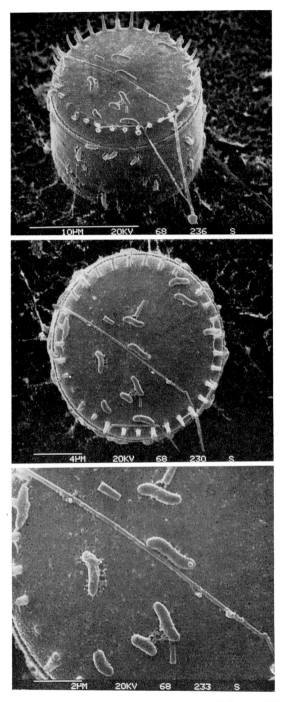

Figure 69   Bacteria aufwuchs on a central diatom (*Thalassiosira* sp.) at three different magnifications. Scanning electron micrographs by U. Palmgren and S. Samtleben

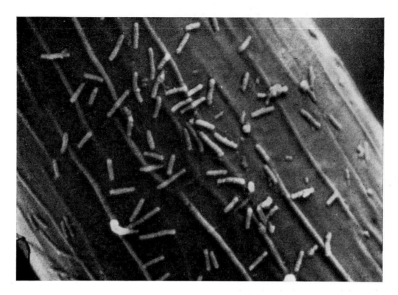

Figure 70   Bacterial aufwuchs on a higher alga ($\times 4250$). Scanning electron micrograph by
U. Palmgren

(*aufwuchs*) (see Figures 68, 69 and 70). From unicellular algae to fish and water mammals, a variety of creatures may serve as a physical substrate for micro-organisms. There are, however, others which — apparently by producing antibiotically active substances — are never colonized by micro-organisms (see Section 9.4). As far as planktonic algae are concerned, observations vary a good deal. Anagnostidis and Overbeck (1966) rarely found surface growth of bacteria on Cyanophyceae or diatoms in the east Holstein lakes. Collins (personal communication), on the other hand, found copious colonization of planktonic algae by bacteria.

Hoppe (1981) observed in the Baltic Sea numerous bacteria, particularly on already spiralized filaments of the cyanophytes *Nodularia spumigena* and *Aphanizomenon flos-aquae*. Stalked bacteria colonized them at first at fairly uniform distances, followed by microcolonies of cocci. Ultimately a more diverse population developed in which rod forms became dominant. An average of 45 bacteria were present on a filament, 75 μm in length. Since the cyanophytes form plankton blooms, these surface growing bacteria can then form a large proportion of the total bacterial numbers.

The peculiar *Leucothrix mucor* grows epiphytically on various marine algae. This bacterium was already discovered in 1844 in the Sound near Copenhagen by the Danish botanist Örstedt. It is a halophilic organism, the filaments of which, consisting of numerous cells, form rosettes and gliding gonidia.

Shiba and Taga (1980) investigated the aufwuchs on different higher algae in Japanese waters. They found saprophyte numbers of 10^4–10^6 per cm^2 on the green algae *Monostroma nitidum* and *Enteromorpha linza*, 10^3–10^4 per cm^2 on the red alga *Porphyrea suborbiculata*, and 10^1–10^4 per cm^2 on the brown alga *Eisenia bicyclis*. Yellow and orange pigmented bacteria of the *Flavobacterium/Cytophaga* group were predominant on green and red algae but not on brown algae. According to Shiba and Taga (1981), extracellular products of *Enteromorpha linza* promoted the growth of this group of bacteria and likewise that of pseudomonads, whereas the growth of vibrios was inhibited.

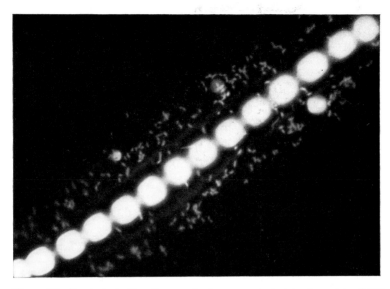

Figure 71   Bacteria in the slime coat of a cyanophyte (*Anabaena*) (× 2 800). R. Zimmermann

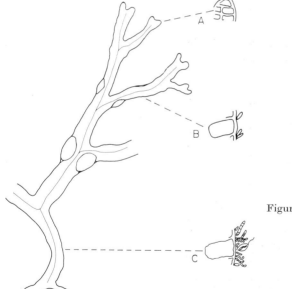

Figure 72   Aufwuchs zones on *Fucus vesi-culosus*. (A) Meristematic zone without aufwuchs. (B) Younger thallus branch with little auf-wuchs. (C) Older thallus branch with dense aufwuchs. (Based on Sieburth, 1968 b)

Micro-organisms are often found in great numbers in the slime capsules of algae (Figure 71). Evidently the slime serves as a nutrient for them (Rieper, 1976; Linley *et al.*, 1981). Fungi have also been found in the slime envelopes of algae; for example, the phycomycete *Amoebochytrium rhizoidioides* lives as a saprophyte in the gelatinous envelope of *Chaetophora elegans* colonies (Sparrow, 1960).

If one tries to reach some conclusions from the far from numerous and, in part, contradictory references to the growth of bacteria and fungi on planktonic algae, the following picture emerges. Vigorously growing algae are usually free from bacterial growth on their surface; while the phytoplankton bloom is building up, one will look in vain for bacteria. On the other hand, populations of algae which are stagnating or declining are increasingly colonized by bacteria. Analogously, in benthos algae the youngest — i.e. apical — parts are free of bacteria (Figure 72), while the older, basal, parts of the same plants show, increasing with the distance from the apex, surface growth of bacteria and fungi (Sieburth, 1968b). At times, this is so dense on some aquatic plants (Figure 72c) that it can be grazed by protozoa and rotatoria and serves as their staple diet.

The aufwuchs obtain nutrients, above all, from plant excretions. Thus, Kirchman *et al.* (1984), for example, established that the seaweed *Zostera marina* provided sufficient photosynthetic products for the growth of the surface microflora. The carbon flow from plant to bacteria, as determined by means of ^{14}C-labelled potassium bicarbonate, was about $0.3\,\mu g$ C h^{-1} cm^{-2}, whilst the highest bacterial production, assessed by the incorporation of [^{3}H] thymidine, amountes to $0.4\,\mu g$ C h^{-1} cm^{-2} (see also Section 11.2). The excretion rate by *Zostera* leaves amounted to 2 per cent of the total carbon fixation.

The absence of bacteria on young, vigorously growing algae (or algal parts) may, to some degree, be due to bacteriostatic or bactericidal substances released by the plants (see Section 8.8); but often the bacteria are repelled by the acid reaction on the algal surface (Figure 73). Thus a pH of 8 in the surrounding sea water may go down to 5 on the surface of the algae; and a bacterium coming from a weakly alkaline (usually optimal) environment into an acid one will tend to leave it again at once (Sieburth, 1968b).

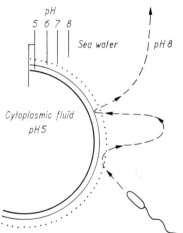

Figure 73   Hypothetical mechanisms of the prevention of bacterial aufwuchs by the acid microzones on the surface of a diatom cell. (Based on Sieburth 1968b, modified)

The surface of many lower animals is also colonized by bacteria and fungi. Knorr and Sonnabend (1964) found that fresh water copepods in Lake Constance carry, at times, numerous bacteria but probably do not play an important role in their spread. A similar situation exists with regard to other crustaceae and worms. Bacteria are also found on the skin of bony fish. Liston and Colwell (1963) found mainly gram-negative non-sporing rods on vertebrates (19 species) and invertebrates (14 species) of the Pacific Ocean. There were only minor taxonomic differences between the bacterial populations of vertebrates and invertebrates. On all animals, *Pseudomonas* and *Achromobacter* species were the most numerous. *Micrococcus* and *Flavobacterium* were represented in slightly greater proportion on the surface of sessile invertebrates than on vertebrates of the same region of the sea. On animals from the tropical parts of the Pacific the numbers of micrococci and (non-faecal) enterobacteria were greater than on those from the North Pacific. Differences were also found with regard to their physiological and biochemical activities. Thus in the tropics mesophilic and proteolytic bacteria predominated, in the cooler regions psychrophilic organisms and organisms which break down carbohydrates. Altogether, the bacteria of vertebrates show greater biochemical activity than those of invertebrates.

Yeasts have also been demonstrated on many aquatic creatures (van Uden and Fell, 1968).

Most micro-organisms living on the surface of plants and animals find their food there. Algae, for example, release dissolved organic substances like amino acids, sugars and organic acids (Gocke, 1969); many are surrounded by slime envelopes, usually sulphuric acid esters of polysaccharides (Round, 1968). Animals shed excreta and rubbed-off bits of tissue which serve as nutrients for the aufwuchs flora on their surface. Some bacteria and fungi, however, use plants and animals only as a physical support and derive their nutrients from the water. This is the case, for example, with some sheathed bacteria and especially with *Caulobacter* and *Hyphomicrobium*.

10.2. Commensals in Animal Organs

Micro-organisms are found as commensals not only growing on the surface but also inside many aquatic animals, particularly in their digestive tracts, where numerous bacteria and yeasts often occur. According to Bertrand and Vacelet (1971) the pore spaces of sponges are colonized by numerous bacteria. In the sponge *Microcionia prolifera*, Madri *et al.* (1971) found representatives of the genera *Pseudomonas, Aeromonas, Vibrio, Achromobacter, Flavobacterium, Corynebacterium* and *Micrococcus*, which likewise occurred in the surrounding water. Their numbers were nearly constant during fortnightly investigations in the months of April to August, 1966, while the saprophyte numbers in the water showed considerable fluctuations. More detailed investigations have been made also on the composition of the gut flora of fish. Mattheis (1964) isolated from the guts of 24 rainbow trout, 7 river trout and 13 carp 219 bacterial strains of which 204 could be identified and, on the basis of their morphological and physiological characteristics, classified as belonging to 32 species, 10 genera and 5 families. In the intestines of carp, genuine aquatic bacteria of the genera *Pseudomonas* and *Aeromonas* predominated; similarly with river trout. The intestinal flora of rainbow trout from pond nurseries was composed of a greater variety of species containing, besides genuine aquatic bacteria, also organisms which are strangers in the aquatic environment but may, under suitable conditions, multiply in water, like members of the genera *Escherichia, Enterobacter* and *Paracolobactrum*. It became evident that the kind of food the fish consumed had considerable influence on their intestinal flora. After prolonged starvation, bacteria, although few in numbers and in species, could still be demonstrated in the empty intestine. Other authors found no bacteria at all in the empty intestines of various fish (Mattheis, 1964). From this it would seem that bacteria get into the fish guts with food. Whether all these bacteria are true commensals or whether some of them also contribute to the breaking down of nutrients, thus making them utilizable for their hosts, is not yet clear.

Though certain strains of *Escherichia coli* are found frequently in the intestines of fresh water fish, the typical faecal types of warm blooded animals do not belong to the normal intestinal flora of fish. Their occurrence always indicates that the animals, before having been caught, were in sewage-loaded water (Wood, 1967).

According to the observations of Ohwada *et al.* (1980), *Vibrio* or *Photobacterium* spp. were the dominant organisms in the gut flora of fishes in the deep sea of the Atlantic whereas species of the genera *Pseudomonas, Achromobacter* and *Flavobacterium* were present in smaller numbers. These organisms were barotolerant forms. Thus, in the deeper regions of the sea, the gut flora is scarcely inhibited by the hydrostatic pressure. Nevertheless saprophyte numbers were lowest in animals from the greatest depths.

Yeasts occur frequently in the digestive organs of fish (van Uden and Fell, 1968); van Uden and Castelo-Branco (1961) state that *Metschnikowi(ell)a zobellii* is the predominant species in the guts of various Pacific fish. The number of viable yeast cells was 25–5700 per g of intestinal contents. *Metschnikowi(ell)a zobellii* was also found in small numbers in the surrounding water. Yeasts have also been demonstrated in the guts of some sea birds; but they seem to be absent from the digestive organs of marine mammals (van Uden and Fell, 1968).

10.3. Parasites

Numerous bacteria and fungi live as parasites in a great variety of aquatic plants and animals and can there cause diseases which may lead to the death of the affected organisms; these are called pathogenic micro-organisms. They are particularly important where certain plants and animals occur in large numbers in a restricted space, for example in plankton blooms, on mussel banks, in fish shoals, etc. In cases of such massed occurrence of one particular species, epidemics have been observed which led to complete destruction of large crops of plants or stocks of animals. Organisms which live by themselves, however, are much less prone to attack by pathogenic bacteria and fungi, as the opportunities for infection are much less. In oligotrophic rivers and lakes, and also in the open sea, conditions for the growth of micro-organisms which cause acute infections in aquatic animals are, throughout, unfavourable; as soon as a fish shows any symptoms of disease it is, as a rule, very quickly eaten by predacious fish or birds. In such nutritionally poor waters only perfectly healthy animals can escape the ubiquitous predators. Even diseased animals in fish shoals soon fall prey to predators in the vicinity. The pathogens thus have very little opportunity to infect other animals (ZoBell, 1946a).

Although parasitic micro-organisms have occasionally been discovered in relatively large numbers of aquatic animals and plants, the course of infection has only been studied in any detail in cases of economically important organisms such as seaweed, oysters, crayfish and any saleable fish.

Hardly anything is known about bacterial diseases in unicellular or multicellular algae. Black rot of *Macrocystis pyrifera* on the coast of California was thought to be due to a bacterium, but this now seems to be a facultative parasite which only invades plants already damaged (Brandt, 1923). Cantacuzene (1930) attributes proliferating tissue of *Chondrus crispus* (Irish moss) and *Sarcorhiza bulbosa* to a bacterial infection.

There are only isolated reports on bacterial pathogens in lower animals. Dimitroff (1926) frequently found spirilla and spirochaetes, presumably parasites, in the digestive tract of oysters. Bacteria have been thought responsible for the death of entire oyster beds on European coasts, but definite evidence for this is lacking. The lobster stocks on the American west coast suffered heavy losses through 'soft shell disease', which seems to be caused by chitin-decomposing bacteria (Hess, 1937). Another infectious disease of the American lobster is caused by *Gaffkya homari* (Hitchner and Snieszko, 1947).

A whole series of bacterial diseases is known that affect commerical fish (Amlacher, 1981; Bullack *et al.*, 1971; Reichenbach-Klinke, 19880). The most important bacterial pathogens are:

Vibrio anguillarum:	vibrioses in salmonids, eel, codfish, pike
Aeromonas salmonicida:	furunculosis in salmonids
Aeromonas punctata:	infectious ascites of the carp, fresh water eel disease, spotty disease in carp, perch and pike
Pseudomonas fluorescens:	ulcers in various fishes
Pseudomonas putida:	red pest in cyprinids
Haemophilus piscium:	ulcer disease in trout and char
Corynebacterium sp.:	kidney disease in salmon and trout
Mycobacterium marinum:	fish tuberculosis
Mycobacterium piscium:	fish tuberculosis
Cytophaga (Flexibacter) columnaris:	mouth mycosis
Cytophaga psychrophila:	bacterial cold water disease

Nocardias and spirochaetes can likewise cause infections in fishes, and also rickettsias, which are very small obligate parasitic bacteria.

The widespread epidemic ascites of carp is caused by the combined action of viruses and Pseudomonadaceae like *Aeromonas punctata* f. *ascitae*, *A. subrubra*, *A. liquefaciens* and *Pseudomonas granulata*. The affected animals show a large distension of the abdomen due to ascites, and also often inflammation of various organs, skin ulcers and destruction of the fins. The disease occurs particularly at the end of winter and during spring and may cause considerable losses in hatcheries. Various pseudomonads cause the so-called spottiness of skin which occurs in some fresh water fish. Red spots appear at first on the skin; inflammation and ulceration follow. A similar disease occurs in pike and eels attacked by *Vibrio anguillarum*; this pathogen is a bacterium of brackish water found in the Baltic and the coastal areas of the North Sea and has often caused heavy losses, particularly amongst eels, in warm summers. The disease, red pest of eels, is sometimes also called salt water eel disease — to distinguish it from the fresh water eel disease, which is caused by *Aeromonas punctata*, *Paracolobactrum anguillimortiferum*, *Pseudomonas fluorescens* and *Ps. putida* in inland waters (Schäperclaus, 1979).

Fish tuberculosis due to infections by mycobacteria occurs fairly frequently both in fresh water as well as in sea water fishes. Such infections are encountered particularly in fishes that are already weakened by unfavourable living conditions, e.g. in aquaria. The marked increase in the practice of fish culture in recent years has led to an increase in bacterial fish diseases. Foremost among these are vibrioses and furunculoses. Correspondingly, treatment has acquired ever greater importance and besides the introduction of antibiotics, the best results are achieved by fish immunization.

The number of parasites amongst aquatic fungi is even greater than amongst bacteria. Many unicellular and multicellular algae are attacked by phycomycetes which either live on the surface of their hosts (epibiotic) and only send fine rhizoids into the cells for food uptake, or grow entirely inside the host cells (endobiotic). Some lower fungi, for example *Polyphagus euglenae*, grow between several individuals of algae (interbiotic) and hold them together by means of their rhizoids (see Section 5.1). There are obligate and facultative parasites. Infection takes place by motile zoospores (see Chapter 5). Generally, only individual algae are infected here and there; but mass occurrence of parasitic fungi has occasionally been observed in plankton blooms. Canter and Lund (1948) found, in the northern English Lake Windermere, that together with the diatom *Asterionella formosa*, its parasite *Rhizophydium planktonicum* multiplied more vigorously in spring, but without any rise in the actual percentage of infected *Asterionella* cells (Figure 74). The fungus may affect the growth rate of the plankton in spring, but it rarely speeds up the decline in the number of cells later. The infection, then, does not alter the maximum production of *Asterionella* cells; this depends, above all, on the concentration of nutrients in the water (Canter and Lund, 1951). Also, various species of the net slime mould *Labyrinthula* (see

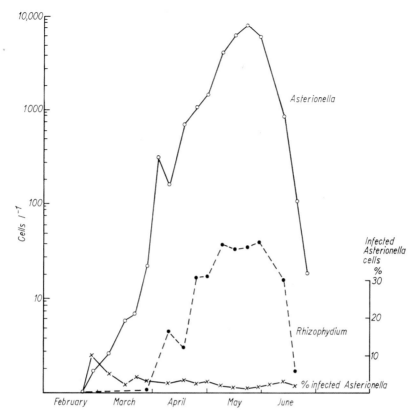

Figure 74   Seasonal changes in the incidence of the diatom *Asterionella formosa* and its parasite *Rhizophydium planktonicum* in the northern part of Lake Windermere, in late winter and spring. (Based on Canter and Lund, 1948, modified)

Chapter 5) live as parasites in higher algae and in seaweed. Schmoller (1960) found *Labyrinthula coenocystis* in tufts of colourless *Cladophora* and regards the myxomycete as the cause of the destruction of the chloroplasts. A disease of the eelgrass *Zostera marina* which has been observed frequently during the last decades is also thought to be due to infection with a *Labyrinthula* (Renn, 1936).

Numerous lower fungi parasitize invertebrates and their eggs and larvae (Sparrow, 1960), but these infections rarely occur in epidemic form. Nevertheless, several phycomycetes have become recognized as causes of dangerous diseases in molluscs and crayfish of economic importance. *Dermocystidium marinum* causes the so-called 'warm season mortality' of oysters on the American Gulf Coast. This disease makes its appearance at high water temperatures of between 25 and 32 °C. Besides the water temperature, the salinity also seems to be important for infection. *Dermocystidium marinum* is a phycomycete whose systematic position is still obscure*. It infects all the tissues of mussels with the exception of the peripheral nerves and the epithelium. The host is injured first by the lytic activity of the fungus and later, in addition, by embolisms in the circulatory system. As a rule, both host and parasite reproduce before the animal dies (Johnson and Sparrow, 1961; Aldermann, 1976).

---

* Recently this organism has been assigned to the Protozoa *(Apicomplexa)* under the name, *Perkinsus marinus* (see Lauckner, 1983).

Various phycomycetes, such as *Lagenidium callinectes* and *Pythium thalassium*, parasitize the eggs of crayfish, but the economic damage they cause does not seem very great (Johnson and Sparrow, 1961).

Members of the genus *Coelomyces* (Blastocladiales) live as parasites in the larvae of aquatic insects (e.g. mosquitoes). They produce a naked, more or less branched, mycelium and numerous cysts within the hosts's body (Sparrow, 1960).

In the region of the sea, asco-and deuteromycetes (Fungi Imperfecti) have been observed, particularly as parasites or as saprophytic inhabitants of higher plants (e.g. of the mangrove tree, *Rhizophora mangle*, and the graminzceous *Spartina townsendii* in brackish waters of river estuaries) as well as of algae and animals. In tropical waters, infections can spread very quickly and the degradation capacity assumes considerable proportions (Kohlmeyer and Kohlmeyer, 1979; Newell, 1976).

Yeasts also probably play a role as pathogens in the case of various marine invertebrates, according to van Uden and Fell (1968) and Kohlmeyer and Kohlmeyer (1979), and some higher ascomycetes and Fungi Imperfecti are parasites of lower animals, particularly fresh water ones.

A number of fungal diseases have become recognized in fish of economic importance (Reichenbach-Klinke, 1966). Lower fungi cause predominantly external mycoses in fresh water fish. Generally, however, only weakened or injured animals are attacked as the spores cannot germinate on the skin of healthy animals. Amongst these pathogens the Saprolegniaceae are important. Diseases due to *Achlya* and *Saprolegnia* species are widespread. They form whitish or brownish lawns like cotton wool on the animals. A species of *Saprolegnia* is the cause of a fatal illness of salmon which occurs frequently in Scotland and Ireland (Willoughby, 1969). Gill rot of rainbow trout may also be caused by Saprolegniaceae, though in other fish this disease usually seems to be caused by *Branchiomyces* species. Here two species may be distinguished: *B. sanguinis*, the causative organism of carp gill rot, and *B. demigrans* which causes gill rot of pike and tench. Hyphae of the latter fungus grow outwards out of the gills. However, the two species also occur in other fresh water fishes. At first the presence of the fungus causes a blockage of the gill vessels and finally leads to the destruction of the gill tissues (Amlacher, 1981).

A very dangerous pathogen, in fresh water and in marine fish, is *Ichthyosporidium hoferi*, a phycomycete, though its systematic position is not entirely clear. It causes the so-called 'Taumelkrankheit', the main symptom of which is loss of muscular co-ordination. Cysts, plasmodia, non-septate hyphae, endoconidia, chlamydospores and spores have been found. Dorier and Degrange (1961) have worked out the various stages of a life cycle. The infection of the fish takes place orally with the intake of material containing the parasite. The digestive juices in stomach and intestines dissolve the cyst walls, thus liberating the infective individuals of the unicellular stage which can pass through the intestinal wall. With the circulating blood they reach the various organs (particularly liver and kidneys). Here cysts are formed which release multinucleate plasmodia. These divide and eventually form new cysts. As the plasmodia hatch, broad hyphae may develop. Old cysts may give rise to chlamydospores or endoconidia. After the host has died, budding hyphae are formed which may grow into rather large hyphal bodies. At the end of the hyphae, infectious plasmodia develop which are likewise taken up by new hosts. Besides liver and kidney, sexual organs, heart, brain, gills and muscles are also affected. Amongst higher fungi, *Penicillium piscium* causes diseases of the internal organs of various fish (Reichenbach-Klinke, 1966).

Viruses also play a role as pathogens of aquatic plants and animals. Thus, virus infections have been observed in different eucaryotic algae (Brown, 1972, Dodds, 1979). Invertebrates and fish also may be attacked by viruses. Various virus diseases of fish are known and both RNA-viruses as well as DNA-viruses may occur. The

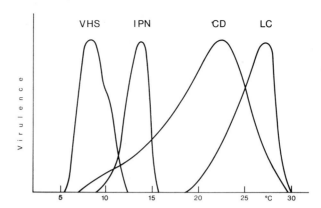

Figure 75   Infectivity of fish pathogenic viruses in relation to the water temperature. VHS, trout pest; IPN, infectious pancreatic necrosis; CD, cauliflower disease; LC, lymphocystis. (Based on Ahne and Wolf, 1980, modified)

former include the causative agents of trout pest (Egtved virus) and of infectious pancreatic necrosis of trout; the latter include lymphocystis in flounders and plaice, cauliflower disease of the eel and also fish pox. The primary cause of infectious dropsy in carp also seems to be a virus (Reichenbach-Klinke, 1966). Mowus. (1963) states that viruses may also be transmitted via intermediate hosts, for example ciliates.

The infectivity of these fish pathogenic viruses, to a large extent, is dependent on the water temperature. It is maximum, for example, in the case of the infective agent of trout pest, at 8 °C, with infectious pancreatic necrosis at 13 °C; with cauliflower disease of eels at 22 °C, and in the case of lymphocystis at 25 °C, Thus, the temperature ranges of infectivity of different viruses are to some extent very different (Figure 75) (Ahne and Wolf, 1980).

10.4. Symbionts

Micro-organisms living in plants or animals may be useful to their hosts. Such symbionts are known in numerous terrestrial animals and also in quite a number of terrestrial plants. Their importance lies above all in the preliminary processing of nutrients. The necessity of this is relatively small in aquatic animals because of their uniform food. However, symbionts play perhaps some role in animals whose food originates on dry land. This is so, for example, with the wood-boring ship's borer (*Limnoria*) and the shipworm (*Teredo*). These animals require the cooperation of fungi and bacteria which partially decompose the wood and so facilitate its colonization by boring organisms. Probably they supply them also with additional nutrients (Kohlmeyer and Kohlmeyer, 1979). A true endosymbiosis as has for long been assumed, however, does not appear to exist.

True endosymbiosis exists in blood leeches (*Hirudo medicinalis*), whose intestinal flora consists of bacteria of the species *Pseudomonas hirudinicola*; these lyse erythrocytes and thus enable the animals to utilize their protein and fat components. If the intestines of the blood leech are made sterile with chloromycetin, the erythrocytes which have been taken up remain intact and cannot be utilized as food. In other kinds of leeches the symbionts live in side pockets of the oesophagus; but as digestion progresses they are found also at the further end of the stomach (Kaestner, 1955; Wilde, 1975). Possibly some inhabitants of the digestive organs of other aquatic

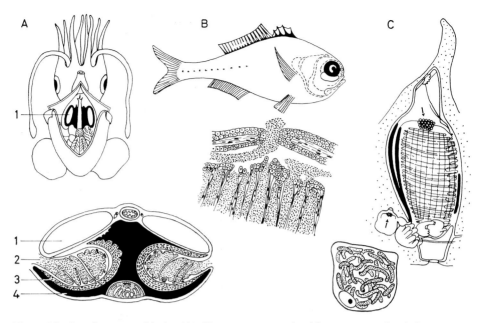

Figure 76   Luminous symbiosis. (A) *Euprymma morsei* with open mantle; below, transverse
section through the region of the luminous organs: 1. lens, 2. orifice of the luminous
organ with bacteria, 3. reflector, 4. ink sac. (B) *Anomalops katoptron*, with luminous
organ under the eye; below, section through the orifice of the luminous organ showing
bacteria in its folds. (C) *Pyrosoma*, with luminous organ (arrows); below, a myceto-
cyte. (Based on Buchner, 1960, modified)

animals are also symbionts (see Section 10.2), but there are no reliable observations of
this. The extremely interesting luminous symbiosis occurs solely in the aquatic
environment. A number of marine animals have complicated luminous organs where
light is produced by luminous bacteria (Figure 76). The luminous organs of the pyro-
somes, which belong to the tunicates, are relatively simple; these lie in pairs above
the branchial region of the intestine, and consist only of an agglomeration of mono-
nuclear so-called mycetocytes which contain the bacteria intracellularly. No other
auxiliary structures are present. In every other case the bacteria are situated extra-
cellularly in cavities that have arisen by invagination of the surface of the body
or of the gut. They are always open to the outside. Some cephalopods have highly
developed auxiliary organs which considerably enhance the luminescence. In *Euprymma morsei*, for example, the luminous organs, arranged in pairs, are sunk into the
sides of the ink bag which serves at the same time as a dimmer. On the underside
they are surrounded by reflectors, and on the outside are very effective lenses which
make the powerful luminescence of these animals possible (Figure 76). In *Loligo*
species, the luminous organs flank the terminal part of the intestine and are provided
with relatively long lenses. The luminosus organs in cuttlefish are a recognition
devide for mating; in some cases they probably help members of a shoal to remain
together. In fish they also, in many cases, function as light traps which help them to
get their food. Thus *Anomalops katoptron* has a very bright bean-shaped luminous
organ under each eye; it emits light — as in *Photoblepharon* — for some time after the
fish has died, so that the luminous organs are used by fisherman of some South Sea
islands as bait. The Japanese fish *Monocentrus japonicus* has a bulb-shaped luminous
organ on its lower lip. It is thought that in some pediculates whose luminous organs

hang from their lips like fishing lines the light is also produced by bacteria. Some other fish, like *Malacocephalus laevis* and several gadiformes, have the luminous organs situated near the anus (Buchner, 1953, 1960). The symbiotic luminous bacteria belong to the genus *Photobacterium* (or *Vibrio* and *Lucibacterium* respectively), which occur also in free water. The animals provide the food and, in some cases, probably also regulate the oxygen supply, thus controlling the bioluminescence. Hastings and Mitchell (1917) found 4000–12 000 millions of luminous bacteria in 1 g of the tissues of the luminous organ of the ponyfish, from which it follows that 10–25 per cent of the luminous organ consists of viable luminescent bacteria.

However, luminescence of animals is not necessarily due to symbiosis with luminescent bacteria. In many cases, the animals themselves luminesce, for example protozoa, crayfish, snails, mussels and deep sea fish. The biochemistry of their luminescence is not much different from that of the luminous bacteria. In bioluminescence the conversion of chemical energy into light is, as it were, a reversal of photosynthesis. The process of luminescence is dependent on the presence of oxygen and is regarded as an aerobic oxidation process which does not lead to production of ATP but to emission of light. In bacterial luminescence the light-emitting substance seems to be a long-chain aldehyde. Besides this 'luciferin', the enzyme luciferase, flavine mononucleotide (FMN), nicotinamide adenine dinucleotide (NAD) and oxygen take part in the process of bioluminescence (see Schlegel, 1981).

A few years ago, a symbiosis was discovered for the first time between chemoautotrophic sulphur bacteria and a sea animal, *Riftia pachyptila*, a pogonophore (Cavanaugh *et al.*, 1981). This involved a worm, without mouth or gut, which lives in the surroundings of hot springs with H_2S-containing water, in the deep sea (see p. 18). Later, symbiotic bacteria were found also in other pogonophores (Figure 77). Cavanaugh (1983) succeeded in detecting chemoautotrophic bacteria, similarly, in mussels from sulphide-containing sediments. In *Solemya velum*, which occurs in the reducing mud of seaweed fields, bacteria were found in the gills, which are very thick and fleshy, in comparison with those of related species from aerobic sediments, whilst the gut is strongly reduced. Similar observations have been made with other mussels in H_2S-containing sediments. It may be presumed, therefore, that symbioses of marine invertebrates with bacteria in corresponding sites are widespread and play an important role in the nutrition of these animals.

Besides endosymbioses in which the larger partner is regarded as the host organism, there are also ectosymbioses in which the partners only form an association. This is the case, for example with *Chlorochromatium aggregatum* in which two different bacteria participate (Figure 78). It grows in barrel-shaped aggregates of cells, each consisting of a relatively large colourless cell with a polar flagellum which is surrounded by 8–16 smaller green cells. The organism grows in mud and in stagnating water with a high content of H_2S, as long as there is sufficient light for assimilation (Breed *et al.*, 1957). The flagellated central cell moves the aggregate along and probably receives its nutrients from the photosynthetic outer cells.

Similar symbiotic relationships also exist between sulphur reducers and photosynthetic sulphur bacteria; an example of this is the association between *Desulphuromonas acetoxidans* and *Chlorobium*. In a central colony of *Desulphuromonas*, acetate is oxidized as a result of reduction of sulphur. Hydrogen sulphide so formed serves as hydrogen donor to the adjacent *Chlorobium* cells. The excreted sulphur diffuses, probably as polysulphide, to the central *Desulphuromonas* colony. Here it is used in turn as an hydrogen acceptor so that it is again reduced to H_2S (Schlegel, 1981). A corresponding syntropic association can develop also between the sulphate reducing *Desulphovibrio* and the purple sulphur bacterium *Chromatium*.

Cyanophytes also have an important role to play as symbionts. Some species together with fungi constitute different lichens (see Chapter 4) in which the symbiosis

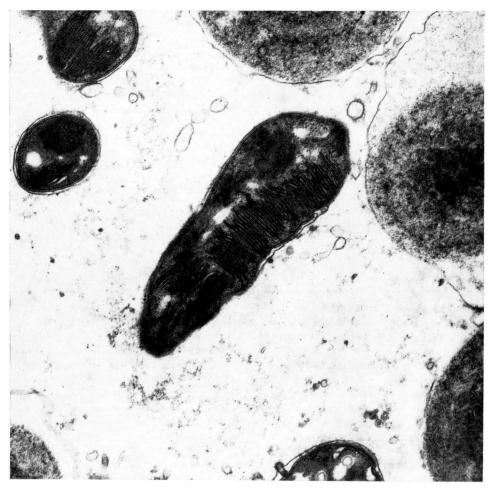

Figure 77   Longitudinal and transverse sections of symbiotic bacteria with lamellar structures characteristic of chemoautotrophs, from the trophosome of *Siboglinum poseidoni* (Pogonophora) ($\times 28\,000$). Preparation and electron micrograph by H. Flügel

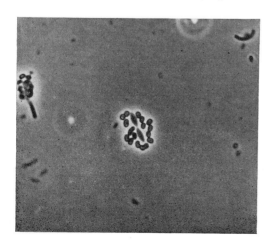

Figure 78   *Chlorochromatium aggregatum* from Lake Cassidy, Michigan, USA ($\times 1300$). (P. Hirsch)

is most clearly defined. The algae supply the organic nutrients which they have acquired through photosynthesis and the fungal hyphae which enclose them like a bark ensure especially uniform humidity. Consequently lichens can maintain themselves even on rocks in the spray region of waters. In some algae, so-called cyanellae have been observed in which cyanophytes might be involved which seem, partially of course, to have lost their independence. In the marine diatom *Rhizosolenia* trichomes with terminal heterocysts occur. This organism has been described as *Richelia intracellularis* (Fogg *et al.*, 1973). The cyanellae certainly play a role in nutrient supply although more detailed investigations have not yet been made. Cyanophytes (*Nostoc*) living in the water fern *Azolla* are important for the nitrogen supply. Not infrequently cyanophytes occur in various sponges, bryozoa and worms. Symbiotic relationships may be assumed in all these cases.

11. *The Role of Bacteria and Fungi in the Cycling of Elements in Waters*

Bacteria and fungi play a decisive role in the cycle of matter in waters. Their part in the primary production of organic material is small but they remineralize it on a large scale as, under the appropriate conditions, they are able to break down virtually all natural organic compounds into the components from which they originated, i.e. carbon dioxide, water and various inorganic salts. The extent to which they are actually responsible for the breakdown of organic matter in waters is, however, an open question (Rittenberg, 1963; Johannes, 1968). Formerly, it was accepted that the remineralization of organic substances in water — as in soil — was essentially due to the activity of bacteria and fungi, while animals had only a subordinate role to play (Waksman, 1933). Today we know that the zooplankton in particular is very important and that colourless algae are also involved. Added to this, there is autolysis. Nevertheless, seen as a whole, the share of bacteria and fungi in remineralization is probably greater than that of all other creatures and autolytic processes together. The suggestion by Johannes (1968) that the zooplankton sets free many more inorganic phosphorus and nitrogen compounds than do bacteria and fungi can hardly be accepted as valid for all waters. The colonization by zooplankton of wide regions of the sea and of numerous inland waters is far too insignificant, while bacteria are practically ubiquitous and are able to multiply quickly whenever food is available (see Section 3.2). In waters with abundant nutrient material, much of the zooplankton feeds on bacteria which, prior to this, have broken down organic matter. Moreover, many substances, for example cellulose, lignin, agar and chitin, can be attacked almost exclusively by bacteria and fungi or with their cooperation. The same is true for hydrocarbons, phenols, etc. The significance of micro-organisms for the cycling of matter in waters is, however, not exhausted by partial or complete breakdown of organic substances; they also produce changes in inorganic compounds, for example fixation of free nitrogen, nitrification and denitrification, sulphur oxidation and sulphate reduction.

The oxygen content of waters is also much affected by the activity of bacteria and fungi. Microbial decomposition involves oxidation processes which lead always to considerable oxygen consumption; under unfavourable conditions this may result in the complete disappearance of oxygen. Niewolak (1974b) found in the Ilawa lakes (Masuria) that oxygen consumption due to bacterial respiration processes amounted to nearly 0.3 mg O_2 per litre per 24 h in summer, whilst in winter the daily average consumption was not more than 0.02 mg O_2 per litre.

In all transformations of matter by micro-organisms, enzymatic processes are involved in which, as a rule, several enzyme reactions are brought into play one after the other.

11.1. Production of Organic Matter

Production of organic substance in waters is, above all, due to cyanophytes and algae, and particularly to the microscopically small forms of the phytoplankton. Higher plants are important only in the regions along the shores.

The bulk of bacteria and all fungi are heterotrophic organisms and therefore not involved in the production of organic material. Only the small group of photo- and chemoautotrophic bacteria can be regarded as primary producers. To judge by the present state of our knowledge, their share in this production must be rather small; both photo- and chemoautotrophic bacteria can become enriched only under very special conditions which are hardly ever present simultaneously in water. Thus, all bacteria capable of photosynthesis are unable to cleave water, and as obligate anaerobes or microaerophils they cannot grow in waters rich in oxygen. They further require sufficient light and suitable hydrogen donors: hydrogen sulphide for the chloro- and purple sulphur bacteria and organic acids or other organic compounds for a few other photosynthetically active organisms. All these requirements do not frequently coincide in waters. But where they do, as in some eutrophic ponds or small lakes, and in lagoons and pools of coastal regions (see Section 11.6), there may be massive growth of photosynthetic bacteria accompanied by production of considerable amounts of organic substance. In August, 1971, at two sites on Lake Wadolek, in north Poland, Czeczuga and Gradski (1973) compared the primary production of *Chlorobium limicola* in the hypolimnion (4–7 m deep) with that of the phytoplankton in the epilimnion (0–3.5 m). The following mean values were obtained:

	Primary production mg C per cm^3 per day		Number of samples
	Site 1	Site 2	
Phytoplankton	22.24	21.66	5
Chlorobium	17.31	8.46	7

The result obtained at the first site shows that the primary production due to photosynthetic bacteria here reaches about three-quarters of that of the phytoplankton.

Table 4   The most important chemoautotrophic bacteria occurring in waters, and their energy and oxygen sources

	H		Autotrophic	
	Donor	Acceptor	Obligatory	Facultative
Nitrosomonas	NH_3	O_2		
Nitrosococcus	NH_3	O_2	+	
Nitrobacter	NO_2^-	O_2	+	
Nitrospina	NO_2^-	O_2	+	
Nitrococcus	NO_2^-	O_2	+	
Thiobacillus	H_2S, S, $S_2O_3^{2-}$	O_2	+	+
Thiobacillus denitrificans	H_2S, S, $S_2O_3^{2-}$	NO_2^-, NO_3^-, O_2	+	
Beggiatoa	H_2S	O_2		+
Thiothrix	H_2S	O_2		+
Sulpholobus	S	O_2		+
Ferrobacillus (Thiobacillus) ferrooxydans	Fe^{2+}	O_2	+	
Gallionella ferruginea	Fe^{2+}	O_2	+	
Leptothrix ochracea	Fe^{2+}, Mn^{2+}	O_2		+
Crenothrix polyspora	Fe^{2+}, Mn^{2+}	O_2		+
Hydrogenomonas	H_2	O_2		+
Micrococcus (Paracoccus) denitrificans	H_2	O_2, NO_2^-, NO_3^-		+

Nor do the chemoautotrophic bacteria (Table 4) seem of great importance for the production of organic material in water. It is true that they have a wider distribution (see also Zavarzin, 1972); nitrifying bacteria, for example, occur in almost all aerobic waters and, in most of them, are able to carry out nitrification (see Section 11.5) and thereby synthesis of organic matter, but their numbers are, almost everywhere, rather small so that only small quantities of organic substance are being produced. Greater numbers of nitrifying bacteria have been observed only rarely in the aquatic environment, if at all, in the surface region of some aerobic sediments. Even under favourable conditions, their performance is rather limited because of the low efficiency of the reaction, which for intstance in *Nitrobacter winogradskyi* is only about 10–11 per cent (Schön, 1964). Other chemoautotrophic bacteria occur widely but are found in large numbers in only a few waters.

Relatively frequent are the sulphur-oxidizing bacteria, particularly members of the genus *Thiobacillus* (see Section 11.6). Occasionally they accumulate where waters containing oxygen and H_2S, respectively, meet; i.e. again in a very limited habitat. The autotrophic iron and manganese bacteria also develop massively only locally; moreover, the oxidation of ferrous or manganous compounds to ferric or manganic compounds, respectively, is associated with only a small gain in energy (see Section 11.8). Starkey (1945) states that 280 g of ferrous iron have to be oxidized by iron bacteria to produce 1 g of cell substance. In eutrophic lakes, bacteria which oxidize hydrogen, carbon monoxide or methane may also play a role as producers:

$$2 H_2 + O_2 \rightarrow 2 H_2O + 138 \text{ kcal}$$
$$2 CO - O_2 \rightarrow 2 CO_2 + 148 \text{ kcal}$$
$$CH_4 + 2 O_2 \rightarrow CO_2 + 2 H_2O + 195 \text{ kcal}$$

From all this it seems that the few bacteria capable of photo- or chemosynthesis contribute in some measure to the production of organic matter in only a few waters. It is, however, not yet clear what is the share of chemoautotrophic organisms with regard to production in the sea; we do not know to what extent chemoautotrophic bacteria are active in the aphotic zone which, after all, constitutes the greatest space of all possible biological habitats.

It must not be overlooked that the energy sources for the chemoautotrophic bacteria originate mostly during remineralization of organic substance (NH_3, NO_2, H_2S, H_2, CH_4). Chemosynthesis in the waters is, then, essentially only a secondary process.

However, bacterial sulphur oxidation in the vicinity of thermal vents seems to be an exception. For, in this case, geothermal energy is involved. Nevertheless, essentially, green plants should be regarded as the primary producers (see Kusnezow *et al.*, 1963).

Secondary production by heterotrophic bacteria and fungi may, however, be considerable, as around 20–60 per cent of the organic nutrients is used for the biosynthesis of microbial substance, whilst 40–80 per cent is involved in energy metabolism. Multiplying the biomass of bacteria present in a body of water (see p. 68) by the number of generations allows us to calculate the annual carbon production of the bacteria. The generation time of aquatic bacterial varies — depending on the kind of organism, the temperature, concentration of nutrients and other factors — between 20 min and many days. In warmer eutrophic waters it is probably, on an average, a few hours; in cool oligotrophic inland waters and in large parts of the open sea it may be several days. ZoBell (1963) calculated, on the basis of 1000 bacteria per ml water and a generation time of 24 h, an annual production of 7.3 mg C per m³. In studies on one of the Masurian lakes, Niewolak (1974a) calculated for a volume of 1446 million m³ an annual bacterial production of 16753 tons of carbon.

According to Overbeck (1979b, 1981, 1982), in the 14 hectare, 29 m deep eutrophic Lake Pluss in East Holstein (Germany), which has a volume of 1.3 million m³ of water,

10*

there is an annual production of 25 tons of algal biomass, 80 per cent of which was already transformed in the uppermost 5 m particularly by bacteria. The glucose turnover, at 2 tons per year, constitutes about 10 per cent of the total heterotrophic activity. Investigations in other Holstein lakes have also yielded analogous results. The proportion of glucose transformed by bacteria here likewise amounted to about 10 per cent.

The proportion of bacterial production in the primary production has been determined in different waters. For example, Meyer-Reil (1977) found an annual bacterial production in the Kiel Bay of about 9 g C per m^2 and in the eutrophic Kiel Fjord of 57 g C per m^2. This corresponds to 15 and 30 per cent, respectively, of the primary production in each case. For the western Baltic Sea, one may assume a total of 20 per cent. A further 15 per cent is converted into bacterial biomass in the sediment of this relatively shallow sea area with an average depth of 20 m (Meyer-Reil and Faubel, 1980). Kuparinen et al. (1984) determined the bacterial production in the Finnish archipelago to be about 10 per cent. The values of 7.5 g C per m^2, or 15 per cent of the primary production, found by Hagström et al. (1979) on the basis of the frequency of dividing cells in the Baltic Sea, to the south of Stockholm, are also of a similar order of magnitude. Jordan and Likens (1980) determined an annual bacterial production of 3–8 g C per m^2 in the oligotrophic Mirror Lake (USA), which is 8–22 per cent of the primary production. In the eutrophic Swedish Lake Norrviken, Bell et al. (1983) estimated a value of about 50 per cent by determination of [^{3}H] thymidine incorporation. Even higher bacterial production, according to Overbeck (1979), is found in the eutrophic Lake Pluss (Holstein, Germany). Here it reached about 90 per cent of the primary production.

Considerable amounts of organic matter, the major part of which can be converted into bacterial biomass, are also produced, under favourable light conditions, by the micro- and macrophytobenthos of shallow waters. According to Meyer-Reil et al. (1980), primary production by the microphytobenthos in sand sediments of the Kiel Bay amounted to 3.7 and bacterial production to 1.8 mg C per m^2 per h, which is nearly 50 per cent. For a typical large algal community on the west coast of the Cape Peninsular (South Africa), Newell and Field (1983) assessed the primary production by phytoplankton as 501, and by the macrophytes as 917, and bacterial production as 130 or 262 g C per m^2 per year, respectively. This is 392 g C per m^2 per year, or about 28 per cent of the total primary production of 1418.

According to Sorokin (1965) and Kusnezow and Romanenko (1966), the proportion of heterotrophic CO_2 fixation in waters would constitute around 6 per cent of the bacterial biomass production. This would, again, permit us to determine the heterotrophic carbon production. However, Overbeck (1972) considers that the assumption of a uniform value of 6 per cent is open to question.

Distinct daily fluctuations of bacterial production were measured by Moriarty and Pollard (1982) in the water and sediment of a seaweed bed in Moreton Bay, Queensland (Australia) by determining the rate of incorporation of [^{3}H] thymidine into the DNA of the bacteria. Rates of cell production increased in the morning by 5–10 times and decreased in the afternoon. However during the night no changes were observed. Total bacterial production was estimated at 43 mg C per m^2 per day.

11.2. Substrate Uptake by Heterotrophic Micro-organisms

Heterotrophic micro-organisms take up organic materials as food and this uptake constitutes a measure of their activity. In pure cultures of bacteria this is quite unequivocal and there exists a positive correlation between the number and biomass of active cells and the consumption of substrate. In mixed cultures and above all in the case of natural aquatic populations, this connection is not always so clear because it is unlikely that all the active bacteria present will take up the proffered nutrients in the same way. Depending on the quality and concentration of the organic materials as well as on various environmental factors such as water temperature, reaction, salt content and hydrostatic pressure, the consuming capacity of the individual micro-organisms will be different. Fundamentally only dissolved substances can be taken up by bacteria and fungi; with protozoa, for example, it is different. Particulate material, therefore, must first be brought into solution by exoenzymes.

One distinguishes between active uptake of material, which proceeds against the concentration gradient and consequently requires energy, and the passive uptake of material due to diffusion and for which no energy is needed. The active uptake of material through the cytoplasmic membrane takes place selectively. Only those substrates for which a suitable transport mechanism is available can be taken up into the cell. In most cases it seems that inducible substrate-specific enzymes, which are termed permeases, are involved. Whilst active material uptake can take place even at very low substrate concentrations with a high transport velocity, passive material uptake is dependent on the substrate concentration and attains a maximum only at very much higher concentrations. Accordingly it plays only a subordinate role in the nutrient uptake of aquatic micro-organisms, since, as a rule, the nutrient concentrations in waters are relatively low (see Section 8.8). It is different with substances foreign to the cell; these can often be passively taken up by cells up to the concentration equilibrium. This is true also for many poisons. This process becomes of importance particularly in polluted waters.

The uptake of CO_2 especially by the photoautotrophs — and of organic nutrients by the heterotrophic micro-organism population of waters — can be measured with the help of ^{14}C- or ^{3}H-labelled substances (see Figure 79) Hoppe, 1978).

The rate of uptake of easily assimilable organic nutrients, such as glucose or acetate, represents a useful measure of the heterotrophic activity of the bacterial population in a body of water. Together with bacterial numbers and biomass, the heterotrophic activity is of great significance in ecological water investigations (Gocke, 1977) and it permits statements on the microbial degradation of organic substances occurring in the water (see Section 11.3).

The rate of uptake is derived from the following equation:

$$v = \frac{f}{t} \cdot (S_n + A)$$

where v is the uptake rate (μg C l^{-1} h^{-1}), f the proportion of the isotope taken up, t the incubation time, S_n the natural substrate concentration and A the concentration of substrate added.

Since the natural substrate concentration S_n is usually difficult to measure, the maximal uptake rate V_{max} is frequently determined (see Parsons and Strickland, 1962; Wright and Hobbie, 1966; Gocke, 1975a, 1977). This is achieved by the addition of a relatively high concentration of the labelled substrate, so that $(S_n + A)$ corresponds nearly to A. The uptake rate is dependent on the substrate concentration and shows a saturation curve. The relationship between uptake rate and substrate concentration is analogous to the velocity of enzymatic reactions, a function of the substrate concentration:

$$v = V_{max} \cdot \frac{S}{K + S}$$

V_{max} = maximal rate of uptake (μg C l^{-1} h^{-1}), K = half saturation constant (μg C l^{-1}), S = substrate concentration (μg C l^{-1}) as the sum of $S_n + A$.

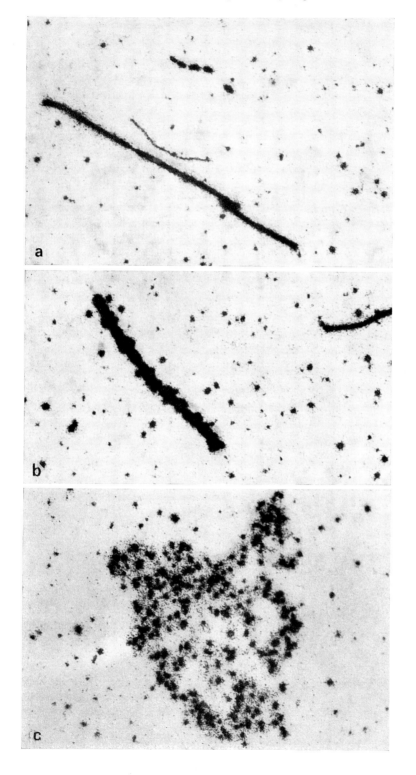

Combining the two equations and replacing K by the transport constant K_t gives a linear relationship, which is known as the Lineweaver–Burk equation:

$$\frac{t}{f} = \frac{K_t + S_n}{V_{max}} + \frac{A}{V_{max}}$$

The maximal uptake rate of one or several substrates is a relative measure for the size of the micro-organism population, which can take up these substances and use them as food. A comparison of the heterotrophic activity of different bodies of water it thus facilitated. Experiments with plankton fractionated according to size (Gocke, 1975 b) have shown that up to 75–90 per cent of the heterotrophic uptake of glucose in the Kiel Fjord is due to bacteria and only a very small part is due to larger plankton fractions. The same might be true also for other organic compounds dissolved in the water. By means of the macroautoradiographical method, Hoppe (1977) established that all saprophytes from quite a large number of samples of different coastal waters could take up glucose. The following list shows that most other compounds tested were also taken up by a large proportion of the saprophytic bacteria:

	Kiel Fjord	Kiel Bay	Number of samples
Xylose	77 %	74 %	19
Galactose	41 %	40 %	12
Lactose	71 %	68 %	19
Sodium acetate	94 %	93 %	18
Glycolic acid	34 %	31 %	12
Uric acid	28 %	29 %	19
Fat	18 %	25 %	19
Phenol	21 %	14 %	20
Riboflavin	34 %	23 %	17

Ammerman and Azam (1982) found that some bacteria take up cyclic nucleotides, especially cyclic AMP.

Investigations carried out in numerous waters in the last two decades showed large differences in heterotrophic activity depending on the season and level of eutrophication (see Figures 80 and 81). In some Masurian lakes of differing degree of eutrophication, Godlewska-Lipowa (1974a–c) found a positive correlation between the degree of eutrophication, total bacterial numbers and the heterotrophic activity of the bacterial flora in the case of glucose, asparagine and peptone, whilst with other substrates for example sucrose, lactose, salicin, cellobiose and cellulose, the connection was by no means so clear. In the western Baltic Sea, Gocke (1977) found a correlation between the maximal uptake velocity of glucose, acetate and asparagine, with the bacterial biomass (Figure 80) and with the saprophyte numbers also in more strongly eutrophicated waters (see Figure 40). In a 15-month investigation in the Kiel Fjord very low values were obtained in the winter months and maxima in spring and late summer. The maximal uptake rate of glucose here fluctuated between 0.016 and 0.663 μg C per l^{-1} h^{-1}. In the less heavily loaded central Kiel Bay, values of between 0.004 and 0.148 μg C l^{-1} h^{-1} were recorded at the same time. From these

Figure 79   Autoradiographic illustrations of active micro-organisms: (a) various bacteria (cocci, rods — singly and in chains — and sheathed bacteria); (b) free bacteria and cyanophytes (*Anabaena*) with bacterial aufwuchs; (c) detritus with heavy bacterial colonization ($\times 1\,200$). (H. G. Hoppe)

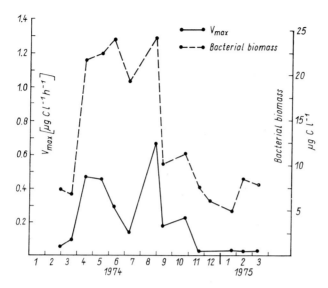

Figure 80   Maximal uptake rate ($V_{\max}$) of glucose and bacterial biomass in the Kiel Fjord during the course of the seasons. (Based on Gocke, 1977, modified)

results a connection is revealed between the primary production of the phytoplankton, bacterial development and heterotrophic activity. Iturriaga and Hoppe (1977) observed that in the summer months from 2 to 21 per cent of the primary production was given off in the form of exudates, which accordingly formed an important source of nutrients for the bacteria. Corresponding observations were made by Straskrabova and Desortova in a retaining dam reservoir in Czechoslovakia. Here, 5–41 per cent of the primary production was assimilated by bacteria and an equally large proportion was respired.

Wolter (1982) determined exudation rates for natural phytoplankton populations in the Kiel Fjord and rates of uptake by bacteria using a modified differential filtration procedure and found a positive correlation between the primary production and exudation. The highest concentration was found in July, 1978, with $239\,\mu$g C per litre, after a 6 hour-incubation with ^{14}C-labelled bicarbonate. This corresponded to 27.6 per cent of the primary-production. The proportion of exudate given off depends in the composition of the phytoplankton population. Thus, 5.1–12.5 per cent of primary production was exuded from *Sceletonema costatum*, 10–40 per cent from nanoflagellates, somewhat over 20 per cent from *Prorocentrum micans*, whilst only very small amounts were exuded from *Gymnodinium* and *Peridinium* species.

Bacteria assimilated up to 90 per cent of all the available exudates. Distinct seasonal fluctuations were revealed thereby. The largest proportion of exudates was taken up by bacteria in June, 1978, when dinoflagellates and nanoflagellates predominated; but, in contrast, only about 10 per cent when diatoms were dominant. Bacteria took up 11.4 per cent of exudates in the sample of 5 July, 1978, which contained only *Sceletonema costatum*.

Bell *et al.* (1947) found that in a *Sceletonema* culture assimilation was stimulated in some bacteria but inhibited in others. Thus the uptake of exudates showed correspondingly large differences:

Pseudomonas	$16 \times 10^{-8}\,\mu$g C per cell per hour
Spirillum	$0.4 \times 10^{-8}\,\mu$g C per cell per hour

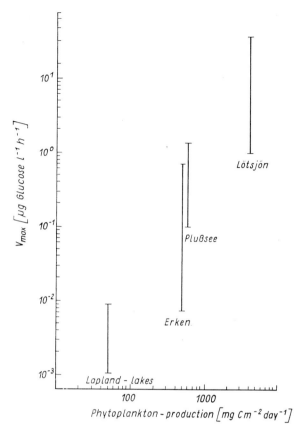

Figure 81   Comparison of the maximal uptake rate (V_{max}) and primary production in northern and central European lakes of differing trophic levels. (Overbeck, 1973)

Bölter (1981) observed that the uptake of exudates corresponded possibly to those of carbohydrates. According to Bauerfeind (1982), bacteria preferentially degrade dissolved organic substances during the remineralization of phytoplankton detritus. Up to 70 per cent of monomeric sugars and up to 60 per cent of dissolved amino acids were taken up daily by bacteria, which utilized about 70 per cent of the substrate by respiration.

Larsson and Hagström (1979) in an investigation in the northern Baltic Sea, some 70 km to the south of Stockholm, in the months March to December, found that the measurable exudates could be completely assimilated by bacteria. Of the annual primary production of 110 g C per m², 45 per cent was exuded and then taken up by bacteria. The bacterial net production amounted to 29 g C per m² and thereby corresponding to rather more than 25 per cent of the estimated primary production.

Investigations by Novitsky (1983) in Halifax Harbour (Canada) showed that the heterotrophic activity in the border region of water and sediment was greater by several powers of ten than in the water above and twice as high as in the sediment immediately below. This high activity is due, above all, to a marked rise in numbers of active bacterial cells and less to an increase in the percentage of active cells or to higher activity per cell. In this layer, especially in shallow waters, a substantial part of the organic matter is decomposed.

11.3. Breakdown of Organic Matter

Heterotrophic bacteria and fungi utilize the organic nutrients that they consume to provide the precursors for the synthesis of their cell substance and the energy for their life processes. The organic material is converted by the micro-organisms into compounds with a smaller energy reserve and finally, under appropriate conditions, into the original mineral substances. The remineralization of organic substrate is the main function of bacteria and fungi in the turnover of matter in waters. The limiting plant nutrients are returned time and again to the cycle of matter and allow new plant growth. Complete mineralization can, as a rule, be attained only in the presence of oxygen, i.e. in aerobic waters; under anaerobic conditions the breakdown often remains incomplete. Substances which are easily decomposed, such as protein, sugar and the like, are decomposed to a large extent by endoxidations; but the more resistant substances such as fats, cellulose and lignin, accumulate and, together with breakdown products, they contribute to the formation of the so-called sea humus.

The rate of decomposition of organic substance varies according to its constituents and the environmental conditions. The extent of the breakdown of dissolved organic nutrients is expressed in the turnover time, in which the total amount of a given substance in the water would be broken down at *in situ* temperatures by the heterotrophic micro-organisms present. The turnover time can be measured by regression analysis using a method in which, after the addition of a given ^{14}C-labelled compound and after, e.g. 3-h incubation at a certain temperature, the substance uptake and the released amounts of CO_2 are determined radiochemically. Determinations of the turnover times for glucose in various lakes and regions of the sea have been made Some examples are given in the following list:

Lake/Region	Turnover time (h)	Reference
Oligotrophic lakes	40–10000	Rhode *et al.*, 1966
		Wetzel *et al.*, 1972
Mesotrophic lakes	80–470	Wetzel, 1968
Eutrophic lakes	0.4–300	Wetzel, 1968
River Inohzawa (Japan)	32	Seki *et al.*, 1975
North Pacific	1300–6000	Seki *et al.*, 1972
English Channel	24–580	Williams and Askew, 1968
Kiel Fjord	2.8–63	Gocke, 1977
Kiel Bay	5.1–523	Gocke, 1977
Shimoda Bay (Japan)	31	Seki *et al.*, 1975
Tokyo Bay	8.7–20	Seki *et al.*, 1975

Seki *et al.* (1975) report the turnover times of different sugars and amino acids for the surface water of some Japanese waters (1 = River Inohzawa, 2 = Shimoda Bay, 3 = Tokyo Bay, 4 = Kuroshio Current):

	1	2	3	4	
Glucose	32	31	8.7	510	h
Galactose	82	50	9.2	1700	h
Aspartic acid	48	36	19	460	h
Glutamic acid	22	12	4.8	150	h
Glycine	16	18	12	280	h
Alanine	23	11	11	160	h
Lysine	96	23	16	333	h
Protein hydrolysate	40	28	13	150	h

At a site in the English Channel (18 nautical miles from the coast), measurements of the turnover times for glucose showed large seasonal differences (Williams and Askew, 1968):

Date	Time (h)
1. 2. 1967	1 580
25. 4. 1967	888
17. 5. 1967	154
21. 6. 1967	144
17. 7. 1967	24

Gocke (1977) found a corresponding seasonal variation with low summer and relatively high winter values in the western Baltic. The results with acetate and amino acids were also similar. In the following example are shown the turnover times for an amino acid mixture (2 μg C per litre) at a site in the more strongly polluted inner Kiel Fjord and the cleaner middle Kiel Bay (Gocke, 1977):

Date	Kiel Fjord 2 m deep		Kiel Bay 10 m deep	
	Temperature (°C)	Turnover time (h)	Temperature (°C)	Turnover time (h)
28. 2. 1974	3.7	32.5	3.0	101.0
21. 3. 1974	4.4	10.0	4.4	238.0
18. 4. 1974	8.6	3.1	6.8	33.5
16. 5. 1974	10.9	2.5	8.7	56.9
11. 6. 1974	12.8	3.6	12.3	24.3
11. 7. 1974	16.7	5.1	15.5	12.0
29. 8. 1974	17.7	2.9	17.5	18.6
12. 9. 1974	15.5	4.6	15.7	26.5
24. 10. 1974	10.5	2.7	10.4	16.6
21. 11. 1974	7.5	28.3	7.5	47.0
10. 12. 1974	6.5	22.2	6.5	119.0
23. 1. 1975	5.3	25.6	5.1	47.6

Meyer-Reil *et al.* (1978) calculated the turnover times for glucose in sand sediments of the western Baltic Sea. Here also there were seasonal differences as the following example from the Falckenstein Beach Station in the Kiel Fjord shows:

Date	Temperature (°C)	Turnover time (h)
10. 11. 1976	7	0.70
18. 1. 1977	1	2.13
14. 4. 1977	4	1.69
2. 5. 1977	9	0.97
4. 7. 1977	19	0.32

On the other hand, the turnover times of corresponding sediments seem to show only relatively minor site differences, as the investigations on the Schleswig-Holstein and south Swedish coasts revealed.

In regions near the surface, especially at summer water temperatures, degradation can take place, therefore, very rapidly. On the other hand, according to the observations of Jannasch *et al.* (1971) food transported into the deep sea was degraded only extremely slowly. For example, apples and sandwiches in the research submarine 'Alvin' which was sunk a depth of 1540 m were still in astonishingly good condition after 10 months, whereas the starch and protein components at the same temperature, 3 °C, in the refrigerator at atmospheric pressure became spoiled in a few weeks. Experiments with different nutrients such as peptone, starch, mannitol, galactose, acetate etc. showed that the microbial degradation in deep sea took place 10–100 times more slowly than in laboratory controls at the same temperature.

Thus, the degradation of acetate, mannitol, glutamate and amino acids, after 8 weeks in 5300 m (i.e. at about 530 atm) proceeded to only 0.15–12.90 per cent of that found at normal pressure. Also in comparative investigations of decompressed and undecompressed water samples from depths of 1600–6000 m, Jannasch and Wirsen (1982) established that at high pressures there was a lower rate of ^{14}C incorporation and a lower $^{14}CO_2$ production than at normal pressures. However, there were distinct differences with different substances. With the microbial decomposition of acetate, the influence of hydrostatic pressure was appreciably smaller than with that of amino acids. On the other hand, Schwarz *et al.* (1976) observed no slowing down of activity at 70 atm *in situ*, compared with 1 atm, in the gut flora of deep sea amphipods. For free-living bacteria (see Chapter 8.3), on the other hand, there is the decisive question of the supply of energy requirements. In deep sea trenches, near to land, allochthonous material (even wood) is certainly still to be found; whereas in distant deep sea basins, only small quantities of energy are present. Accordingly, bacterial activity here is extraordinarily low. The complete utilization of the energy supply for the life of organisms in the deep sea is thereby hindered (see Morita, 1979 b).

As a rule, independent of the site, sugars and protein are decomposed first, then starch, fats and finally high molecular weight compounds such as chitin, cellulose, lignin, etc.

With the decomposition of organic substances there are corresponding permanent changes in the microflora, which may be recognized by microscopic observation. Thus, Olah (1972) found four different bacterial populations within 14 days during the degradation of finely divided reeds (*Phragmites communis*) in water from an inland lake. Within less than 12 h, large rods were dominant in the liquid phase and they used the dissolved nutrients as food. After about 4 days they were almost completely replaced by small cocci and after 9 days large cocci, which also colonized the reed particles, became dominant. Soon afterwards myxobacteria (*Cytophaga*) developed on the detritus and on the large cocci and after 14 days attained a proportion of over 50 per cent of the total bacterial numbers. As cellulose decomposers, *Cytophaga* play an important role (see below). In many inland and coastal waters fungi and lower animals also participate in such nutrient-conditioned successions. Thus, on mangrove seedlings (Newell, 1976) and *Spartina alterniflora*, a grass of brackish water areas (Gessner, 1977), the occurrence of different fungi one after the other was observed.

Particularly numerous in waters are proteolytic bacteria which can utilize proteins as food. Many species of *Pseudomonas* and other eubacteria and also various fungi, can carry out proteolysis. In the marine habitat their proportion of the whole microflora is probably even greater than in inland waters. ZoBell and Upham (1944) found that of 60 strains of marine bacteria all could release ammonia from peptone and 47 could liquefy gelatine. The micro-organisms first hydrolyse the proteins by means of proteolytic exoenzymes. The polypeptides and oligopeptides taken up by the cells are then broken down to amino acids by peptidases. The amino acids are then either used for the synthesis of cell material or deaminized with the liberation of ammonia. This process is, therefore, called ammonification (see Section 11.5). In an anaerobic environment, the amino acids may be decarboxylated with the formation of primary amines and carbon dioxide. The carbon skeleton proper is then broken down further in various ways, depending on its structure and the micro-organisms involved (Karlson, 1980).

With animal (and human) excreta *urea* reaches the water, where bacteria, by hydrolytic deamination, break it down to ammonia and carbon dioxide. In many urea-cleaving bacteria, a repression of urease formation by ammonium ions takes place so that no more ammonia is produced than that required for protein synthesis. In only few bacteria — as for example *Proteus vulgaris* which is common in effluents — does the enzyme formation suffer no repression of this kind. Therefore the urea present is split completely in a short time into ammonia and carbon dioxide and so can bring about an increase in the pH.

Steinmann (1976) found urea concentrations in the western Baltic Sea of between 0.02 and 9.04 μg N per litre, the highest values being recorded in the relatively heavily polluted fjords. In winter the number of urea-decomposing bacteria was distinctly higher than in the summer months. In the Kiel Fjord up to 2500 per ml of water were counted. However, urea can also be taken up by fungi (Sguros *et al.*, 1973), cyanophytes (Fogg *et al.*, 1973) and by various algae (Guillard, 1963) and used as a nitrogen source.

Uric acid (2,6,8-trihydroxypurine), which is formed particularly by birds and reptiles during protein breakdown, can be decomposed by a series of bacteria which possess the enzyme uricase. Steinmann's investigations (1974) showed that the proportion of uric acid-degrading micro-organisms in the microflora of lakes and ponds with an abundant bird life is distinctly greater than in those which are much less frequented by birds. Thus, in the Russee 5000 uric acid degrading bacteria were found per ml of water — this amounts to 25 per cent of the saprophytes. In the Schulensee, which is less populated by birds, the proportion was only 6 per cent. Of the uric acid decomposers isolated from the inland and coastal waters in the surroundings of Kiel (Germany), most were representatives of the genera *Pseudomonas, Vibrio, Alcaligenes* and *Nocardia*. In sediments *Streptomyces* play a big role. There are fresh water and brackish water forms among them and they can all cleave urea also.

Many eubacteria as well as actinomycetes and numerous fungi are able to degrade simple *sugars*. In aerobic media, hexoses are first split into three-carbon compounds and these are converted to pyruvic acid and then, after decarboxylation via the tricarboxylic acid cycle, are finally oxidized to carbon dioxide (for details, see Karlson, 1980).

Often in an anaerobic environment only fermentations are possible, which represent incomplete oxidations. According to conditions, alcohols, organic acids (like lactic acid, butyric acid, propionic acid, acetic acid or formic acid), hydrogen and carbon dioxide may be produced.

Some disaccharides which are widely present in the sea, like sucrose, lactose and maltose, can, unlike the simple sugars, be broken down by only a few bacterial species (ZoBell, 1946a).

The sugar alcohol, mannitol, is excreted into the water particularly by brown algae. This is especially important in the case of large forms of the genera, *Laminaria* and *Ecklonia* (Lucas *et al.*, 1981). However, it can also be produced by bacteria which ferment fructose and, besides mannitol, produce also lactate, acetate and CO_2. Mannitol is a nutrient for some bacteria and yeasts (e.g. *Rhodotorula* and *Torulopsis*) and therefore, as a rule, it is quickly decomposed, the oxidation proceeding via D-fructose. Consequently, in areas of brown algal stands, usually an increase in numbers of mannitol decomposers may also be observed (Koop *et al.*, 1982). In anaerobic media, *Clostridium pasteurianum* can ferment mannitol to butyric acid (Schlegel, 1981).

Starch, an important reserve substance of many plants, is found mainly in inland waters, but is also present in the marine habitat. It is a polysaccharide, composed of amylose and amylopectin. Numerous bacteria and fungi are able to hydrolyse starch by means of exoenzymes (amylases). The resulting product is the disaccharide maltose, which is then hydrolysed to glucose by the enzyme maltase. In aerobic waters, the starch breakdown occurs rapidly, mainly due to pseudomonads, various *Bacillus* species, actinomycetes and higher fungi. In anaerobic sediments, starch is decomposed chiefly by various *Clostridium* species.

Cellulose is the substance from which the skeleton of most plants is constructed. It is therefore particularly abundant in inland waters, but it also plays an important role in the sea. In the sediments of ponds and lakes it may constitute a large fraction of the organic matter present. Organisms which decompose cellulose are therefore very important in water — as they are in the soil. In an aerobic environment, cellulose is broken down mainly by myxobacteria and higher fungi (ascomycetes, Fungi Imperfecti). According to Bärlocher and Kendrick (1976), aquatic hyphomycetes stand foremost in the degradation of leaf litter in flowing waters. They are able to penetrate quickly into the interior of leaves by the formation of hyphae and so represent the pathfinders for bacteria, protozoa and invertebrates.

Amongst bacteria, *Cytophaga* and *Sporocytophaga* species are predominantly active in the decomposition of cellulose. Ostertag (1950) states that they are responsible for almost the entire cellulose breakdown in the River Elbe; fungi do not seem of great importance here (von Brandt, 1953). Cellulose-decomposing bacteria are always abundantly present in the River Elbe but their activity is much reduced with the lower water temperatures in winter (Figure 82). Whereas at temperatures of more than 12 °C the tearing strength of cotton threads hung up in the water had diminished by 80–90 per cent after 7 days, it remained almost unchanged during the cold season, and even after 2–3 weeks of exposure had not declined (Rheinheimer, 1960). Similar results were obtained by von Brandt and Klust (1950) in lakes and coastal waters. One can say then, that only in the warm season — from May to October — is there any important cellulose decomposition in the rivers and it proceeds here more rapidly

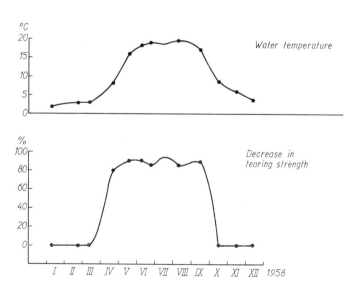

Figure 82   Microbial decomposition of cellulose in the Elbe near Zollenspieker (km 598), expressed as the percentage decrease in tearing strength of a cotton thread after 7 days' exposure in the water (below), and the average water temperature for every 10 days (above). The curves show the great extent to which cellulose decomposition depends on water temperature

than in stagnant waters, but in these it does not die down completely even in winter. Maciejowska (1968) likewise found in the Gulf of Danzig a vigorous increase in cellulose decomposition in summer and detected the highest values above the bottom. At a site near the coast, northwards of Gdynia (Poland), the decrease in tearing strength of cotton threads after an 18-days exposure in August amounted to 11.0 per cent on the surface, 31.5 per cent in moderate depths of water and to 47.4 per cent above the bottom. Besides myxobacteria, various organisms of the order of Pseudomonadales are involved in the breakdown of cellulose in lakes, for instance *Vibrio* and *Cellvibrio* species (Kadota, 1951; Kusnezow, 1959) and some actinomycetes. Lower and higher fungi also take part here in cellulose decomposition. In coastal waters (for example of the North and Baltic Seas) a maximum of cellulose decomposers was observed in autumn (Lehnberg, 1972). That may be connected with the decay of many algae in this season and at the same time a big increase in the proportion of *Cytophaga* organisms also occurs.

Cellulose is hydrolysed by exoenzymes (depolymerases). Some organisms are able, by means of cellulases, to carry the breakdown as far as the disaccharide cellobiose or even to glucose (Norkrans, 1967). In myxobacteria, however, no extracellular cellulases have, so far, been demonstrated. The thin flexible cells position themselves close against the cellulose fibres and parallel to them. Apparently they hydrolyse them while they are in the closest possible contact and immediately take up the products of this reaction (Schlegel, 1976).

Anaerobic cellulose breakdown is important, particularly in mud. It is carried out mainly by clostridia. During anaerobic cellulose fermentation, ethanol, formic acid, acetic acid, lactic acid, hydrogen and carbon dioxide are produced. Apostol (1977) found quite large numbers of anaerobic cellulose decomposers both in various sewages as well as in the water of the lower Danube; a substantial rise in numbers was shown in the summer months in the Danube.

Xylans (hemicelluloses) are present in many plants and have the dual role of providing mechanical strength and as food reserves; they occur in most waters. They can be attacked by a greater variety of micro-organisms than cellulose, by bacteria as well as fungi. Kusnezow (1959) states that hemicelluloses are decomposed anaerobically by *Clostridium tertium* and in the process hydrogen, carbon dioxide and probably also fatty acids are produced.

Pectins (polygalacturonides) are also decomposed by numerous bacteria, which first lyse them by means of a propectinase and then cleave them with pectinases (pectin methylesterases). The breakdown products are pectic acid and methanol. With regard to quantity, pectins do not, of course, play an important part in waters. But in former times, retting of flax was done in streams, and organisms which decomposed pectin accumulated. In an anaerobic environment clostridia, especially *Clostridium pectinovorum*, carry out this reaction.

Agar is a product of numerous algae and the most important are marine red algae of the genera *Acanthopeltis, Ahnfeltia, Ceramium, Gelidium, Gracilaria* and *Pterocladia* (Levring *et al.*, 1969). They occur in large masses in many regions of the seas and are the basis for the commercial production of agar. As might be expected, a number of bacteria are found in the same regions which can lyse agar by means of a specific enzyme (agarase). These are predominantly members of the genera *Pseudomonas, Vibrio, Achromobacter, Flavobacterium, Agarbacterium, Bacillus* and *Cytophaga*. These bacteria are found mainly as a surface growth on a wide variety of algae, and can be identified and isolated readily by making impression cultures on agar media, where the colonies of organisms which break down the agar sink into the medium and form little craters on the plates. In sediments containing much algal debris, agar-decomposing bacteria can accumulate and their numbers may amount to several per cent of the total microbial counts. In the open water they are found near places where large quantities of algae occur. Thus, in the western Baltic only small numbers of them are found, whereas they abound in the North Sea near Helgoland which has large stocks of algae.

Alginic acid (polyuronic acid) is produced by various brown algae. The raw material for the industrial exploitation of derivatives of polyuronic acid is supplied mainly by *Macrocystis pyrifera*, several *Laminaria* species and *Ascophyllum nodosum*. Near the sites where brown algae occur, those derivatives are found in the water and the sediments, and may be broken down there by some bacteria. Particularly common is *Alginomonas alginovorus* and *A. alginica*, less frequently *A. fucicola* (Waksman *et al.*, 1934). The last two are able to break down agar also.

Chitin (polyacetylglucosamine), which provides the skeletal structure of lower animals and fungi, occurs widely in waters and their sediments. The bulk of it is provided by crustaceae, whose outer skeletons consist predominantly of chitin. This substance is broken down very actively by some bacteria and fungi. Amongst the first, members of the genera *Pseudomonas* and *Vibrio* predominate; less common are those of *Beneckea* or actinomycetes (*Micromonospora*). Chitin-lysing bacteria were found by Rheinheimer and Soemintapoera (unpublished) regularly in the Baltic and the North Sea, both in the water and in the sediments. Thus, the water of the western Baltic and its fjords contained between five and several thousand chitino-clastic bacteria per ml, i.e. 0.1–6.8 per cent of saprophytes. In aerobic sediments there were between a thousand and several tens of thousands per g dry weight, or 0.4–8.8 per cent of total saprophyte counts. They are particularly abundant on the shells of dead crayfish. These, throughout, are *Pseudomonas* or *Vibrio* species. Some of them have a worldwide distribution, for example *Pseudomonas cryothasia* and *Vibrio algosus* are found in the North Sea and the Baltic as well as in American and in Japanese waters (ZoBell and Upham, 1944; Campbell and Williams, 1951; Seki and Taga, 1965).

Of the psychrophilic bacterial strains isolated by Morita (1975), many were chitin decomposers which still showed vigorous chitinase activity at temperatures around 0 °C. In view of the prodigious amounts of chitin that are produced in the sea by copepods alone, great importance may be attached to these psychrophilic chitin degraders. Since animals with chitinous skeletons form the food of many fishes, chitin-degrading bacteria are also present in their digestive organs. For example, 1 ml of the stomach juice of *Leptocottus armatus* contained 15 billion chitin degraders.

Chitin-lysing fungi have also been found in waters. An obligate chitin-decomposing organism is, for example, the phycomycete *Karlingiomyces asterocystis* (Murray and Lovett, 1966).

Various other chytridineae, as well as thraustochytriae, likewise play a role in the decomposition of chitin, especially in sediments of inland and coastal waters (Ulken, 1979, 1981).

The microbial breakdown of chitin is carried out by exoenzymes (chitinases). In the process only little N-acetylglucosamine is produced, together with much chitobiose and chitotriose which are then broken down further by chitobiase (Schlegel, 1981).

Lignin, an important constituent of wood, is abundantly present in many inland and coastal waters, but only very little is found in the oceans. Lignin is broken down much more slowly than cellulose and can, therefore, accumulate to a considerable extent in the sediments of smaller inland lakes. Thus the organic material of the surface mud of lakes was found to contain 34–78 per cent of lignin-humus complexes (Kusnezow, 1959). Lignin is not just one compound but is composed of various units. These are phenylpropane compounds, particularly coniferyl alcohol, sinapal alcohol and cumaral alcohol. The phenylpropane units are linked together in various ways by ether-and C—C bridges which are very resistant to enzymes.

The lignin in waters is broken down mainly by higher fungi. Numerous lignin-decomposing ascomycetes and Fungi Imperfecti are always found on submerged wood. These, together with cellulose-lysing organisms, contribute to the destruction of wooden buildings in water (see Section 16.4). Many lignin-decomposing fungi have also been isolated from the sea. Bacteria which can utilize lignin as food have been found in water. In an anaerobic environment lignin is very resistant to microbial attack.

Fats (fatty acid esters of glycerol) are produced by plants and animals and hence are present in almost all waters and their sediments. They are broken down by a number of bacteria and fungi which produce lipases. Lipoclastic bacteria are widely distributed in waters. In lakes, quite often the fats produced, especially by the plankton accumulate in the form of an oil film on the surface so that fat-degrading bacteria can become enriched here. In the sea they are able to attack all sorts of oils and fats (ZoBell, 1946a). ZoBell and Upham (1944) isolated from the sea 13 species of lipolytic bacteria; members of the genera *Pseudomonas*, *Vibrio*, *Sarcina*, *Serratia* and *Bacillus* were particularly active. Hydrolysis begins with the formation of glycerol and fatty acids. Glycerol is particularly important as a source of energy and is, therefore, usually quickly oxidized.

Waxes (esters of higher monovalent alcohols with higher carboxylic acids) can also be broken down by micro-organisms, but usually much more slowly than fats.

Fatty acids are produced not only as a result of hydrolysis of fats and waxes but also during various fermentation processes. Thus, anaerobic decomposition of cellulose may yield considerable quantities of fatty acids and the mud of lakes always contains fatty acid salts (Kusnezow, 1959). Under suitable conditions these may be broken down by bacteria and fungi. In anaerobic environments there is the formation of methane and carbon dioxide. Some fatty acids are, of course, only incompletely oxidized. A small group of obligate anaerobic bacteria are capable of this methane fermentation. Included are several species of the genera, *Methanobacterium*, *Methanospirillum*, *Methanogenium*, *Methanosarcina* and *Methanococcus*, all belonging to the Archaebacteria. The hydrogen acceptor is either CO_2 (in *Methanosarcina barkeri*, also CO) or —CH_3 groups. Some of these organisms are able to reduce CO_2, using molecular hydrogen, to methane.

Polybasic acids likewise occur in waters. Smith and Oremland (1983) found 0.1 to 0.7 mmol l^{-1} oxalate in sediments of different North American lakes and of San Francisco Bay. However, only about 3 per cent were present in the dissolved form. Oxalate is contained in plants and, together with plant detritus, gains access to the bottom of waters. In all the sediments tested, labelled oxalic acid was degraded anaerobically to CO_2.

Hydrocarbons get into waters in a variety of ways. Methane is produced in the sediments — particularly of lakes and ponds — during the anaerobic breakdown of fatty acids and is present abundantly in so-called marsh gas (see Section 8.9).

In experiments with sediment from the eutrophic, forest-encompassed Lake Pluss in east Holstein, methanol, which may possibly be formed in the decomposition of pectins, was detected as a precursor of methane production. In this case, peptone proved to be a preferred nitrogen source (Naguib, 1982).

Acetate and hydrogen were found to be the most important precursors of methane formation in the sediment of the eutrophic Wintergreen Lake in Michigan (USA) and here, methane producers are the largest consumers of hydrogen (Lovley and Klug, 1983). In contrast, in the sediments in the tidal regions of the coast of Maine (USA), 35.1–61.1 per cent of the total methane production proceeds via trimethylamine (King *et al.*, 1983). In the Baltic Sea, Lein *et al.* (1981) found 10–1000 methane-producing bacteria in 1 g of moist sediment. The rate of methane production from CO_2 varied from 0.2×10^{-3} to 27.3×10^{-3} cm^3 per kg per day, and from acetate from

0.015×10^{-3} to 0.93×10^{-3} cm^3 per kg per day. Methane production diminished from the shallow Riga Bay to the Gotland deep. Organic carbon consumption for methane formation was between 0.14 and 7.9 mg C per kg moist sediment per year. In contrast, the amount consumed in sulphate reduction was up to 492.7 mg C per kg per year.

According to the investigations of Winfrey and Ward (1983), sulphate reduction in different anaerobic sediments of the tidal zone was 100–1000 times higher than methane production. However, after the disappearance of sulphate, the rate of methane formation rose steeply. Oremland and Polcin (1982) established that sulphate ions inhibited methane formation strongly when hydrogen or acetate served as the substrate; but not when methanol, trimethylamine or methionine were utilized. Sulphate reduction was promoted by hydrogen and/or acetate whereas the other compounds mentioned had no such influence. Accordingly, with hydrogen and acetate, methane producers succumb to the competition of sulphate-reducing bacteria. However, these two physiological groups do not compete for methanol, trimethylamine or methionine; therefore, methane formation and sulphate reduction can proceed side by side in anaerobic sediments.

Ethane, propane and butane are produced in small amounts during methane formation; these hydrocarbons may also gain access to some waters due to the penetration of natural gas. Long-chain aliphatic and aromatic hydrocarbons are produced by plants and are contained in petroleum, which in oil-bearing regions gets into the water by natural means, but elsewhere mainly through the agency of ships (see Section 15.3). Rubber is a plant product, and bituminous substances often develop in sediments (Kusnezow, 1959).

Under suitable conditions, almost all hydrocarbons can be broken down by microorganisms (Fuhs, 1961). From inland waters as well as from the marine habitat, numerous bacteria and also a certain number of fungi have been isolated which can oxidize either individual hydrocarbons or a variety of them (Kusnezow, 1959; Fuhs, 1961; ZoBell, 1964 b).

Methane is, as a rule, oxidized by bacteria which do not attack long-chain hydrocarbons. *Methanomonas* (*Pseudomonas*) *methanica* can utilize methane (or methanol) as its only source of carbon (see Section 11.1). Higher hydrocarbons are oxidized only along with methane; by themselves they cannot be utilized, nor any other organic substances (sugar, alcohols or fatty acids). The oxidation of methane is probably carried out via methanol, formaldehyde and formic acid to carbon dioxide (Schlegel, 1981).

As well as *Methanomonas methanica*, there are — particularly in lakes with active methane formation — other bacteria able to oxidize methane (Overbeck and Ohle, 1964; Anagnostidis and Overbeck, 1966). Besides pseudomonads, some actinomycetes are very active methane oxidizers. Romanenko (1959) found up to 400 methane oxidizers in 1 ml of water in the retaining dam reservoir of Rybinsk (USSR). The proportion of the oxygen consumed by the bacterial oxidation of methane can reach 40–60 per cent of the total oxygen present in eutrophic lakes.

Iversen and Blackburn (1981) determined the rates of methane oxidation in the anaerobic sediment of a Danish fjord and found that these increased from 3.34 to 17.04 μmol CH$_4$ per m^2 per day during the time from April to August, with a simultaneous rise in temperature from 4 to 21 °C.

Short-chain aliphatic hydrocarbons such as ethane, propane and butane, can be oxidized by bacteria of various orders, especially from the genera *Pseudomonas*, *Flavobacterium* (Eubacteria) and *Nocardia* (Actinomycetes). Evidently there are more bacteria which can oxidize propane than those that can break down ethane or butane.

Long-chain aliphatic hydrocarbons can be degraded by a greater number of bacteria, particularly pseudomonads, micrococci, corynebacteria and nocardias. Also, some yeasts of the genus *Candida* are able to oxidize these hydrocarbons. Interestingly enough, the greater the number of carbon atoms in the chain, the more different species of micro-organisms there are which can decompose these hydrocarbons. Some oligocarbophilic bacteria are able to live on the water surface on petroleum vapour alone (Hirsch and Engel, 1956; Hirsch, 1958, 1960).

Microbial oxidation of the higher paraffins starts at the terminal carbon atom and leads to the formation of higher fatty acids via paraffinic alcohols. This process requires the presence of molecular oxygen. Schöberl (1967) found in the water of Hamburg harbour a strain related to *Pseudomonas erythra* Fuller et Norman which could utilize aliphatic hydrocarbons from hexane to liquid paraffin. Octane is normally broken down via *n*-octyl alcohol, caprylic aldehyde and caprylic acid. As a rule, pseudomonads oxidize aliphatic hydrocarbons completely, whereas micrococci and nocardiae accumulate intermediate products.

Aromatic hydrocarbons are broken down by some bacteria, by yeasts and other fungi. Amongst bacteria, predominantly members of the genera *Pseudomonas, Vibrio, Spirillum, Flavobacterium, Achromobacter, Bacillus* and *Nocardia* are able to oxidize aromatic hydrocarbons (Fuhs, 1961). The immediate products of the breakdown are aromatic ring compounds which carry almost exclusively hydroxy groups. These rings are further cleaved open to aliphatic acids. This requires molecular oxygen whose incorporation is catalysed by oxygenases (Schlegel, 1981).

Different hydrocarbons are oxidized by bacteria without being able to grow on them. For growth, they require other organic substances which may also be hydrocarbons. Co-oxidations of this kind are known between cycloalkanes and chlorinated hydrocarbons (Perry, 1979).

A small group of bacteria and fungi can attack phenols (Iturriaga and Rheinheimer, 1972), provided the concentration does not exceed about $2 \, g \, l^{-1}$. Bacterial phenol degradation can be markedly influenced by other nutrients. Thus Pawlaczyk (1965) found that glucose caused an inhibition of phenol decomposition by *Pseudomonas fluorescens*; but urea, on the other hand, was stimulatory.

11.4. Carbon Cycle

The element carbon, which forms the basis of all organic matter, undergoes a constant cycle in nature, at the center of which stands carbon dioxide (see Section 8.9). The air of the earth's atmosphere contains about 0.032 per cent by volume of CO_2 — altogether 2.3×10^{12} tons. In the sea approximately 50 times the amount is in solution, for the most part in the form of bicarbonate. Gas exchange with the atmosphere is, however, small and is estimated at 0.1×10^{12} tons per year.

All living creatures participate in the cycling of carbon: the vegetable organisms capable of photo- or chemosynthesis bring about the synthesis of organic matter through reduction of CO_2 — heterotrophic plants and animals its degradation through the oxidation of organic matter to CO_2. The presence of carbon dioxide, therefore, is the prerequisite for life on earth and correspondingly also in the complete ecosystem of the waters (Chapter 12). It is guaranteed by the functioning of the carbon cycle. Where this is not the case, far reaching disturbances of the ecosystem in question may be caused. In waters the proportion of micro-organisms participating in the carbon cycle is greater than in terrestrial habitats. The production of organic matter is effected above all by the unicellular algae and cyanophytes of the phytoplankton — the degradation chiefly by eubacteria. In contrast, on land, organic matter is largely produced by higher plants; in its degradation, actinomycetes, fungi and lower animals play a much greater role than in the aquatic environment.

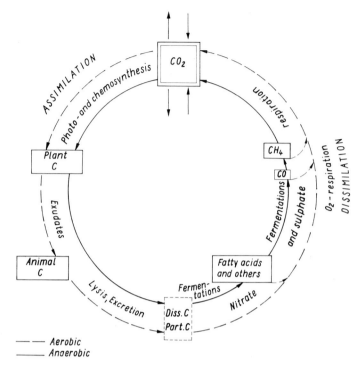

Figure 83   The carbon cycle

The carbon cycle (Figure 83) may be divided into assimilation — i.e. synthesis and transformation of organic material into the multitude of natural carbon compounds — and dissimilation, which is the stepwise breakdown of all these substances by the respiration of the heterotrophic plants and animals.

Plants and animals give off exudates and excretions into the surrounding water and dead organisms are lysed. The dead particulate organic matter can be converted into soluble organic substances and, under suitable conditions, these can also in part be transformed again into particulate matter (see Section 13.1).

The cycling of carbon is influenced to a high degree by the oxygen content of the waters. It proceeds most rapidly in the presence of molecular oxygen. When oxygen is lacking the breakdown of carbon compounds is at least slowed down. As long as bound oxygen is still available in the form of nitrite, nitrate or sulphate, it can still go to completion through nitrate and sulphate respiration. This is, in principle, also possible by intramolecular respiration — however, with fermentation processes there is often an accumulation of intermediate products and only part of the organic material present is broken down to CO_2. In this way, in addition to other compounds, organic acids are often accumulated and then under suitable conditions a methane fermentation becomes possible (see Section 11.3). In these processes a small amount of carbon monoxide is also formed. If oxygen becomes available all these intermediate products can be oxidized to CO_2 (Figure 83).

Of course, the carbon cycle in waters depends also on further factors — as for example on the pH and on the presence of nitrogen and phosphorus compounds.

Whenever through unfavourable conditions the carbon cycle is disturbed, the enrichment of intermediate products is facilitated and they can be deposited with the sediments (see Chapter 13). In the course of geological time these organic deposits can assume enormous proportions and they can undergo further transformations. In this way the formation of oil, natural gas, brown coal and peat deposits has been brought about (see Chapter 14).

The majority of natural organic compounds consists only of the three elements carbon, hydrogen and oxygen. Their quantitative proportion is greater in the organic matter produced on land than in that formed in waters. However, there is a strong enrichment of nitrogen in humus formation in soils, so that here the C/N ratio is displaced from 40 : 1 to 10 : 1.

Most natural organic substances can undergo fermentations in an anaerobic medium. However, there is also a series of substances which are stable under anaerobic conditions and can not be attacked by micro-organisms since with them intramolecular respiration is not possible. This is true, for example, of the hydrocarbons which do not contain any oxygen atoms. Likewise, neither can higher fatty acids ($>C_5$), steroids, carotenoids, porphyrins and terpenes be fermented.

11.5. Nitrogen Cycle

Nitrogen, as a constituent of protein, is a vital element. It is taken up by green plants mainly in the form of ammonia or nitrate. These compounds, however, occur in most waters in only very small quantities, so that the nitrogen is the limiting factor for plant life in rivers, lakes and the sea. Hence the production of organic substance depends to a large degree on the ammonia and nitrate concentrations. In the euphotic zone almost all the nitrogen is bound in the proteins of living creatures; but bacteria, particularly the putrefying bacteria, keep on liberating ammonia (see Section 11.2) which thus becomes available to the green plants as a source of nitrogen, either directly or after microbial oxidation to nitrate. Some micro-organisms are able to bind free nitrogen and thus harness it for the use of the living organisms in the environment. Others, by reducing nitrate, release nitrogen which is then lost from the system. It follows that the nitrogen cycle occupies an important position in the biological processes taking place in waters.

The *fixation of molecular nitrogen* is carried out in water — as in soil — by various bacteria and cyanophytes. Among the former, particularly *Azotobacter agile* and *A. chroococcum*, as well as various species of the genus *Azomonas*, play an important role in aerobic environments. They are frequently detected in rivers and lakes. *Azomonas agilis* and *A. insignis* have been isolated hitherto only from waters and do not appear to occur in soil (Becking, 1981).

In anaerobic sediments, *Clostridium pasteurianum* and some related organisms are important nitrogen fixers. More recently it has been established that, besides *Azotobacter* and *Clostridium* species, *Desulphovibrio* (Herbert *et al.*, 1977) as well as some oligonitrophilic bacteria from the genera *Pseudomonas*, *Aeromonas*, *Vibrio*, *Achromobacter*, *Flavobacterium* and *Corynebacterium* are also able to fix free nitrogen (Niewolak, 1973; Niewolak and Korycka, 1972) — even though their importance might be less. According to Pschenin (1963) a *Spirillum* is the most intensive nitrogen fixer in the water of the Black Sea.

The obligate aerobic *Azotobacter* species can utilize many organic substances as energy sources. They require approximately 50 g of glucose to bind 1 g of nitrogen. Deufel (1965) found that the *Azotobacter* counts in Lake Constance increased from a few cells to 1000–3000 per litre of water between 1958 and 1962; he attributes this to progressive eutrophication with the consequent increase in the concentration of organic carbon compounds. A similar finding was reported by Niewolak (1972) also in the Ilawa lakes of northern Poland. Here a maximum of 90000 *Azotobacter* cells per litre of water was observed.

According to Niewolak (1971a), *Azotobacter* possesses a high resistance to sunlight and ultraviolet radiation. Consequently members of this genus are abundant in the hyponeuston of inland lakes and appear to play here a greater role in nitrogen fixation than in deeper zones. In the water and sediment of the Ilawa lakes, their numbers as a rule were greater in summer than in winter — maxima were found in May or June and in September and November. *Azotobacter* cells were found preferentially on the larger planktonic organisms and only to a small extent isolated in the free water.

In contrast to the small numbers of nitrogen-fixing heterotrophic species, practically all photoautotrophic bacteria are capable of fixing molecular nitrogen (Wetzel, 1975). This is true above all of eutrophic lakes at the time of the summer stagnation, when large amounts of purple or chlorobacteria develop in the metalimnion (see Section 7.2). The heterocyst-forming cyanophytes (Chapter 4) are even more important for nitrogen fixation than the bacteria. It has been known for some years that in the aerobic environment nitrogen fixation takes place in the heterocysts (Thomas and David, 1972). It runs roughly parallel to the assimilation and shows a distinct daily rhythm — although even at night it does not quite come to a standstill (Hübel and Hübel, 1974a). The enzyme, nitrogenase, which catalyses nitrogen fixation, is apparently present in all the cells because under anaerobic conditions the vegetative cells also bind nitrogen (Gorkom and Donze, 1971). That is also true for cyanophytes which do not form heterocysts — and likewise for filamentous as well as unicellular forms.

As yet relatively little is known about the extent of microbial nitrogen fixation in waters. Only since ^{15}N isotopes were employed and particularly since it was possible to measure nitrogenase activity by means of the acetylene reduction method (Stewart et al., 1967) have some reliable data become available. Horne and Goldmann (1972) found a total of 460 tons of fixed nitrogen in the 176.7 km² large Clear Lake in California for the year 1970. Rinne et al. (1977) estimated the amount of nitrogen produced in the northern and central parts of the Baltic Sea by fixation in the heterocysts of *Aphanizomenom flos-aquae*, *Nodularia spumigena* and *Anabaena lemmermannii* at about 100000 tons. According to Hübel and Hübel (1974b), in the shallow coastal waters of the middle Baltic Sea (bays of Darss and North Rügen) a considerable fixation of nitrogen takes place due to the benthic cyanophytes.

The importance of the heterotrophic bacteria, on the other hand, might be very small, especially in water. This is already clear from the low cell numbers given by Kusnezow (1970) for lakes of differing degrees of trophism:

	Water (ml⁻¹)		Sediment (g⁻¹)	
	Azotobacter	*Clostridium*	*Azotobacter*	*Clostridium*
Oligotrophic lakes	0	0	0–10	0–10
Mesotrophic lakes	0–10	0–10	0–10	0–1000
Eutrophic lakes	0–10	1–20	0–10	100–10000

Bacterial nitrogen fixation plays a greater role in the sediments of eutrophic waters and has a corresponding effect here on the nitrogen cycle also.

Ammonification (see Section 11.3) is of particular importance for aquatic life as it continuously returns ammonia to the cycle of matter. Numerous protein-decomposing (proteolytic) bacteria and many fungi can perform it. Their numbers, of course, differ from water to water and, moreover, are subject to considerable fluctuations. Wherever proteins become available they multiply very rapidly, as the generation time of most proteolytic bacteria is relatively short. The optimal temperature for ammonification is generally 30–35 °C and this temperature is only rarely attained in waters of the temperate climatic zone. The water temperature of the Elbe, for example, remains always below 25 °C. In winter ammonification is slowed down very much — corresponding to the diminished activity of the proteolytic bacteria. But even at 0 °C it does not stop completely and hence continues under a thick layer of ice. However, the quantity of proteolytic bacteria in sewage-loaded rivers, lakes and coastal waters is often considerably greater in winter than in summer. Investigations in the lower Elbe and its tributaries have shown over and over again that in these waters the intensity of ammonification and the number of proteolytic bacteria (MPN) are higher in winter than in summer (Rheinheimer, 1965a). The maximum was found regularly from December to March. The number of proteolytic bacteria in the lower Elbe varied, according to the site of sampling and the season, between a few thousand and more than one million per ml of water. The greater number of proteolytic bacteria present during winter in the river water offsets to a large degree the lower activity, so that also during the coldest time of the year ammonification in the river is quite considerable. The ammonia content of the Elbe water over the 100 km between Schnackenburg and Tesperhude remains approximately steady all the year round and may occasionally even increase somewhat during the winter months. The marked increase of proteolytic bacteria during the cold season is, however, limited to sewage-loaded waters and will not occur normally in clean rivers, lakes and areas of the sea.

Niewolak (1965) states that a relatively intense bacterial protein decomposition takes place in the Ilawa lakes throughout the year. However, ammonification reaches the highest values in summer and early autumn both in the surface water as well as in the bottom water and sediment, whereas in winter there is a decline although not a very pronounced one.

Ammonia which is liberated during protein breakdown serves as a source of nitrogen for numerous plant organisms, heterotrophic as well as autotrophic ones. It also provides energy for nitrite bacteria which, in the presence of oxygen, oxidize ammonia to nitrite which is then, as a rule, oxidized further to nitrate by another group of nitrifying bacteria. This extremely important process for the cycling of matter in nature was already investigated by Winogradsky (1890, 1891) at the end of the previous century. It is termed *nitrification*:

$$NH_4^+ + 1\frac{1}{2}O_2 \rightarrow NO_2^- + H_2O + 2H^+ + 76 \text{ kcal} (= 317.7 \text{ kJ})$$
$$NO_2^- + \frac{1}{2}O_2 \quad \rightarrow NO_3^- + 24 \text{ kcal} (= 100.3 \text{ kJ})$$

The first stage is called *nitritation* and the second *nitratation*.

The chemoautotrophic nitrifying bacteria need the energy obtained by nitrification to reduce carbon dioxide and thus to build up organic substance (see Section 11.1). With nitrate the final step of the mineralization of organic nitrogen compounds has been reached. It is taken to the sea via streams and rivers where it is of great importance as a source of nitrogen for the phytoplankton.

In nitritation, N_2O is also formed, especially when there is a low oxygen concentration. According to Goreau et al. (1980), in pure cultures of *Nitrosomonas* the proportion of N_2O increases when there is a drop in the oxygen content from 7.0 to 0.18 mg l^{-1}, from 0.3 to 10 per cent of the nitrite production, although the growth rate of the bacteria diminished by less than 30 per cent. At the same time, nitrite formation declined from 3.6×10^{-6} to 0.5×10^{-6} mmol per cell per day.

In inland waters the same nitrifying bacteria are found as in soil, i.e. particularly the nitrite bacterium *Nitrosomonas europaea* and the nitrate bacterium *Nitrobacter winogradskyi* (Engel, 1960). Besides the chemoautotrophic bacteria, heterotrophs also are able, to a small extent, to carry out nitrification, without acquiring any importance in nature (Engel, 1958b). Nevertheless, in a few eutrophic lakes heterotrophic nitrification seems to be more extensive than the autotrophic (Gode, 1970, personal communication).

Witzel and Overbeck (1979) isolated from the Kleine Plöner See in Holstein (Germany) an *Arthrobacter* strain which, in contrast to most other heterotrophic nitrifiers studied until now, possesses the capacity for the relatively vigorous formation of nitrite. The organism can grow with very varied carbon compounds. However, acetate, malate, citrate or ethanol, as well as Mg^{2+} ions, are required for nitritation. Glucose, tryptose and yeast extract inhibit nitritation but not growth. Nitrite formation by *Arthrobacter* appears, therefore, to be connected with acetate metabolism.

In spring water and in very clean streams and lakes hardly any nitrifying bacteria are found, but in eutrophic lakes and rivers they can always be demonstrated and, at least temporarily, there may be vigorous nitrification. In the lower Elbe near Bleckede (km 550), in 1963 and 1964, the number (MPN) of nitrifying bacteria was 1–100 per ml of water (Rheinheimer, 1965c). The number of nitrite bacteria was mostly higher than that of the nitrate bacteria and the difference was greatest (by a factor of 10, at least) during the warm season. The minima were normally found in winter or early spring, the maxima in summer (Figure 84). The number of nitrifying bacteria thus follows a clear-cut seasonal rhythm which is exactly the reverse of that of the proteolytic bacteria (see above).

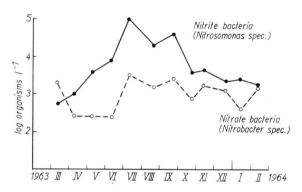

Figure 84   Counts of nitrifying bacteria in the water of the Elbe near Bleckede (km 550)

Investigations in the lower Elbe, carried out over a number of years, permitted Rheinheimer (1965a) to detect a far-reaching dependence of the numbers of nitrite bacteria on the water temperature. Only at a threshold of 10–15 °C is there an increase of some significance. This is reached in the Elbe normally in May. The nitrite bacteria increase rapidly in numbers, and in summer there is not infrequently a response to quite small temperature fluctuations of 2–3 °C (Figure 85). In October they diminish again considerably. In practice nitrite bacteria can multiply in rivers only in summer, and only then is there any demonstrable oxidation of ammonia.

The behaviour of the nitrate bacteria in the Elbe is somewhat different. Their numbers are usually also higher in winter but the seasonal rhythm is not so pronounced. The growth of nitrate bacteria seems to be inhibited in river water. This may be due to their greater sensitivity to suboptimal oxygen and nutrient supplies (Schöberl and Engel, 1964). Also, light and poisons inhibit *Nitrobacter* more strongly than *Nitrosomonas* (Ulken, 1963; Bock, 1965). The better growth of nitrite bacteria is followed by a vigorous increase of the nitrite concentration, up to 2 mg NO_2^- per litre in the Elbe estuary during the summer months (Lucht, 1964; Rheinheimer, 1965a). The peak could be found, as a rule, between Blankenese and Stadersand; further downstream the nitrite concentration declined rapidly (Figure 86), while the

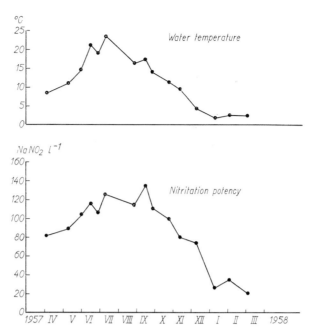

Figure 85 Nitrite-forming potency as a relative measure of numbers of nitrite bacteria, and the water temperature of the Elbe near Bleckede (km 550). The curves show that tempera- ture fluctuations of only a few degrees Centigrade affect the nitrite-forming capacity

nitrate concentration rose correspondingly. It seems that the nitrite bacteria multiply when the river water has become polluted in the Hamburg area while the growth of nitrate bacteria stagnates and increases only 10–20 km downstream of the city.

Ritter (1977) also made similar observations with respect to the Rhine and Neckar. Nitrite and nitrate bacteria occurred in the same numerical ratio only in water samples with low carbon content. When the level of pollution was more severe the proportion of nitrate bacteria was always smaller. In the River Olawa, in Silesia, Pawlaczyk and Solski (1967) found an inhibition of nitrification in times of heavy pollution — in particular during the sugar season. In the relatively heavily loaded River Passaic in New Jersey (USA), Matulewich and Finstein (1978) counted far more nitrifiers in the sediment than in the water:

	Nitrite bacteria MPN per ml		Nitrate bacteria MPN per ml		Number of samples
	Range	Mean	Range	Mean	
Water	5.6–2780	395	0–702	30	41
Sediment surface	1280–1620000	214000	116–42800	7422	23

The highest numbers of nitrifiers, however, were found on water plants and on rocks. Here, on all substrates, again there were always more nitrite bacteria detected than nitrate formers.

In lakes, conditions are somewhat different as the nitrifying bacteria are found not so much in the open water as in the upper sediment zones. The ammonia available is often oxidized much more slowly because of the lack of oxygen in the hypolimnion

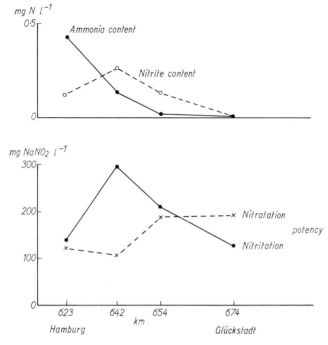

Figure 86   Nitritation and nitratation potencies as relative expression of the numbers of nitrite and nitrate bacteria, and the contents of ammonia nitrogen and nitrite nitrogen of the Elbe water at four sites between Hamburg and Glückstadt on August 29, 1958. The nitrite peak coincides with the maximum of nitritation potency and the minimum of nitratation potency

during the stagnation periods, and nitrification to any extent occurs only during the circulation periods in autumn and spring, i.e. at times when the activity of the nitrifying bacteria is not very great because of the relatively low water temperatures. Consequently, those are the times when heterotrophic nitrification assumes increased significance. In the relatively strongly eutrophicated Ilawa lakes, Niewolak (1965) observed a quite similar seasonal course of nitrification, however, to that described for the rivers. He found maximum nitrite formation in July and the maximum nitrate formation in August — on the other hand, the minima were in the winter months December to February.

Relatively little is known, so far, about nitrification processes in the sea. Some decades ago, nitrifying organisms were isolated from the Baltic and the North Sea (Brandt, 1902; Thomsen, 1910; Liebert, 1915), but these were found mainly in the sediments. Later, nitrification was also demonstrated in various coastal waters (Vargues and Brisou, 1963; Rheinheimer, 1967). The search for nitrifiers in the oceans was unsuccessful for a long time, although the occurrence of microbial ammonia oxidation in the sea had been suspected (ZoBell, 1946a). Some years ago, however, Watson (1963, 1965) succeeded in isolating the chemoautotrophic nitrite bacterium *Nitrosococcus oceanus* from the Atlantic. It could be demonstrated even at a depth of 1500 m. More recently, Watson and Waterbury (1971) also found two nitrate bacteria, *Nitrospina gracilis* and *Nitrococcus mobilis*.

Using immunofluorescence methods, Ward (1982) counted 1000 to 10000 cells of the ammonia-oxidizing bacteria, *Nitrosococcus oceanus* and *Nitrosomonas marina* per litre of water in the open ocean, and 100000 to 1000000 in coastal waters. These are distinctly higher values than have previously been assumed. They allow sufficiently high nitrification rates to maintain the cycle of matter in equilibrium. In shrimp aqua cultures of Tahiti (Polynesia), Bianchi and Bianchi (1982) found about 10×10^6 nitrifying bacteria per litre of water at temperatures of 25–28 °C and were able to determine a doubling time of 7 h.

Investigations by Swerinski (1981) in the Kiel Bay showed that here nitrification in the aerobic sediments is greater than in the water. This also applies to the cold seasons. In the winter months an average nitrification rate of $3 \times 10^{-3} \mu mol$ NO_3^- per cm^3 per hour was observed in the uppermost zone (1 cm) of the sandy sediment. In that season, nitrate production ran to $9 \mu mol$ NO_3^- per cm^3. If this quantity of nitrate reached the water layer (average depth, 18 m), it would lead to a concentration of $5 \mu mol$ NO_3^- per litre towards the end of winter. Before the start of the first phytoplankton bloom, $5-10 \mu mol$ NO_3^- per litre of water were measured. The values, therefore, are in good agreement. Nitrification by chemoautotrophic bacteria, therefore, plays a decisive role in the Kiel Bay and similar coastal waters in the nitrogen cycle.

In the water of the Baltic Sea, in summer, a nitrite maximum was found in places below the thermocline. In the experiments of Enoksson (1980), nitritation certainly took place; nitratation, however, was suppressed.

According to Olson (1981), light has a inhibiting influence on the activity of nitrite bacteria and nitrate bacteria, but it stimulates nitrate assimilation of the phytoplankton. The limiting values for nitration lie between 0.2 and 2.0 per cent of the light on the surface; they occur, therefore, within the region of the compensation depth with about 1 per cent. Consequently, the nitrite bacteria in the photic zone are unable to compete with the phytoplankton for ammonium. Since nitratation is even more sensitive to the effect of light (see Section 8.1), there is then an accumulation of nitrite below the thermocline.

Figure 87   The two pathways of nitrate respiration: denitrification (above) and nitrate ammonification (below)

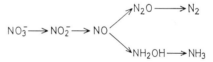

Whilst nitrification is possible only in the presence of oxygen (though very small concentrations will suffice), in an anaerobic environment — if organic hydrogen donors are present — *denitrification*, i.e. dissimilatory reduction of nitrate through nitrite to nitric and nitrous oxide (NO and N_2O) and molecular nitrogen (Figure 87), is carried out by many facultative anaerobic bacteria. They use, under anaerobic conditions, nitrate or nitrite respectively as hydrogen acceptor. This represents an anaerobic respiration, also called nitrate respiration. In most habitats, however, more organisms can reduce nitrate to nitrite than nitrite to free nitrogen. Consequently, wherever there is lively denitrification, abundant nitrite is produced in the first instance (Rheinheimer, 1964). Bacteria capable of denitrification can be demonstrated in practically all waters. As they are facultatively anaerobic, they occur also in waters rich in O_2 and here use oxygen for respiration. Only if oxygen is in short supply does the enzyme system switch over to nitrate respiration; this, however, requires some time as the particular nitrate reductase is synthesized only under anaerobic conditions. The switch from nitrate to oxygen respiration, on the other hand, is instantaneous. It seems that nitrate respiration is a kind of emergency device which comes into effect when no more oxygen is available. The energy yield, though, is only about 10 per cent less than when oxygen is available as hydrogen acceptor.

In some waters denitrification processes may be demonstrated although oxygen is still present; these, however, are always waters which contain relatively much detritus, and the flora growing on the surface of debris particles can create microzones free of oxygen where denitrification can proceed (Jannasch, 1960). According to Ottow (personal communication) denitrification can occur also in the presence of oxygen when abundant easily assimilable carbon compounds are available.

As, in the absence of oxygen, the denitrifiers have an advantage over the obligate aerobes, a shift of species within the microflora may occur in favour of the former. In rivers like the Elbe, their numbers are usually higher in winter than in the summer months (Rheinheimer, 1965a). In this, they behave like the proteolytic bacteria with whom some are, indeed, identical as many facultatively anaerobic protein degraders can also carry out denitrification. In the lower Elbe the counts (MPN) of denitrifiers were found to vary between several hundred to several tens of thousands per ml of water.

In the open water of our rivers, though, denitrification is generally rare, as the oxygen content of the water is usually too high. Even if there is quite a lot of detritus, demonstrable nitrate reduction is possible only if the oxygen concentration drops considerably below 50 per cent of saturation level. It is of real importance only in severe winters, when the rivers carry an ice cover for some time and no gas exchange with the atmosphere is possible. This was the case with the Elbe in the cold winter 1962–63 (Rheinheimer, 1964). The river upstream of Hamburg carried for almost 3 months an ice cover 40–50 cm thick — and downstream of Hamburg there was, for all that time, continuous drifting of ice. Under the ice, high oxygen consumption, mainly due to bacteria, took place, so that during January — or at the latest the beginning of February — the oxygen deficit at various locations had reached 80 per cent or more. As the water contained sufficient debris and dissolved organic substance, vigorous denitrification had already started in January, particularly upstream from Hamburg. The nitrate concentration declined and the nitrite content reached a particularly high level for winter, up to 1 mg NO_2 per litre. In February, no more nitrate could be shown to be present at all, and the nitrite concentration diminished as reduction went on to free nitrogen. At that time, the inorganically bound nitrogen was present almost exclusively as ammonia, in a large part of the river.

In clean rivers, not loaded with sewage, there is much less oxygen consumption even under ice, hence not much loss of nitrogen through denitrification takes place except in slack water zones.

Denitrification is much more common in stagnant waters, for example in the hypolimnion of eutrophic lakes during summer and winter stagnation. Niewolak (1965) found the most vigorous denitrification in the Ilawa lakes in the summer months and a severe decline in the winter, if the lake carried an ice cover. Particularly at the sediment surface, vigorous bacterial nitrate reduction takes place with liberation of molecular nitrogen. By contrast, this does not happen in mud since nitrate is absent (Kusnezow, 1959).

Active denitrification may occur in anaerobic zones of the sea and is made evident by the decline of the nitrate concentration and the temporary increase of nitrite. Here too the process is particularly active on the surface of the sediment. Heitzer and Ottow (1976) counted several millions of denitrifiers per g in samples from the sediment surface from deep water in the Red Sea. In the sediment at 1–2 m deep, however, only about 10–25 per g were found. A taxonomic investigation showed that, for the most part, the organisms were pseudomonads.

Investigations by Rönner (1983) showed that, in the deep water of the Baltic Sea, denitrification occurred if the oxygen content fell to below 0.2 ml l⁻¹. The limiting factor is nitrate. Denitrification rates determined by the use of a mathematical model indicate that, in the Baltic Sea, denitrification results up to 80–90 per cent in the sediment but only to 10–20 per cent in the oxygen-impoverished deep water. On the basis of the denitrification rates measured, it was calculated that about 500 000 tons of nitrogen are evolved annually into the atmosphere from the Baltic Sea. This corresponds to almost the estimated nitrogen addition in this sea area. Sörensen (1978a), using the acetylene reduction method, determined the maximal denitrification rate in sediments off the Danish coast as 35 nmol N per cm³ per day at 2.5 °C and 0.99 nmol N per m² per day for the total sediment. In the oceans denitrification may be of no great significance.

In stagnant waters *nitrate ammonification* also seems to play a role; this represents a reversal of nitrification, i.e. nitrate is reduced to ammonia, via nitrite (Figure 87). It seems that far fewer bacteria are able to carry out this reaction than can carry out denitrification; for example various *Bacillus* species and several strains of *Escherichia coli* are capable of it. The process was first discovered in a small pond in Bohemia (Denk, 1950). In ponds and small lakes the deeper parts often contain much ammonia which cannot be attributed to normal ammonification (see above) alone. In sewage purification plants also, nitrate ammonification is quite important. In anaerobic zones of the sea, for example in the Black Sea, its occurrence has been demonstrated.

Sörensen (1978b), using the nitrogen isotope ¹⁵N, investigated denitrification and nitrate ammonification in marine sediment and found that the capacity for reduction of nitrate to free nitrogen decreased rapidly with increasing depth in the sediment, whereas that for reduction of nitrate to ammonium was still relatively large, even in deeper zones. From the results obtained, it was concluded that nitrate ammonification is as important for nitrate turnover in marine sediments as denitrification.

Koike and Hattori (1978) likewise found vigorous nitrate ammonification in some sediments off the Japanese coast. On the whole, however, it is probably much rarer in nature than denitrification; but it must be admitted that the process of nitrate ammonification has, so far, been insufficiently investigated.

The *nitrogen cycle* in water is shown in a simplified manner in the scheme illustrated in Figure 88. This allows to appreciate the importance of oxygen; its presence determines whether nitrate is produced or destroyed in a water, and whether the nitrogen balance ends up positive or negative.

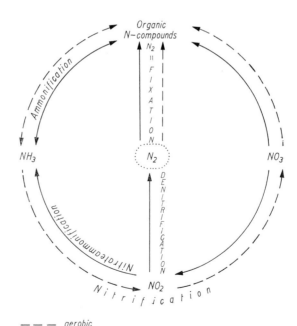

Figure 88　The nitrogen cycle

In detritus-rich water and especially in sediments, nitrification and denitrification processes can even occur side by side.

Blackburn and Henriksen (1983) studied the nitrogen cycle in different sediment types in the Kattegat and Skagerrak in November, 1978, and July, 1979. Rates of production and uptake of ammonium were dependent on changes in the N/C ratio on the sediment surface. Thus the activity in July was greater than in November. Nitrate flow from the sediment to the water was correlated with the nitrification rate and the season. The transfer rates of inorganic and molecular nitrogen amounted to 44–66 per cent of the net rates of mineralization of organically bound nitrogen. Nitrate and ammonium from the sediment was sufficient to cover 30–82 per cent of the nitrogen requirement of the phytoplankton. A stimulating influence by bioturbation was observed by Henriksen et al. (1983). On the basis of quantitative investigations of sediments off the Belgian North Sea coast, Billen (1978) gives examples of the nitrogen flow through ammonification, nitrification and denitrification in areas adjacent to (1) and distant from (2) the coast:

$$(1)\quad \text{Org. N} \rightarrow NH_4^+\ 27.0 \xrightarrow{\ 9.4\ } NO_3^-\ 17.6 \xrightarrow{\ 11.5\ } N_2\ 6.1$$

$$(2)\quad \text{Org. N} \rightarrow NH_4^+\ 12.5 \xrightarrow{\ 4.6\ } NO_3^-\ 7.9 \xrightarrow{\ 6.0\ } N_2\ 1.9$$

The figures refer to g of N per m^2 in the yearly mean. They show that the proportion of denitrification is distinctly higher in the nutrient-rich coastal sediments than in the nutrient poorer areas distant from the coast, where the oxygen supply is better.

Nishio et al. (1983) also came to similar conclusions in their studies of sediments from the Tama estuary and the Odawa Bay in Japan.

11.6. Sulphur Cycle

During protein degradation, besides ammonia, small quantities of hydrogen sulphide are liberated originating mainly from the sulphur-containing amino acids cystine (or cysteine) and methionine. Some proteolytic bacteria can carry out this reaction, splitting off sulphydryl groups by means of desulphurases.

Hydrogen sulphide is not stable in an aerobic environment and is oxidized either chemically or by some bacteria and fungi. Microbial oxidation proceeds through several intermediate products down to sulphate, which represents the terminal step of mineralization of organic sulphur compounds, and serves as an important source of the vital element, sulphur, for green plants. This process is also called sulphurication.

$$2\,H_2S + O_2 \rightarrow S_2 + 2\,H_2O + 80\ \text{kcal}\ (=334.4\ \text{kJ})$$
$$S_2 + 3\,O_2 + 2\,H_2O \rightarrow 2\,H_2SO_4 + 240\ \text{kcal}\ (=1003.2\ \text{kJ})$$

Some chemoautotrophic bacteria are able to oxidize hydrogen sulphide and other oxidizable sulphur compounds, like sulphur itself, thiosulphate and sulphite, and to utilize the energy gained for the reduction of carbon dioxide. Amongst them are, above all, members of the genus *Thiobacillus*, and probably also the filamentous sulphur bacteria of the genera *Beggiatoa* and *Thiothrix*. Besides the obligate and facultative chemoautotrophs, also a number of heterotrophic micro-organisms, bacteria as well as fungi, are able to oxidize sulphur compounds. However, as far as our present knowledge goes, they do not seem to be important in that respect. The photoautotrophic purple sulphur and chlorobacteria also oxidize reduced sulphur compounds to sulphur or sulphate to gain hydrogen for the reduction of CO_2 (see Section 11.1).

In waters *Thiobacillus* species seem to be the most important sulphur oxidizers. They have a very wide distribution and their presence can be demonstrated in most rivers, lakes and coastal waters (Kusnezow, 1959; Rheinheimer, 1965a; Bansemir, 1970). Under suitable conditions they become enriched wherever H_2S is produced. Besides aerobic species, there is also a facultative anaerobic form (*Thiobacillus denitrificans*) which can carry out nitrate respiration. Its particular importance lies in its ability to oxidize sulphur compounds in the anaerobic environment of the aphotic zone, provided nitrate is present. As hydrogen sulphide is very rapidly oxidized chemically in the presence of oxygen, it can accumulate only in an anaerobic environment. In eutrophic lakes there is, therefore, usually a sharp interface between H_2S-and O_2-containing water. This interface, particularly if it lies in the aphotic zone, is the main location of the aerobic thiobacilli; if it lies in the euphotic zone, particularly the purple sulphur and (or) chlorobacteria multiply, and the thiobacilli play only a subordinate role. They can cause daily fluctuations in the hydrogen sulphide concentration of the water. Thus, Parkin and Brock (1981), in the Knaack Lake in Wisconsin (USA) during the summer months, observed a dense population of green sulphur bacteria, which exhibited vigorous photosynthetic activity during the day and completely oxidized the hydrogen sulphide present to sulphate. In the dark, accumulation of hydrogen sulphide could then again occur. Conditions in coastal waters are quite similar. Here, too, the purple and green sulphur bacteria occur in vast numbers in the shallow lagoons or coastal lakes, while the thiobacilli grow preferentially in the aphotic zone.

Thiobacilli are almost always present in rivers also. In the lower Elbe they have been found all the year round, though no H_2S could be demonstrated. Their numbers (MPN) were between 20 and 600 per ml of water. Mütze and Engle (1960) showed the presence of *Thiobacillus denitrificans*, *T. thiooxydans* and *T. thioparus*. The quantity of thiobacilli in the river water depends to a high degree on the volume of water carried by the river; it decreases considerably, for example, in the stretch of the river between Schnackenburg and Hamburg. It seems that the majority of thiobacilli present in the Elbe are carried in with sewage but do not multiply in the river. This is supported also by the regular finding of *Thiobacillus thiooxydans* which, however, is quite inactive at the weakly alkaline reaction ($\pm$ pH 7.5) of the Elbe water; its proper location is in extremely acid waters with a pH of less than 5, and it can grow even at pH 1. Altogether, although thiobacilli seem to occur in most waters, only the presence of H_2S allows them to take part in the sulphur cycle; if this condition is met they multiply vigorously and contribute to the oxidation of hydrogen sulphide. *Thiobacillus thiooxydans* is probably also important in oxidizing colloidal sulphur of volcanic lakes (Kusnezow, 1959).

On the surface of sediments containing hydrogen sulphide, *Beggiatoa* and *Thiothrix* species grow and may form a white cover like a spider's web. Sorokina (1938), investigating this by the method of slide culture in the Beloye Lake, established that the surface layer of the mud contained predominantly such filamentous sulphur bacteria. Within 6 days, 3000 bacterial filaments grew per cm^2. From this it appears that the hydrogen sulphide is already oxidized to sulphur and sulphate in the zone of water near the bottom (Kusnezow, 1959).

Achromatium and *Thiovulum*, which have egg-shaped or spherical cells with sulphur granules and partly $CaCO_3$ inclusions also, play a similar role. *A. volutans* can occasionally form, on the margins of beach ponds of the Baltic coast, thick white deposits which are found over a layer of purple bacteria.

According to the observations of Jörgensen and Revsbech (1983), *Beggiatoa* populations, by formation of mats on the sediment, and *Thiovulum* by forming a veil in the water, create a stable environment with an optimal substrate supply. By the gliding movement of *Beggiatoa* or the flickering motion of *Thiovulum*, the individual cells of these sulphur bacteria can satisfy their requirement for oxygen and for hydrogen sulphide also.

Oxidation of hydrogen sulphide, not seldom by bacteria, also takes place in the sediments. For example, Kepkay and Novitsky (1980) in the 40-cm horizon of the sediment of Halifax Harbour, Canada, found vigorous production of organic matter and also a maximum of dissolved sulphate, simultaneously with a pH minimum. This was clearly due to the activity of chemoautotrophic sulphur oxidizers.

Besides the relatively small quantities of H_2S released by protein decomposition in all surface waters, it can also be produced in an anaerobic environment by bacterial reduction of sulphate. This may occur anywhere, and certainly occurs in the water of lakes, rivers and the sea (see p. 19). A prerequisite for bacterial sulphate reduction is the absence of oxygen and the presence of organic matter which supplies hydrogen for reduction. As hydrogen donors, organic acids and alcohols are used predominantly, the terminal product frequently being acetic acid. Molecular hydrogen can also be utilized. Kadota and Miyoshi (1963) state that, in addition, several nucleotides and amino acids are required for the stimulation of growth.

As during bacterial sulphate reduction H_2S escapes into the air and is thus lost from the substrate, this process is called *desulphurication*. It may be compared to denitrification (see Section 11.5). The sulphate reducers are, however, usually obligate anaerobes and cannot grow in an aerobic environment. They are thus entirely dependent on sulphate respiration and, unlike the denitrifiers, cannot utilize oxygen for respiration. The dissimilatory sulphate reduction starts with sulphate activation which requires relatively much energy. An ATP sulphurylase exchanges a phosphate group of ATP for sulphate. The adenosine-5-phosphosulphate thus produced is then reduced, with the formation of sulphite and the splitting off of AMP. Further reduction of sulphite is apparently carried out with the participation of cytochrome c_3 (Schlegel, 1981). The most widely distributed sulphate reducer is *Desulphovibrio desulphuricans*. This organism is found almost everywhere in the mud of lakes and ponds and can often be demonstrated in water also. In the sea, the closely related *D. aestuarii* is found. The *Desulphovibrio* species are frequently associated with the sporer *Clostridium* (*Desulphotomaculum*) *nigrificans* (Stüven, 1960), which also reduces sulphate. This rod-shaped bacterium has curved spores which somewhat resemble the comma-shaped *Desulphovibrio* cells; they used, therefore, to be frequently confused. As *Clostridium nigrificans* has the very high temperature optimum of 55 °C, and *Desulphovibrio*, by contrast, no longer grows at that temperature, these two organisms can be separated with certainty. Besides these obligatory anaerobic sulphate reducers, a facultative anaerobic bacterium is said by Sturm (1948) to exist and to reduce sulphate vigorously. The organism was called *Pseudomonas zelinskii* and is supposed to occur widely in the mud of the Black Sea but also at other sites.

Although only a few kinds of bacteria are capable of desulphurication, it plays an important role in nature; particularly in the mud of waters, as all the prerequisites for vigorous sulphate reduction are present, conditions are usually anaerobic, and organic substances which can be utilized as hydrogen donors are generally abundant. In marine sediments great significance should thereby be attached, above all, to acetate (Winfrey and Ward, 1983). According to Cappenberg *et al.* (1984), sulfate-reducing bacteria oxidize lactate to acetate in the near-surface of the sediment of the Vechten lake (Netherlands) and the acetate is utilized by methane producers in the underlying zone.

Desulphurication processes are also found very often in the lower oxygen-free zones of eutrophic lakes and ponds, and in enclosed parts of the sea. Considerable quantities of H_2S may thus be produced. In fact, most of the hydrogen sulphide present in the Black Sea seems to have originated by desulphurication. Opinions differ, though, as to what proportion precisely has been produced in this way. Daniltshenko and Tschigirin (1926) attribute 99.4–99.6 per cent of the hydrogen sulphide to desulphurication and only 0.4–0.6 per cent to protein putrefaction, but Kriss (1961) argues that there is no evidence either way for the origin of hydrogen sulphide in the Black Sea. Gunkel and Oppenheimer (1963) conclude from experiments on H_2S formation in sediments on the coast of the Gulf of Mexico and the North Sea near Helgoland that a considerable proportion is due to protein putrefaction. Nonetheless, in many waters with small amounts of protein much more hydrogen sulphide seems to be produced by desulphurication than by protein decomposition. This is supported also by the investigations of Bansemir (1970) in the western Baltic. Jörgensen (1977) calculated that only 3 per cent of the sulphide in the sediments of the Limfjord (Denmark) came from organic sulphur compounds.

In experiments with sediments from the intertidal region of the Irish Sea, Nedwell and Floodgate (1972) found a connection between the origin of the hydrogen sulphide and temperature. Thus, at 5–10 °C the main part of the H_2S was liberated from organic sulphur compounds, whilst at 20–30 °C a greater proportion was formed by sulphate reduction. This might reflect the conditions in summer and winter in the natural situation. Kusnezow (1959) states that only in lakes rich in sulphate can a high concentration of H_2S be reached. In such a lake, 1 g of mud may contain up to 10 million desulphurizing bacteria. In the following table Kusnezow (1959) has compiled the figures in thousands for H_2S-producing bacteria per g of mud in some fresh water and estuary lakes of the USSR:

	Malotshni (estuary lake)	Kujalnicki (estuary lake)	Slavianski (fresh water lake)	Golanpristanskoye (fresh water lake)
Proteolytic bacteria	100–1 000	10–1 000	1 000	100–1 000
Desulphurizing bacteria	10–100	10–10 000	1 000–10 000	100–1 000

Jörgensen (1977), in July 1975, counted from 20 000 to 130 000 sulphate reducers per cm^3 in the mud of Limfjord at seven sites. The highest numbers were recorded usually at a depth of 3.5 cm in the sediment. Here the average values were 93 000 compared with 56 000 on the surface and 43 000 at 7 cm deep.

In the water of various lakes, however, relatively few desulphurizers could be found (up to several thousand per litre), though some were found even in the oxygen-rich zones. This was also confirmed for the western Baltic (Bansemir, 1970). They seem to reach a resting stage here which is not very sensitive to oxygen.

In the eutrophic north American Mendota Lake, by employing radioactively labelled $^{35}SO_4^{2-}$, high rates of sulphate reduction from 50 to 600 μmol SO_4^{2-} cm^3 per (according to season) were found in the topmost sediment zone. In the water, on the other hand, they were less by three orders of magnitude. Correspondingly, less than 18 per cent of the total sulphate reduction took place during the summer stagnation in the hypolimnion (Ingvorsen et al., 1981).

In Limfjord, Jörgensen (1977) determined sulphate reduction rates by means of tracer techniques and parallel determinations of different sulphur compounds. He found very high values on the sediment surface with 25–200 nmol SO_4^{2-} per cm^3 per day. Activity could also still be detected in 1.5 m depth of sediment. Sulphate reduction showed marked seasonal fluctuations with clear peaks in the summer months (Figure 89). Of the hydrogen sulphide formed, 10 per cent was precipitated by metallic ions in the oxygen-free sediment, on the other hand 90 per cent was oxidized to sulphate again at the surface. For sulphate, the turnover times amounted to 4–5 months and for hydrogen sulphide 1–5 days.

Recently it has been established that *Desulphuromonas acetoxidans* can use, as a hydrogen acceptor, elementary sulphur which is thereby reduced to H_2S. This organism therefore, is capable of sulphur respiration (see Schlegel, 1981).

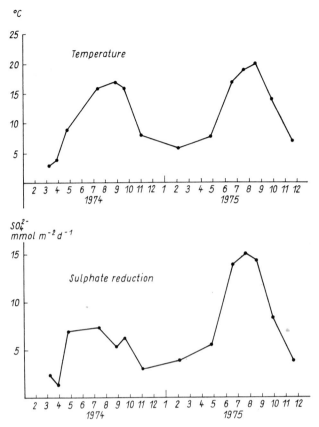

Figure 89   Annual curves of sulphate reduction rates and of temperature in the upper 10 cm of the sediment in Limfjord. (Based on Jörgensen, 1977, modified)

The *sulphur cycle* — like the nitrogen cycle — is influenced strongly by the oxygen conditions (Figure 90). While under aerobic conditions the organic sulphur compounds are mineralized to sulphate, in an anaerobic environment the result is production of H_2S and loss of sulphur.

11.7. Phosphorus Cycle

Like the inorganic nitrogen compounds, phosphate is the limiting factor for plant life in many waters. Phosphorus, particularly as a constituent of nucleic acids, is a vital element for all organisms. In addition, it is present in phospholipids, phosphorylated sugars, phytin, etc. A number of bacteria are able to store phosphoric acid in the form of polyphosphates in volutin granules. A shortage of phosphorus in the water may limit the decomposition of organic matter by bacteria and fungi. The turnover time, thereby, is often very short and at summer water temperatures ranges from a few minutes to a few days (Campbell, 1981).

Many bacteria and fungi, by breaking down organic phosphorus compounds, are able to release phosphate from them and thus return it to the cycle of matter.

Phosphorus exchange between water and sediment in relation to the oxygen content is of greater importance. Under aerobic conditions, phosphorus is precipitated

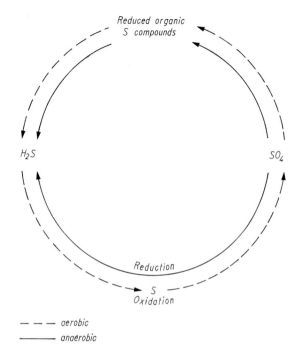

Figure 90   The sulphur cycle

    – – – aerobic
    ———— anaérobic

mainly in the form of iron and aluminium phosphates. After oxygen has disappeared, e.g. during the summer and winter stagnation periods of eutrophic lakes, iron becomes converted to the divalent form and it forms soluble ferrous phosphate, so that the phosphate concentration in the water increases (Kusnezow, 1970). Since bacterial sulphate reduction often occurs in anaerobic environments, a close connection between hydrogen formation and phosphate concentration may frequently be observed. Iron sulphide is thereby precipitated. This process takes place, above all, in sulphate-rich waters in coastal regions (e.g. in beach ponds and estuaries).

Because waters sometimes contain considerable amounts of phosphorus in solid form — as bones and as $Ca_3(PO_4)_2$ — it is important for the phosphorus cycle that numerous bacteria are able to solubilize tricalcium phosphate. This is done, most frequently, by production of organic acids. The solubility of $Ca_3(PO_4)_2$ is assisted also by the formation of ammonium compounds. In various Polish and Lithuanian lakes, ponds and reservoirs, members particularly of the genera *Pseudomonas*, *Aeromonas*, *Escherichia*, *Bacillus* and *Micrococcus* were found to participate in the decomposition of tricalcium phosphate (Niewolak, 1971b; Paluch and Szulicka, 1967; Gak, 1959). Some bacteria, however, may take up small quantities of undissolved tricalcium phosphate (ZoBell, 1946a).

It was assumed formerly that there is microbial reduction of phosphates to hydrogen phosphide, but this could not be confirmed (Näveke, personal communication); nor is there any bacterial oxidation of reduced phosphorus compounds.

The cycle of phosphorus is, then, much easier to survey than those of nitrogen and sulphur. As a rule, phosphorus is taken up by plants as pyrophosphate; this is changed into organic phosphorus compounds and, from these, phosphates are released mainly due to the activity of micro-organisms.

Heinen (1962, 1965) states that in bacteria, as in plants, phosphoric acid may be replaced by silicic acid, which partly takes over the former's function. In this process carbohydrate-silicic acid esters are formed and the enzymatic characteristics of the organisms altered, but reversibly. Glucose respiration is possible with both phosphate and silicate, but the switchover of the enzyme system requires several hours in either direction.

It follows that bacteria may also affect silicon turnover of waters.

11.8. Iron and Manganese Cycles

Iron is ubiquitous on earth and is present in all waters, even if — as for example in the sea — often in very small concentrations. It belongs to the elements necessary for life and is required by all micro-organisms as a component of important enzymes like cytochromes.

There is also a group of bacteria which can oxidize ferrous to ferric compounds on a large scale.

$$Fe^{2+} \rightarrow Fe^{3+} + 11.5 \text{ kcal} (= 48.1 \text{ kJ})$$

These iron bacteria use the energy thus gained for the reduction of carbon dioxide. They are, therefore, chemoautotrophs. Most species, however, seem to be only facultatively autotrophic and can utilize organic nutrients. As the energy gain during iron oxidation is only small, a relatively large turnover is necessary which may lead to extensive deposits of ferric hydroxide $Fe(OH)_3$ (see Section 11.1).

Some of these bacteria may utilize manganese instead of iron, and in quite an analogous manner manganous compounds are oxidized to manganic compounds. This process is rarer in nature, as manganese is not so common.

Particularly interesting is *Thiobacillus* (*Ferrobacillus*) *ferrooxydans*, which oxidizes ferrous to ferric compounds at acid reactions. This acidophilic organism is obligatory chemoautotrophic and its importance consists, above all, in its ability to oxidize ferrous compounds in an acid environment where auto-oxidation is not possible. The chemoautotrophy of the peculiar *Gallionella ferruginea* has for long been disputed, but may, however, be justified, according to more recent results. The organism could be cultured in purely mineral nutrient solution for years, but did not grow in organic media (Hanert, 1981). Iron can not be replaced here by manganese. *Gallionella ferruginea* develops best in cool waters.

The most widely distributed iron bacteria, *Leptothrix ochracea* and *Crenothrix polyspora*, seem to be facultative chemoautotrophs (Starkey, 1945) and, accordingly, can also use organic substances like peptone. Both belong to the filamentous bacteria, the Chlamydobacteriales, where the individual cells remain connected to form a filament surrounded by a thick gelatinous sheath. These sheaths are coloured an ochreous yellow or dark brown due to deposits of iron or manganese compounds. The sheaths get thicker and thicker with iron or manganese encrustations and the bacterial filaments may creep out from them and form fresh envelopes. Often, therefore, empty sheaths are found amongst the living filaments.

While in these two species the filaments are always unbranched, so-called false branching may occur in *Clonothrix fusca* (Breed *et al.*, 1957). Occasionally a cell is squeezed out of the filament but remains loosely attached to the mother filament and grows into a side branch. The cell filaments of *Leptothrix ochracea* swim freely, the two other species are firmly attached to a base by means of gummy pads. Multiplication is by egg-shaped or spherical so-called gonidia, which are produced at the end of the filaments, often in great numbers.

In iron-manganese concretions from the Baltic Sea, Ghiorse and Hirsch (1982) found stalked bacteria which encrust the ironoxides and manganese oxides and so contribute to the formation of metallic concretions. In this case, *Pedomicrobium*-like strains are involved.

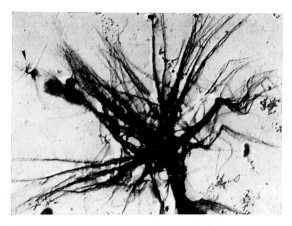
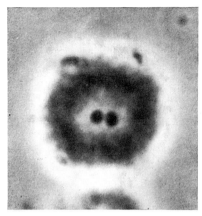

a   b

Figure 91   Micro-organisms with incrustations of iron and manganese hydroxide from lakes
(a) *Metallogenium* from Lake Pluss with buds and daughter colonies (arrow) ($\times 8\,600$)
Electron micrograph by W. D. Schmidt. (b) *Siderocapsa geminata* ($\times 1\,600$). (W. D
Schmidt)

Metallogenium presents a very peculiar structure (Figure 91a). This forms numerous
fibrillae, often arranged in star formation and possibly reproduces by budding. Manganese hydroxide is deposited on the fibrillae thus often causing them to clump
together (Schmidt, personal communication). Possibly, they are mycoplasmal, that
is to say, bacteria which have no cell wall. Recent investigations by Schmidt (1979)
showed that *Metallogenium* is very possibly not an organism, but consists of non-
living organic matter. In lakes, dissolved manganese compounds are converted by
Metallogenium into particulate hydroxides. With its appearance, therefore, a decrease
in dissolved and an increase in particulate manganese ensues.

Besides these morpholgically very characteristic organisms, some iron-oxidizing
eubacteria, belonging for the most part to the family of Siderocapsaceae, seem to
exist; they also include, besides autotrophs, heterotrophic bacteria, which have
only the iron-encrusted mucilaginous capsules in common with the iron bacteria
proper (Breed *et al.*, 1957). Schmidt (personal communication) regularly found
Siderocapsa geminata (Figure 91b), which can use iron and manganese compounds,
in Lake Pluss (east Holstein). In summer it occurred in the hypolimnion, but attained
its highest numbers immediately after the full circulation in November and December
at a low oxygen concentration. Not infrequently the bacteria show pleomorphism —
then hydroxy formation in the capsule no longer occurs. Some fungi are also able to
oxidize iron and manganese (Schweisfurth, 1969).

Iron bacteria are common in fresh water, and are often particularly numerous
in wells and springs, where frequently their masses can be seen even with the naked
eye. They are also sometimes found in great quantities in marshy streams, bogs and
ponds. They may do damage to water mains: if the water contains small quantities
of ferrous salts and iron bacteria get a foothold in the pipes they may, in time, by
constant precipitation of ferric hydroxide, block them completely. In a similar way
manganese precipitation can also be brought about. Tyler and Marshall (1967), for
example, found a heavy deposition of manganese in the pipeline of a hydroelectric
station, which they attributed especially to the activities of hyphomicrobia.

Iron bacteria normally require for their growth ferrous salts, oxygen and carbon
dioxide. When these are present and the reaction is alkaline, the purely inorganic

oxidation of ferrous to ferric iron also takes place easily. Both the biological and the chemical oxidation of iron may thus occur side by side. Iron is present in water usually in the form of dissolved $Fe(HCO_3)_2$. In oligotrophic waters with plenty of oxygen and a weakly alkaline reaction it is precipitated mostly as ferric hydroxide or iron phosphate; these form the brown iron deposits frequently found in springs and their streams. The water, as it emerges, becomes enriched with oxygen and gives up some of the dissolved carbon dioxide. In waters which flow slowly or which are stagnant, one not infrequently sees a fine iridescent skin of $Fe(OH)_3$, and the aquatic plants are often covered with a thick layer of colloidal ferric hydroxide. In eutrophic lakes also, iron precipitates occur when, during the circulation periods, the water is oxygenated. In the following stagnation periods the iron is again reduced and goes into solution, so that relatively much dissolved iron is found in the hypolimnion (Kusnezow, 1959).

According to Gorlenko et al. (1983), a close connection exists between the concentration of iron oxides and manganese oxides and the number of iron bacteria in the water. In Lake Glubokoje in the Soviet Union (Moscow region), in the course of a year, close relationships between these two parameters were found. Transport of reduced iron from the sediment into the water of the lake resulted in an increase of iron bacteria. Therefore their number rose during the times of circulation by many times and vigorously declined again with the subsequent precipitation and sedimentation of the oxidized iron.

Investigations by Hanert (1974) of drainage pipes in north-western German lowland soils, showed that here iron(III) precipitation (ochre formation) occurs extensively, due to direct and indirect bacteriological processes. According to Khrutskaya (1970), in groundwaters of the Soviet Union, the proportion of bacteriological ochre formation amounts to 81–98 per cent and that of chemical ochre formation to 2–19 per cent.

Sörensen (1982) found that the reduction of Fe^{3+} to Fe^{2+} in marine sediments is connected with the activity of nitrate-reducing facultative anaerobic bacteria. In the surface zone of the sediment, Fe^{3+} reduction appears to result, mainly due to micro-organisms. This process might be important for mineralization when little nitrate is available.

While iron precipitates in acid waters must always be attributed to the activity of iron bacteria, this is only partially so in neutral or alkaline waters. This coexistence of biological and chemical iron precipitation is surely the reason why the role of iron bacteria in nature's cycle of matter is still an open question. This is true also for lake or bog ore (see Section 14.3). Considered then as a whole, we still know rather little about the iron bacteria, although they are a group of organisms which have aroused the interest of eminent scientists over and over again (Ehrenberg, Winogradsky, Cholodny, and others).

Micro-organisms are also capable of forming organic iron and manganese complexes (chelates). Various fungi can even synthesize numerous different complexes (Winkelmann, personal communication). Organic metal compounds and complexes are broken down again by micro-organisms (see above). In waters this takes place above all by bacterial action. These decompositions also play a role in the cycling of iron and manganese in waters, which is greater than previously assumed, especially in some eutrophic lakes.

11.9. Interactions of Cycles

The cycles of individual elements are parts of the great cycle of matter in nature. Accordingly, the participating micro-organisms often cause, simultaneously, transformations of quite different compounds and so participate in several individual cycles.

The desulphurizers, for example, reduce sulphate to hydrogen sulphide in order to oxidize carbon compounds. The thiobacilli oxidize, in turn, the hydrogen sulphide to gain energy for reduction of carbon dioxide and thereby for the synthesis of carbon compounds. *Thiobacillus denitrificans* can produce, in this way, the required oxygen by the reduction of nitrate. With that, this organism is already participating very actively in three cycles: those of carbon, sulphur and nitrogen. This is true also, even if in a quite different manner, of proteolytic bacteria which can liberate carbon dioxide, hydrogen sulphide and ammonia in the decomposition of proteins and so likewise play an important role in the cycles of the elements cited. The phosphorus cycle or that of iron may be equally associated with cycles of other elements through the activity of micro-organisms.

Thus, the quantitative transformations caused by bacteria and fungi and involving different elements are of decisive significance for the understanding of aquatic ecosystems in their mutual relationships. Nevertheless our knowledge is still very incomplete. Now, for example, it is known that, for the synthesis of 1 g of cell dry matter, *Thiobacillus ferrooxidans* oxidizes 156 g of Fe^{2+}, *Thiobacillus neapolitanus* 30 g of $S_2O_3^{2-}$, *Nitrosomonas* 30 g of NH_3 and *Alcaligenes eutrophus* 0.5 g of H_2 (Schlegel 1981). Since the bacterial cell contains about 50 per cent carbon, 20 per cent oxygen, 14 per cent nitrogen, 8 per cent hydrogen, 3 per cent phosphorus, 1 per cent sulphur, 1 per cent potassium, 0.5 per cent calcium, 0.5 per cent magnesium and 0.2 per cent iron, one can easily recognize which manifold transformations are involved when, for example, in the nitrogen cycle, ammonia is oxidized to nitrite by *Nitrosomonas*. The cycles of all the elements taking part in the structure of the cell are thereby implicated to a greater or lesser extent.

Whereas under defined growth conditions in pure cultures reactions may be quantified, this is scarcely possible in natural environments because of the numerous subsidiary and opposing biochemical processes that are occurring. Thus, under suitable conditions, nitrogen fixation, ammonification, nitrification and denitrification can proceed simultaneously in the water and sediment, whereby organic matter is built up by nitrifiers and broken down by micro-organisms participating in other processes. With the aid of various isotopes, it is possible to quantify the nitrogen and carbon flows. However, on account of the expensive methodology, this could only seldom be done. Mostly, in investigations of this kind, one limits oneself to the cycles of individual elements, particularly nitrogen and sulphur (see Sections 11.5 and 11.6). The elucidation of the quantitative relations between cycles, therefore, is one of the most important tasks for the ecologist in future years.

CHAPTER 12

12. Bacteria and Fungi in the Ecosystems of Waters

The important role played by bacteria and fungi in the cycling of matter in waters already indicates their significance in ecosystems.

An ecosystem represents an activity structure of living creatures which although certainly open to influences from outside is to a certain degree capable of self-regulation. So we can consider, for example, a lake, a river or a region of the sea as an ecosystem that, in turn, is a component of a larger ecosystem — from landscapes and seas up to the total biosphere. In the symposium volume *Ecosystem Research* published in 1973, Ellenberg attempts a classification of the earth's ecosystems from the functional point of view. In this, both aquatic as well as terrestrial ecosystems are considered in a five part hierarchy.

The fact that ecosystems are open and have no sharp boundaries explains why they are subject to a variety of external influences. This is true for small and apparently relatively sharply limited ecosystems such as small lakes or bogs, as well as for whole landscapes. The equilibrium of such a system is always dynamic because all the factors affecting it, as for example light, temperature and nutrient concentrations, are themselves subject constantly to more or less large fluctuations. In addition there are the interventions of man.

The ecosystem is the fundamental functional unit in ecology since it embraces equally the organisms and their environments which are mutually influenced in many different ways (Odum, 1980).

Bacteria and fungi are also to be counted among the essential components of nearly all ecosystems. In the model of a complete ecosystem given by Ellenberg (1973) (Figure 92), heterotrophic micro-organisms have their place among the 'decomposers' and here they are essentially the 'mineralizers'. But they are also found among the 'symbionts', the 'parasites' and the 'superparasites'. An ecosystem can only be termed complete when there are sufficient autotrophic organisms present which gain the greatest part of the energy required through photosynthesis — thus converting into chemical energy part of the light energy provided by the sun. In waters it is primarily the algae of the phytoplankton that take care of this. Only exceptionally do photosynthetic bacteria play here a greater role (see Section 11.1).

According to Winogradsky (1925), micro-organisms in ecosystems can be classified into two groups, the autochthonous and the zymogenous. Autochthonous micro-organisms are indigenous and present in the ecosystem in question constantly without any particular nutrients being introduced from outside. In part, they are specialists which are largely adapted genetically to the conditions of their habitat. This is the case, *inter alia*, with most true water bacteria. Zymogenous or allochthonous micro-organisms, as a rule, belong to the ecosystem only transiently and are dependent on the addition of nutrients. Frequently they are widely distributed forms (ubiquiters) which require high concentrations of easily assimilable nutrients. They are found therefore particularly in eutrophicated waters.

To elucidate the many-sided activity of micro-organisms in ecosystems, the investigation of so-called microcosms is suggested. These are, for example, food chains or parts there of, such as the relationship between individual predators and their

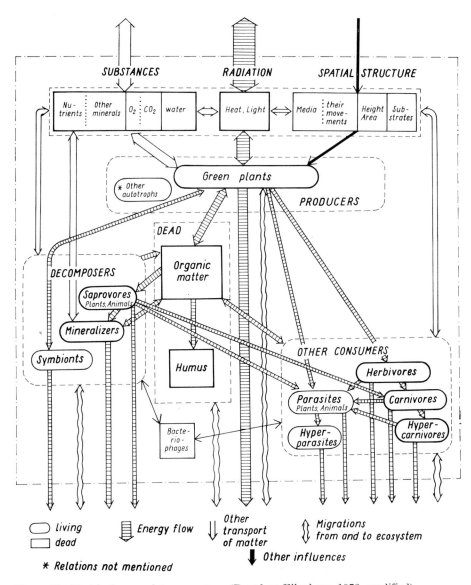

Figure 92   Model of a complete ecosystem. (Based on Ellenberg, 1973, modified)

prey organisms, matter cycling or the microbial degradation of individual compounds, energy flow, etc. (see Giesy, 1980). Each microcosm then represents a little mosaic stone which can be assembled with others into a picture and so facilitate a glimpse into the often complicated functions of micro-organisms in the ecosystems.

12.1. Bacteria and Fungi in Food Chains

Food chains occupy a dominant position within natural ecosystems. Considerable practical significance is attributed to them — for example, in waters, for fishing. Since on transfer from one member of the food chain to the next higher one, 80–95 per cent of the chemical energy bound in the living creatures may be lost, it is important in view of the worldwide shortage of high-value foodstuffs to fish not only the terminal members of the food chain but to secure the earliest possible members also. Thus much higher protein yields may be obtained if plankton feeders are fished in place of predator fishes. The creation of the fundamental principles for this is the special concern of fish biological research.

In all waters, bacteria and fungi have important functions in the food chain. They take up dissolved organic substances, which for the most part are released into the water by the primary producers — i.e. by the phytoplankton. In addition there are other substances derived from animal organisms and, in inland and coastal waters, from the land also (allochthonous substances). They are converted by bacteria usually very rapidly into particulate material and then a substantial part is consumed by other organisms. They are thus taken up into the food chain and in this the capacity of bacteria to take up very low concentrations of organic compounds is of great importance (see Section 8.8). In this way, in fact, such organic substances are brought into the food chain that otherwise could scarcely be utilized on account of their low concentration. Bacteria and also fungi are used as food by widely differing animals (see Section 9.2), which can represent different members of the food chain (Figure 93). In the first place, they are used as food by the herbivorous zooplankton, which are reckoned as primary consumers. But carnivorous zooplankton, which are included among the secondary consumers, also take up bacteria and fungi with their usual food and use them, in a sense, as supplements for their nutrition. With the increasing size of the animals, the importance of this supplement, however, becomes less and less. Consequently, for the final consumers it is of little or no consequence.

In the sea, according to Azam et al. (1983), bacteria can utilize 10–50 per cent of the carbon bound by the primary producers as food (see Section 11.2). The number

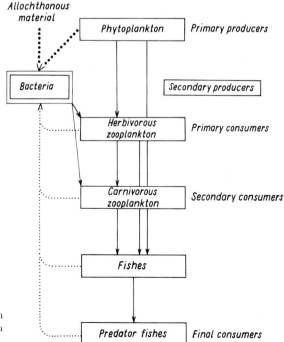

Figure 93   The position of bacteria in the food chain of the open sea

of free-living bacteria in the water is controlled probably by the heterotrophic flagellates belonging to the nanoplankton, which on their part, together with autotrophic flagellates serve as food for members of the microzooplankton. These, at 10–80 μm, correspond to the size of most phytoplankton algae and with these represent the food for the next larger fraction of the zooplankton. In this way, a part of the organic substance which the phytoplankton gives off by exudation and lysis is ed back into the direct food chain.

The transfer of carbon compounds by bacteria is of special importance when such phytoplankters dominate that are not consumed or only in very small amounts by the zooplankton. For example, this is the case with some filamentous cyanophytes, like *Nodularia spumigena* and *Aphanizomenon flos-aquae*, which regularly form blooms in the Baltic Sea in summertime (see Section 7.3). The older filaments concentrate, when the sea is calm, in the area of the water surface in the form of flocks, which may consist of many thousands of *Nodularia* filaments. For the most part, these are colonized by bacteria. In the initial phase of flock formation there is a large increase in the bacterial biomass. The *Nodularia* filaments, which are at first straight, often form spirals. On both the straight and the already spiral filaments, mostly colonies of micrococci at first develop and then rods. Later other bacterial forms come in and finally a *Nocardia* species can dominate the bacterial population. *Nodularia* flocks represent a specific microcosm, which contains a multiple of the biomass of the surrounding water. Here the most varied aufwuchs organisms may be found: in addition to bacteria there are fungi, algae (*Navicula salina*, *Chlorella*) and protozoa (*Vorticella marina*, *Amoeba*). Also present in the flocks are numerous free-living bacteria, heterotrophic flagellates, ciliates, rotatoriae and crustaceans. The flocks consist of both living and dead material (decayed cells, faecal pellets, etc.) and they can remain intact for some weeks. The decomposition of organic matter is due largely to bacteria. The latter are consumed by flagellates and ciliates, which, in turn, serve as food for larger zooplankters (rotatoriae, crustaceans). In this way, the carbon fixed by cyanophytes is transferred into the food chain via the bacteria and so ultimately made available to fishes.

Caron *et al.* (1982) made similar observations in marine snow and *Rhizosolenia* mats of the Sargasso Sea. Here, populations of bacteria and bacteria-feeding protozoa also developed which were up to 10^4 times greater than those in the water.

According to Suberkropp *et al.* (1983), fungi play an important part in streams and rivers in the degradation of leaves after the autumn leaf-fall in that they convert these into a more usable food for detritus-feeding invertebrates. The fungi have the capacity to depolymerize cellulose, xylans and pectin. The material thereby becomes softer and can be taken up more easily by animals. The fungi likewise serve as food for animals, e.g. trichoptera larvae. Of course, different fungi are evidently consumed more or less readily. Through the activity of fungi and, to a lesser extent bacteria also, considerable amounts of terrestrial plant material are brought in this way into the food chains of waters.

Bacteria and fungi are also of some importance as suppliers of CO_2 to the phytoplankton. This is especially true when, in the case of algal blooms, the atmospheric carbon dioxide becomes the limiting factor for algal growth (Lange, 1970).

The special position of the heterotrophic bacteria and fungi becomes still clearer when their nutrient suppliers are considered. These may be found among all members of the food chain because all organisms release, with their excrement, organic nutrients, which are consumed preferentially by bacteria. In addition, in lakes and coastal waters they are able to utilize allochthonous substances. This takes place above all where rivers discharge large amounts of organic matter. In river lakes and particularly in artificial lakes the proportion of allochthonous substances can be even larger than that produced in these waters by the assimilation activity of primary producers.

Thus, in the artificial lake of Rybinsk, Kusnezow and Romanenko (1966) estimated for the year 1964 a biomass production by bacteria of 117 388 tons as against a primary production of 99 750 tons by the phytoplankton.

12.2. The Role of Bacteria and Fungi in Energy Flow

An essential aim of ecosystem research is the understanding of the relevant energy flow because nearly all the components of the system are linked through the transfer of energy. In a complete ecosystem this begins as radiation energy supplied by the sun and is next bound as chemical energy by the primary producers in the form of cell substance and then part of it is passed on via the food chain.

As far as we know at present, however, even under the most favourable conditions only a few per cent of the solar energy irradiated into waters is transformed into chemical energy by the primary producers. In his investigation of the Silver Springs in Florida (USA), Odum (1957) found an absorption of 41 000 cal per cm^2 of water surface per year (see Figure 94), of which only 2081 were used by algae for photosynthesis — i.e. about 5 per cent. Of these 2081 cal, however, 1198 were lost through respiration, so that the net production amounted to only 883 cal. Finally 506 cal per cm^2 per year fell to the 'decomposers' in the form of dead organic material of which roughly 90 per cent was respired. If one supposes that the 'decomposers' in the lake in question were, in the first place, bacteria, then it follows that approximately half of the net production of the green plants was decomposed directly by bacteria. The proportion of other members of the food chain, on the other hand, can have been only a comparatively small amount, as may be seen from the scheme in Figure 94. This example can certainly not be applied directly to other ecosystems,

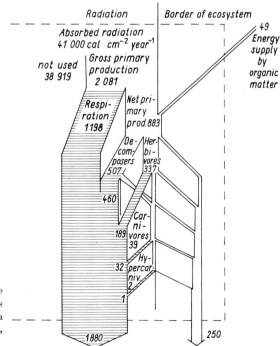

Figure 94   Energy flow through the ecosystem of Silver Springs in Florida (USA). (Based on Odum, 1957, and Ellenberg, 1973, modified)

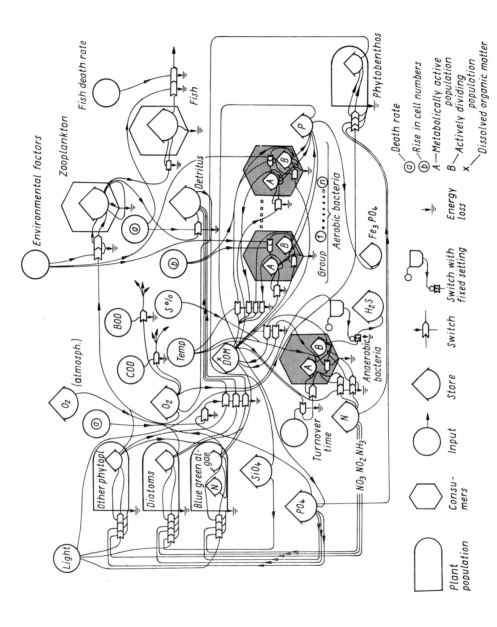

Figure 95   Model of the ecosystem of the Kiel Bay using Odum's symbols, with special reference to micro-organisms. (Based on Bölter et al., 1977, modified)

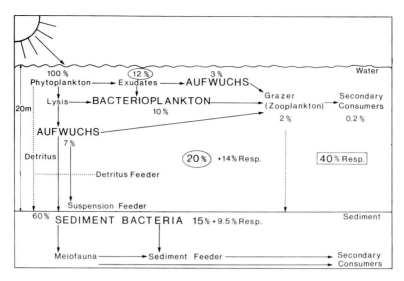

Figure 96   The role of bacteria in the energy flow in the food chain of the Kiel Bay

but it points out the way of energy flow with its most important sites, which, nevertheless, are basically everywhere the same in all details, and it gives some idea of the size ratios which make clear the prominent position of bacteria. In eutrophic lakes and regions of the sea, fungi will certainly play a greater role within the ecosystem than is possible in a spring lake like Silver Springs, although even then bacteria might have the greater part in energy transport. The great importance of bacteria was also confirmed by the results of ecosystem research in the Baltic Sea (Jansson, 1972).

Figure 95 shows a model of the ecosystem of the Kiel Bay with special reference to micro-organisms (Bölter *et al.*, 1977). It was based on an extensive microbiological-ecological investigation of the Kiel Bay in the years 1974–75 (Rheinheimer, 1977c). The Odum symbols, which express the energy relations, have been used for this model.

On the basis of the present existing data, reliable quantitative information on energy flows in waters is only rarely possible. Figure 96 gives an abstracted presentation of the relative energy flow through initial members of the food chain with particular reference to bacteria, based on extensive studies in the Kiel Bay. Accordingly, about 20 per cent of the carbon produced by the phytoplankton is converted in the water into bacterial carbon, whereas up to 60 per cent sinks to the bottom of this shallow coastal water with its average water depth of 20 m, because, at the time of the spring blooms, scarcely any herbivorous zooplankton is present here. About 15 per cent of the primary production is transformed by sediment bacteria into biomass. Therefore, in the Kiel Bay, a considerable part of the energy flows through bacteria into the food chain (Rheinheimer, 1981).

If sufficient zooplankton is present (for example, during the summer blooms), the major part of the primary production enters directly into the food chain of the pelagial. In this case the bacteria might, on the whole, play a lesser role in the energy flow. In the temperate climatic zone, correspondingly, the spring bloom has a different influence on bacterial activity than the summer bloom (see Williams, 1981) and the energy flow proceeds in different ways.

CHAPTER 13

13. *Micro-organisms and Sedimentation*

Bacteria and fungi affect the formation of sediments in various waters and may, by their activity, alter them. Hence they play an important part in the origin of sedimentary rocks. Although interest in geomicrobiology has been growing over the last years, our knowledge in this field is still fragmentary. However, the importance of micro-organisms for many geological processes can hardly be over-estimated.

13.1. The Role of Micro-organisms in the Formation of Sediments

Bacteria, and to a lesser extent fungi also, colonize suspended particles, whether they are of organic or of mineral origin (see Section 8.4). This may change their size and shape and consequently their sedimentation rate. In turn, this affects the stratification of the sediments.

Colonization by micro-organisms may destroy, partly or completely, the suspended particles if they are used for food or, for instance, dissolved by acid.

The growth of micro-organisms not infrequently also causes an enlargement of the particles, particularly by aggregation of several small particles into bigger ones. Fungi are thus able by means of their rhizoids or hyphae to hold together a number of particles and eventually unite them. Copious slime production by bacteria leads to similar results (ZoBell, 1946a). Recent research points to bacteria with fimbriae or pili (see Chapter 3) as the cause of aggregation of tiny suspended particles. These bacteria attach themselves to the particles by means of their fimbriae and thus connect several particles with one another. Näveke (personal communication) reports that *Vibrio extorquens* sticks to soil particles so firmly with its fimbriae that the organisms are not flushed out by rain water. Electron micrographs suggest that fimbriae play an important part in the compacting of soil and thus affect its structure. They probably have a similar function in sandy beaches and sandbanks. They also certainly affect the layering of some sediments. The number of bacteria which posses such fimbriae is likely to be considerably greater than has formerly been assumed. They are widespread amongst gram-negative bacteria and are found particularly in strains freshly isolated from their natural environment (Brinton, 1965). However, there are various types of fimbriae, which doubtless differ in their functions (Moll and Ahrens, 1970).

Micro-organisms are even able to trigger off sedimentation processes through their metabolic activities. Thus the precipitation of lime in tropical shallows is said to be due to changes in the pH following various bacterial reactions (Bavendamm, 1932; Greenfield, 1963; Moritz, 1980b). Such a pH shift towards the alkaline region occurs, for example, during ammonification, denitrification and desulphurication (see Sections 11.5 and 11.6) and during breakdown of organic calcium salts. The most important microbial processes which may be associated with precipitation of lime may be represented as follows:

1. $(NH_4)_2CO_3 + CaSO_4 \rightarrow CaCO_3 + (NH_4)_2SO_4$

   Ammonium carbonate is produced by ammonification during putrefaction of protein.

2. $Ca(NO_3)_2 + 3H_2 + C \rightarrow CaCO_3 + 3H_2O + N_2$

3. $CaSO_4 + CH_3COOH \rightarrow CaCO_3 + H_2S + H_2O + CO_2$

4. $Ca(COOCH_3)_2 + 4O_2 \rightarrow CaCO_3 + 3CO_2 + 3H_2O$

Precipitation of calcium carbonate takes place at a pH of around 9.4 (Wood, 1967), but as sea water is very effectively buffered, such a shift of reaction is not possible in oceans which are poor in nutrients. It can occur only in those regions of the sea which are densely colonized by micro-organisms and favour their activities by high water temperatures. This is the case in some tropical shallows, for example the Bahama Bank. Vigorous assimilation activity of algae and seaweed have, in such regions, raised the pH so far that a relatively small increase in alkalinity due to bacterial metabolic processes may lead to precipitation of chalk. This happens most easily at the sediment surface, by denitrification and desulphurication; but under suitable conditions it may also occur in the water. Possibly, such bacterial processes have played an important part in the formation of unstratified calcareous rock, often thousands of metres thick.

The calcium carbonate particles adsorb organic and inorganic nutrients, enabling micro-organisms to grow on their surfaces and thereby enlarge the particles.

In inland lakes, calcite crystals can be formed by the CO_2 produced in the bacterial decomposition of organic substances in the mud (Kusnezow, 1966).

Mats particularly of cyanophytes, but also of eucaryotic algae, photosynthetic and various heterotrophic bacteria, represent a special form of sedimentation at the bottom of shallow waters (Golobic, 1976); from these, in geological times, the so-called stromatoliths have developed. They consist of limestones with a characteristic form and surface. A small part of the frequently considerable production of these mats is converted into carbonate. According to Krumbein *et al.* (1977), heterotrophic bacteria also take part in this process. Investigations of the hypersaline Solar Lake (Sinai) showed that here more than 99 per cent of the primary production is remineralized and less than 1 per cent is transformed into carbonate. The growth of stromatiolithic chalk, therefore, takes place very slowly and amounts to only a few centimeters in a hundred years.

Sieburth (1965) states that in sea water with much protein, organic particles may be formed through the activity of bacteria. Thus, in sea water, enriched with peptone or casein and inoculated with mixed cultures, organic aggregates come into existence. Sieburth (1968a) suggests the following mechanisms for the formation of such organic particles by bacteria:

Protein substrate $\xrightarrow[\text{ammonification}]{\text{Bacterial}}$ NH_3 $\longrightarrow$

Microzonal pH rise $\longrightarrow$ Inorganic microparticles $+$

Dissolved organic substance $\longrightarrow$ Particulate organic aggregates

A large part of the organic deposits, however, is produced by the phytoplankton. Investigations by Iturriaga (1977) into the bacterial activity of the aufwuchs on sunken material in the Kiel Bay showed that these can be very rapidly colonized and broken down by micro-organisms, particularly in the summer months. This is shown, among other things, by the pronounced decrease in 'biocarbon' proportion with increasing depth of water:

g per m² per year				Percentage of total carbon		
Depth (m)	Dry matter	Organic substance	C	Bio-C	Phytopl.-C	Bact.-C
10	223.8	41.1	19.6	47	17	1.85
18	569.6	101.8	45.2	23	14	1.52

Total bacterial numbers of the deposits varied from 200×10^6 to 600×10^6 per mg of organic matter. The heterotrophic activity (see Section 11.2) declined in winter to about half at a depth of 10 m and to a third at 18 m. Degradation of particulate organic matter takes place, therefore, to a considerable extent in the uppermost zones of the water column where the velocity is strongly influenced by the temperature. Accordingly, the deposits suffer a more or less large change in their composition even in shallow waters on their way from the site of formation to the bottom.

Bacteria, heterotrophs as well as autotrophs, also participate in the precipitation of iron and manganese compounds (see Section 14.3).

13.2. Transformation of Sediments by the Activity of Micro-organisms

There are multifarious activities of bacteria and fungi in most sediments of inland waters and of the sea (see Sections 7.4 and 7.5). As a rule, these are most vigorous in the upper, i.e. newest, zones, which are still relatively rich in organic substances capable of being broken down. Here, therefore, the changes brought about by micro-organisms in composition and structure are the most marked. But bacteria live and act in the deeper layers also, though in much smaller numbers.

Due to bacterial activity in sediments, the total amount of organic matter gradually diminishes and its composition is changed. As the compounds which are more easily attacked are broken down first, the proportion of substances which decompose with difficulty increases in the deeper layers. In inland lakes, these are predominantly lignin-humus complexes (Kusnezow, 1959). Marked changes in pH and in redox potential are often associated with these chemical processes.

In the sediments of eutrophic waters, rich in nutrients, the vigorous microbial consumption of oxygen leads frequently to the formation of anaerobic zones or pockets, where fermentation as well as denitrification and desulphurication occur. The latter reaction particularly has a lasting effect on the sediment diagenesis. The hydrogen sulphide released during sulphate reduction combines, in the presence of iron, to form FeS, and offensive black iron sulphide mud is produced. This can accumulate to such an extent in sewage-loaded ponds and harbour basins that they have to be dredged regularly.

In regions bordering on aerobic zones, sulphur-oxidizing bacteria, particularly thiobacilli, or, on the sediment surface *Beggiatoa* and *Thiothrix* species also, may again contribute to the oxidation of sulphides (see Section 11.6). A cycling of sulphur often takes place between sediment and the water above it, but, as a rule, it is incomplete since only part of the sulphur reaches the water again. The remainder is left in the sediment in the form of metallic sulphides. Jörgensen (1977) found in the Limfjord that 9.8 mmol of S per m² per day entered the sediment in the form of sulphate and organic compounds and only 8.8 mmol of S returned to the water again (Figure 97).

The formation of hydrogen, methane, nitrogen and carbon dioxide affects the structure of sediments to a greater or lesser degree. This is particularly the case where — as in some eutrophic lakes — bubbles rise up (Ohle, 1958). With them, ma-

13*

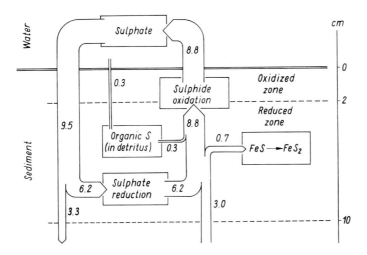

Figure 97   Cycling of sulphur between water and sediment of the Limfjord. The numbers represent the mean transport rates in mmol S per m² per day. (Based on Jörgensen, 1977,
modified)

terial from the surface of the sediment may be carried back into the water repeatedly
and undergo changes there.

The diagenesis of limnic and marine sediments is, to a considerable extent, affected
by microbial acid or alkali production. Particularly in anaerobic sediments, fermentation processes may lead to the production of various organic acids besides carbon
dioxide (see Section 11.4), which may form insoluble calcium salts.

Whilst in oligotrophic lakes the organic material can be broken down to a considerable extent, this is not possible in eutrophic and dystrophic lakes. In the former,
the so-called sapropel often occurs. Here the microbial decomposition of organic
matter is clearly inhibited at a depth of over 1 m only (Kusnezow, 1959) and may
fade out completely at greater depths, although the lower mud layers often contain
more hemicelluloses, celluloses and nitrogenous compounds than the regions nearer
the surface. The cause of this may be the lack of nutrients which are easily assimilated, like protein and sugars.

Bioturbation — i.e. the mixing of different zones by sediment-inhabiting animals —
may have an influence on bacterial activities. There is also, thereby, an exchange of
pore water with the body of water to be found above the sediment surface. Bioturbation can facilitate frequent alterations in the composition and concentration of
nutrients and the admission of oxygen is possible. Not infrequently this has led to an
increase in activity and growth rates of micro-organisms and hence an acceleration
in the decomposition of matter in the sediment. This influence is most pronounced
in soft bottom (Yingst and Rhoads, 1980).

In most sea sediments bacterial activity is not completely absent even in the deepest
parts. Although the saprophyte numbers here are very small (see Section 7.5), their
activity over long periods of time may affect the diagenesis of the sediments. Thus,
in a deep sea sediment of the Arabian Sea from 3300 m there were still, in the 4-m
horizon, 1500 saprophytic bacteria per g wet weight. This sediment is about 100000
years' old, and the colonization by bacteria has probably been constant for a long
time. Although the activity of micro-organisms here is small, it should not be overlooked.

14. The Role of Micro-organisms in the Origin of Mineral Resources

The formation of valuable resources of the earth is closely connected with the actions of micro-organisms in sediments. Bacteria and fungi have contributed to the formation of peat, coal and petroleum, and of certain ore and sulphur deposits.

14.1. Peat, Brown Coal and Petroleum

We are best informed about the function of micro-organisms in the formation of *peat*, as this process is still going on in many parts of the world where it can be studied. The various kinds of peat all have a high content of lignin and of humic acids. It has been shown that the top layers of peat have, as a rule, a rich bacterial and fungal population, though the numbers of micro-organisms diminish rapidly with increasing depth. In a bog region of the USSR, Kurbatowa-Belikowa (1954) found 700 million bacteria per g dry peat in the surface layer, but at a depth of 25–50 cm there were only 25 million. The numbers then varied very little down to 6 m. Throughout one can distinguish an oxidative surface zone with a very lively microbial activity, an intermediate zone which may be oxidative or reductive according to the weather and climatic conditions, and a predominantly reductive deep zone with relatively little microbial activity (Kusnezow *et al.*, 1963). In the formation of peat, higher fungi particularly (and above all Fungi Imperfecti) are of great importance as they very quickly decompose the moss thalli. Neofitova (1953) states that sphagnum tissue was destroyed to a large degree after only 2 months. Further decomposition is thought to be due mainly to bacteria. In the surface zone, the presence of easily assimilated organic substance, abundant supply of oxygen and, in summer, relatively high temperatures cause the rapid conversion of plant material to peat. This seems to be accomplished within a few seasons. On the other hand, the lack of inorganic nutrients and the high content of humic acids prevent the further breakdown of peat.

It is likely that micro-organisms also played an important role in the formation of *brown coal*. Like peat, this is of plant origin. It is formed predominantly of wood from tertiary forests. The frequent occurrence of fragments of various bog cypresses (*Taxodium*) suggests that some of the deposits originated in bog regions. The microbial decomposition of the wood resulted in a material consisting mainly of lignin and humic acids. This process probably ran a similar course to that which can be observed nowadays in the formation of forest peat, which is due mainly to the activity of higher fungi.

Less clear is the role of micro-organisms in the formation of *petroleum*. If we assume that it is of biogenic origin, we can be certain that micro-organisms have contributed to it. According to prevailing theory, the material from which petroleum originated is of plant and animal nature, and is supposed to consist first and foremost of marine plankton (Beerstecher, 1954). The dead organisms undergo rapid changes due to the activity of bacteria and fungi. ZoBell (1943) believes that, at present, the organic substances in sediments are being gradually changed into marine humus whose chemical composition resembles more that of petroleum than that of marine plankton.

	Marine organisms	Marine humus	Petroleum
Carbon (%)	32–45	52–58	82–87
Hydrogen (%)	5–9	6–10	11–14
Oxygen (%)	25–30	12–20	0.1–5
Nitrogen (%)	9–15	0.8–3	0.1–1.5

During the anaerobic breakdown of humus material, bacteria produce methane and small quantities of liquid hydrocarbons (see Section 11.3). Thus, Jankowski and ZoBell (1944) found hydrocarbons in *Desulphovibrio* cells which had been cultivated in caprylic acid. As the hydrocarbons are more stable than the bacterial cytoplasm, it is possible that they gradually accumulated in the sediments. In this case hydrocarbons would have been synthesized in the bacteria from organic material of predominantly vegetable origin, and would have been released on the death of the bacteria. Also, bacteria which are capable of activating molecular hydrogen can convert unsaturated into saturated hydrocarbons. Possibly, micro-organisms may also have had a share in the accumulation of hydrocarbons, for example by decomposing the organic residues or by dissolving calcareous rocks.

It must, however, be admitted that in recent sediments no accumulation of hydrocarbons has been found anywhere which would allow us to conclude that oil was being formed (Kusnezow et al., 1963), so that all our ideas on the formation of petroleum are, for the time being, still hypotheses.

14.2. Sulphur

More than 90 per cent of the sulphur present on earth is supposed to be of sedimentary origin and barely 10 per cent volcanic. In the origin of the former, micro-organisms have played an important part. In the formation of sulphur deposits, various sulphur-oxidizing bacteria, photoautotrophic purple and green sulphur bacteria as well as chemosynthetic thiobacilli have probably been involved. In waters rich in hydrogen sulphide they oxidize H_2S to elementary sulphur, which, in consequence, can accumulate. A prerequisite for this, however, is a high original concentration of hydrogen sulphide, or else not much sulphur is produced as the oxidation is carried further to sulphate. The hydrogen sulphide may have a variety of sources. It comes either from volcanic springs or it is produced during bacterial sulphate reduction or it may be released in small quantities during protein putrefaction.

Butlin and Postgate (1954) studied the formation of sulphur in a lake of brackish water in Cyrenaica (Libya). On the bottom of the lake they found a sulphur layer 15–20 cm thick, even though the inhabitants of the surrounding country exploited it regularly. The water showed a milky turbidity due to its content of elementary sulphur. Butlin and Postgate found an abundance of sulphate-reducing bacteria in the mud of this lake and attributed the H_2S formation to their activity. They observed in the water massive development of green and purple sulphur bacteria which oxidize hydrogen sulphide to elementary sulphur (see Section 11.6). These were mainly members of the genera *Chlorobium* and *Chromatium*. The former precipitate the sulphur into the substrate, whilst the latter store sulphur granules within their cells.

Similar conditions prevail in Lake Sernoje in the USSR (Kusnezow et al., 1963). Its water contains 83–85 mg H_2S per litre. Besides *Chlorobium* and *Chromatium*, numerous thiobacilli were also found. By means of tracer experiments, adding $Na_2^{35}S$ to the water samples, it could be established that, in this lake, the bulk of the sulphur is produced by thiobacilli. These oxidize 30 per cent of the sulphide to sulphur within

24 h; 17 per cent is oxidized chemically to sulphur and a further 8 per cent by photo-synthetic bacteria to sulphate.

Bacterial sulphur production may also occur in groundwater of regions where rocks carry mineral oil. Such groundwater often contains sulphate. In the deepest parts this is reduced by *Desulphovibrio desulphuricans* to hydrogen sulphide, hydrocarbons supplying the necessary energy. Wherever water containing H_2S comes into contact with water carrying oxygen, thiobacilli, particularly *Thiobacillus thioparus*, oxidize the hydrogen sulphide to sulphur. Large numbers of thiobacilli are frequently found in the vicinity of deposits of finely crystalline sulphur in such border areas. By means of isotope experiments it was established that 50 per cent of the sulphides in water samples were oxidized to elementary sulphur within 24 h, i.e. 30 per cent chemically and 20 per cent biologically, by thiobacilli (Kusnezow *et al.*, 1963).

In the biological formation of sulphur, thiobacilli play the most important overall role, as far as our present knowledge goes; filamentous sulphur bacteria and the photosynthetic organisms are of importance only on a local scale. Thus purple sulphur and chlorobacteria show mass development almost solely in shallow waters at present times, and it is unlikely that they behaved differently in earlier epochs. Heterotrophic bacteria are also able to oxidize sulphides, but it is not known how far they have participated in the origin of sulphur deposits.

As the desulphurizing bacteria prefer the light isotope of sulphate sulphur (^{32}S), bacterial sulphate reduction leads to a partial separation of the sulphur isotopes. Marine sediments contain a greater proportion of light ^{32}S than would correspond to the terrestrial mean, while the sulphate present in the sea is enriched with heavy ^{34}S. There is than a deficit ($^{34}S \simeq -15$ per thousand) in the sediment sulphur, matched by a surplus ($^{34}S \simeq +20$ per thousand) in the sulphate sulphur of sea water. This permits us to establish the biogenic origin of sulphides and sulphur deposits on the basis of isotope distribution (Kaplan and Rittenberg, 1964).

14.3. Ores

In the formation of various iron and manganese ores, bacteria have apparently played an important part. The participation of iron bacteria in the formation of lake iron has long been a matter of dispute, as no bacteria could be observed either in the iron-containing ooze of the lakes or in iron concretions. Perfiljew (1926) first succeeded in demonstrating bacteria in lake iron ore. In the iron ooze of dystrophic lakes he found an organism which he named *Sphaerothrix latens*. It very much resembles *Gallionella ferruginea* and may even be identical with it (Breed *et al.*, 1957). The organism grows in concentric colonies consisting of a great number of filaments encrusted with ferric hydroxide. Perfiljew suspected that the iron manganese concretions in lakes represent giant colonies of *Sphaerothrix latens*. The iron manganese deposit is due to the oxidation of ferrous and manganous compounds (see Section 11.8).

Later it was found that *Metallogenium personatum* can bring about iron-manganese precipitations. This occurs in many waters and is thought to be largely responsible for the formation of lake iron ore in the Karelian lakes (Perfiljew and Gabe, 1969). The organism can cause the deposition of iron as well as of manganese compounds.

As with lake iron ore, bacteria have probably also participated in the formation of bog iron ore (limonite). Wherever bog iron ore is being formed, filamentous iron bacteria of the genera *Leptothrix* and *Crenothrix* are present (Kusnezow *et al.*, 1963) which are able to oxidize ferrous to ferric compounds (see Section 11.8). Dubinina and Derjugina (1972) found that the iron hydroxide of weakly acid peats (pH 6.05 to 6.25) in the neighbourhood of the Punnus-Järvi Lake consists largely of the mineralized envelopes of filamentous bacteria. Electron microscopic analysis revealed that *Gallionella ferruginea, Gallionella filamenta, Metallogenium personatum, Leptothrix ochracea, Toxothrix trichogenes* and others were involved. The microflora of an acid shallow peat (pH 2.8–3.5), on the other hand, contained only rods. In addition to those of normal size, very small cells were also found. Chemical and microbiological investigations showed that the oxidation of divalent iron, the concentration of which here reached 12 mg l^{-1}, is attributable to *Thiobacillus ferrooxydans*.

In mine water, *Thiobacillus ferrooxydans* can cause an accumulation of iron. This acidophilic organism is able to oxidize pyrites (FeS_2); in the process sulphuric acid and a rust-like ooze are formed. *Thiobacillus ferrooxydans* is the only organism which can oxidize ferrous to ferric compounds in an acid environment.

The formation of marine iron and manganese concretions was likewise originally attributed to the activity of chemoautotrophic iron bacteria (Butkewitsch, 1928). But this mode of formation of manganese nodules, which are widely distributed in the sea, is today held to be improbable (Ottow, 1983). The speed of growth of manganese nodules can be estimated at 1 mm in 1 million years, thus a manganese nodule of 4 cm in diameter would be aged about 20 million years. During this time, with a concentration of $2\,\mu g$ Mn^{2+} per litre and a streaming velocity of between 2 and 25 cm per sec, 2500–31500 tons of manganese would have flowed through a 1-mm thick layer over 1 m² of the sea bottom. In 20 million years, however, only about 10 kg of manganese nodules, with an average manganese content of 16.5 per cent, were formed per m²; less than a millionth part of the amount available. The energy gain thereby through oxidation of Mn^{2+} to Mn^{4+} would amount to only 3286 kJ (679 kcal), which is somewhat less than by the oxidation of 1 mol of glucose. Since this small energy gain over such a long time period does not suffice for the development of chemo-autotrophic bacteria, some other method of formation of manganese nodules must be presumed. Probably the formation depends on a combination of chemical oxidations and microbiological precipitations by heterotrophic bacteria.

According to Ehrlich (1972), the manganese tubercles harbour a characteristic microflora that includes both Mn^{2+} oxidizers as well as MnO_2 reducers. A prerequisite for bacterial Mn^{2+} oxidation is that it must first be adsorbed onto the tubercles; the presence of oxygen and small amounts of organic substance are also required. It is an enzymatic process and thereby the incorporation of other elements, such as iron, copper, cobalt and nickel, is also affected. Adsorption of Mn^{2+} and the subsequent bacterial oxidation to MnO_2 causes the growth of the tubercles. These two processes, according to Ehrlich (1972), may be represented as follows:

$$H_2MnO_3 + Mn^{2+} \rightarrow MnMnO_3 + 2H^+$$

$$MnMnO_3 + 2H_2O + {}^1/_2O_2 \xrightarrow[\text{peptone or NaHOC}_3]{\text{bacteria}} (H_2MnO_3)_2$$

In the presence of abundant organic substances, the bacterial reduction of the MnO_2 can occur and thereby the dissolution of the manganese tubercles; this is also an enzymatic process and the solution of the trace elements, copper, cobalt and nickel, can be brought about, but not that of iron.

15. Micro-organisms and Water Pollution

The microflora of waters is affected by refuse and sewage in many ways. With domestic sewage particularly, many micro-organisms get into rivers, lakes and coastal waters. Added to these are great quantities of organic and inorganic nutrients which cause mass development of bacteria and fungi. On the other hand, the microflora is not infrequently inhibited or even largely destroyed by poisonous substances. With sewage, pathogenic bacteria and fungi are also carried into waters and may give rise to epidemics.

By the breakdown of organic refuse, micro-organisms contribute decisively to the natural self-purification of waters. They have a similar function in the purification of sewage. During these processes the concentration of organic nutrients is diminished so much that eventually the bacterial content of the water decreases correspondingly.

15.1. The Microflora of Sewage

In many instances sewage has a characteristic microflora. Particularly rich in bacteria is domestic sewage, which consists largely of faeces, waste water and remains of food. Thus the Kiel urban sewage examined monthly in 1968–69 showed saprophyte numbers of between 3 and 16 million per ml. The organisms found were mostly putrefying bacteria, for example, *Pseudomonas fluorescens*, *P. aeruginosa*, *Proteus vulgaris*, *Bacillus subtilis*, *B. cereus*, *Aerobacter cloacae*, *Zoogloea ramigera* and others. In addition, there are, in varying numbers, representatives of many other physiological groups, particularly those which break down sugar, starch, fat, urea and cellulose. Relatively high is the proportion of coliform bacteria, which are an important indicator of the pollution of the water with faecal material. In the sewage of the city of Kiel the figures for *Escherichia coli* vary between several tens of thousands and several hundreds of thousands per ml. Frequently *Aerobacter aerogenes* is also found, which belongs, like *Escherichia*, to the Enterobacteriaceae, along with *Streptococcus faecalis*, also an inhabitant of the human intestine. Moreover, numerous bacteriophages are present.

In sewage rich in organic matter sheathed bacteria are important, particularly *Sphaerotilus natans*, which is often wrongly called 'sewage fungus'. It is a typical sewage organism and often covers the bottom of strongly polluted waters with a thick lawn, visible to the naked eye. In rivers, bacterial filaments and bundles are torn off and masses of them drift in the water. This is referred to as 'mycelial drifting'. Massive development of *Sphaerotilus natans* may occur at temperatures between 5 and 20 °C and a pH of 6–9, if the oxygen supply is good (Scheuring and Höhnl, 1956). As nutrients, mainly carbohydrates, certain organic acids, proteinaceous substances and their component units are utilized. *Sphaerotilus natans* grows not only in domestic sewage but also in sewage from cellulose works and from various food industries. Particularly during the spring and autumn months, when the water temperature is about 10 °C, *Sphaerotilus natans* shows mass development in suitably polluted flowing waters. It already grows optimally at that temperature (Figure 98)

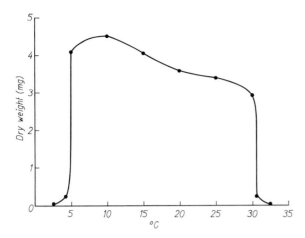

Figure 98   Growth of *Sphaerotilus natans* in relation to temperature. (Based on Scheuring and Höhnl, 1956, modified)

and has, therefore, an advantage over most competitors from sewage. Its vigorous growth is associated with considerable consumption of oxygen; hence a critical situation often develops if great quantities of *Sphaerotilus* settle in areas of stagnant water; this may very quickly result in complete disappearance of oxygen. The masses of *Sphaerotilus* eventually die and undergo putrefaction, and amongst other products H_2S is formed. The formation of hydrogen sulphide is usually enhanced by bacterial sulphate reduction, as sewage often contains abundant desulphurizing bacteria — above all *Desulphovibrio desulphuricans* — which find here very favourable conditions with a rich supply of nutrients. Many sewage waters also contain sulphur-oxidizing bacteria, particularly of the genera *Thiobacillus*, *Thiothrix* and *Beggiatoa*, which can multiply rapidly when H_2S is being produced. Denitrifiers, like *Thiobacillus denitrificans*, *Micrococcus denitrificans* and many others, as also methane producers and 'Knallgas' bacteria, are present in quantity in domestic and in some industrial sewage; also various iron bacteria like *Leptothrix ochracea* and *Thiobacillus ferrooxydans*. In sewage waters containing oil, bacteria which break down hydrocarbons accumulate; these are mainly members of the genera *Pseudomonas* and *Nocardia*.

In agricultural waste and sewage, especially those which contain excrement from domestic animals, large amounts of myxobacteria are generally found. Particularly the members of the genera *Myxococcus*, *Cystobacter* and *Polyangium* are involved. These, therefore, can serve as indicators for the corresponding pollution of groundwater, rivers and lakes (Lecianova, 1981). In the course of a year, 20–410 myxobacteria per ml of water were counted in the Danube at Bratislava. The values in winter and spring, at 20–50 per ml, were relatively low. In summer they rose greatly and reached a maximum in October (Miklošovičová, 1981).

In predominantly organic sewage numerous fungi occur besides bacteria. Thus urban sewage is usually rich in yeasts and yeast-like fungi. In the sewage of the city of Kiel, for example, between 4000 and 200000 yeast cells per litre were counted. Most frequent were members of the genus *Saccharomyces*. *Candida*, *Cryptococcus*, *Rhodotorula* and others were present in smaller numbers (Hoppe, 1970). Very high yeast counts are often found in the sewage of certain food and beverage industries.

Domestic sewage, as a rule, also contains large numbers of fungal spores and hyphae. Sewage fungi in the strict sense are *Leptomitus lacteus* and *Fusarium aquaeductuum*. In waters heavily loaded with sewage they may show massive development and

— similar to *Sphaerotilus natans* — cause the dreaded mycelial drifting. These fungi grow best in waters polluted with sulphite lye from the cellulose industry. Their main foods are carbohydrates, of which there are 35–75 g in 1 litre of spent sulphite lye. They are chiefly glucose, mannose and xylose, in varying quantities according to the kind of wood used in the particular works (Scheuring and Zehender, 1962). Genuine sewage fungi grow at pH 3–9. Unlike *Sphaerotilus natans*, they are capable of massive growth in relatively acid waters.

15.2. Pathogens in Waters

Human pathogenic bacteria and fungi get into waters mainly with domestic sewage. For the most part, they cannot grow there permanently and eventually die off in inland waters as well as in the sea; but, depending on the kind of water and the prevailing conditions, various pathogens can survive for a time. They may remain virulent, hence sewage-loaded lakes, rivers and sea regions may be dangerous sources of infection. Particularly frequent in polluted waters are pathogenic intestinal organisms like *Salmonella typhi* and *S. paratyphi* which cause enteric fever. Salmonella infections may, however, be caused not only by the water as such, but by eating oysters and other shellfish from sewage-loaded waters. Less frequent are shigellae (cause of dysentery). In tropical countries *Vibrio comma*, the causative agent of cholera, occurs epidemically and is commonly spread by water contamination. Before the supply of drinking water was regulated, cholera epidemics also occurred in European cities; this last happened in Hamburg in 1892. *Vibrio cholerae* was found in coastal waters where there was no faecal contamination and evidently is a natural occurrence in brackish water areas (Colwell, 1979). Often tubercle bacteria (*Mycobacterium tuberculosis*) also occur. The spores of pathogenic clostridia, particularly of those causing gas gangrene like *C. perfringens*, *C. novyi* and *C. septicum*, can nearly always be demonstrated in sewage-loaded waters. They remain alive in the sediments for a relatively long time and even may become enriched there (Bonde, 1967). The spores of the causative agent of anthrax (*Bacillus anthracis*) are also very resistant. They survive the sewage purification process and can even penetrate through the filter beds into the outflow. Occasionally, infections due to halophilic vibrios from sea water have also been observed. For example, *Vibrio alginolyticus* can cause inflammation of the middle ear (otitis media) (von Graevenitz and Carrington, 1973). Various marine strains of *Vibrio parahaemolyticus* cause intestinal disorders (enteritis) in humans.

Pathogenic bacteria are found at times in mussels and other filtering organisms, which come from sewage-loaded waters. Hence typhoid and cholera epidemics (for example, in Naples in 1973) have been caused repeatedly through the enjoyment of raw oysters and sea mussels. In regions at risk, therefore, mussels intended for raw consumption are submitted to 'self-purification' in clean water — if need be, sterilized by chlorination and subsequently dechlorinated again — which should last for 48 h. In this time, the surviving bacteria are excreted again (Mason, 1972).

As regards pathogenic fungi, the yeast-like *Candida albicans* in particular and related kinds may be found in sewage-loaded waters. Also, more highly developed dermatophytes, belonging to the Fungi Imperfecti and causing skin infections, may often be found. *Trichophyton* species (Figure 99) occur in the sand of bathing beaches.

In addition to bacteria and fungi, domestic sewage contains numerous human pathogenic viruses (Primavesi, 1970) which remain virulent in that environment for some time. Polluted water is said to have repeatedly caused infection with poliomyelitis viruses. In the sewage of cities, polio viruses can be demonstrated almost

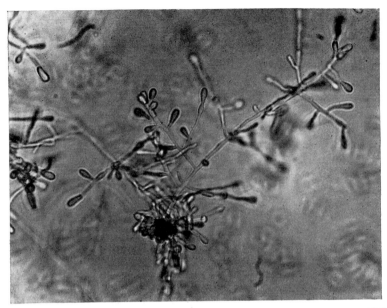

Figure 99   *Trichophyton* sp. from the sand of a bathing beach on the Kiel Fjord ($\times$ 350).
           (J. Schneider)

every summer. Much more frequent are viral intestinal infections following ingestion of polluted water, again particularly in summer. Such infections occur time and again in polluted outdoor swimming baths. They are frequently caused by Coxsackie or Echo viruses which are regularly found in sewage during the warm season of the year. Certain hepatitis viruses may also be transmitted by polluted water; such cases occur occasionally after the consumption of raw mussels obtained from sewage-loaded fresh water (Jawetz *et al.*, 1963).

The survival time for most pathogenic bacteria is greater in fresh water lakes and rivers than in the sea, as sea water is bactericidal for non-marine bacteria (see Section 8.6). Frequently, survival times are greater in sediments than in free water but, again, in marine sediments they are shorter than in those of inland waters. Viruses also behave quite similarly. Therefore, danger of infection during bathing exists normally only in the immediate vicinity of sewage inlets. However, the danger of infection through drifting particles of little pieces of meat or fish dropped by birds extends over quite a distance, as the pathogens in such proteinaceous material are, to a large degree, protected against the bactericidal effect of sea water, and may even multiply.

In inland waters and also in the sea, there seems to be a connection between the survival times of intestinal bacteria and the size of the bacterial population present in the water. According to Mitchell (1972), they are inversely proportional. An example of this is given in Figure 100 which shows that in autoclaved sea water the survival time of *Escherichia coli* decreases with increasing numbers of added marine bacteria.

Investigations in the North Sea by Moebus (1972) showed that annual fluctuations of anti-bacterial activity of the sea water against test bacteria (e.g. *Escherichia coli* and *Staphylococcus aureus*) are correlated with the development of phytoplankton and are especially pronounced with the break-up of diatom blooms. The inactivation of these bacteria is influenced by the growth of the natural bacterial population in sea

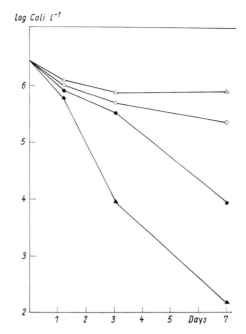

Figure 100   Survival times of *Escherichia coli* in autoclaved sea water after addition of varying amounts of marine bacteria. △, autoclaved sea water; ○, autoclaved sea water + 10 marine bacteria per ml; ●, autoclaved sea water + 1000 marine bacteria per ml; ▲, autoclaved sea water + 100000 marine bacteria per ml. (Based on Mitchell, 1972)

water, partly by nutrient competition. Inactivation of viruses also undergoes an acceleration with increasing bacterial numbers in the water. It could be demonstrated that bacteria produce anti-viral substances which are active both against RNA as well as against DNA viruses. On the other hand, the virus-inactivating abiotic factor of sea water (see above) is mitigated by the presence of bacteria which clearly have the function of protective colloids. Likewise, cyanophytes can form anti-viral substances. Thus, for example, Walter *et al.* (1981) established, in model experiments, that *Microcystis aeruginosa* caused inactivation of adenoviruses, probably by metabolic products.

In oligotrophic waters, according to Zaske *et al.* (1980), cell wall damage in *Escherichia coli* can be brought about within a short time.

Besides human pathogenic organisms, sewage also contains animal and plant pathogens. They may cause disease not only of aquatic animals but also of livestock and wildlife. The so-called brucelloses, which are caused by bacteria of the genus *Brucella*, are examples of this. They give rise to inflamations of different organs of the affected animals and to abortions. By sprinkling with surface water, plant diseases may be transmitted to vegetables and fruit. Greater still is the danger of infection when sewage is used for irrigation.

Waterfalls of sewage-loaded rivers represent a certain danger in mountainous countries, because with the fine water droplets bacteria and viruses are also flung into the air and, particularly in narrow valleys, can cause infections of man and animals.

15.3. The Role of Micro-organisms in the Self-purification of Waters

Most rivers and lakes and some areas of the sea are almost continuously exposed to pollution by refuse and sewage; hence the natural self-purification of waters is extremely important. This process keeps removing pollutants from the water so that, for instance, rivers are already reasonably clean again some kilometres below the location of the sewage inlet. Physical and chemical processes like sedimentation and oxidation play an important part, but the decisive role must be attributed to biological processes. Numerous living creatures take part in this, from birds and fish down to micro-organisms. Where unpurified domestic sewage emerges, numerous seagulls and other sea birds collect, and fish pick up the coarsest pieces. In general they can, however, make use of only a very small fraction of the polluting material as food. Somewhat greater is the importance of lower animals, particularly various insect larvae, worms and protozoa which take up smaller particles. The decisive role is played by bacteria and fungi. They can break down organic compounds, both those which are present in solid form and those in solution; under the most favourable conditions the breakdown is carried through to the original building components of carbon dioxide, water and a few inorganic salts (see Section 11.3). They are thus able to effect complete remineralization of many organic pollutants. Protein, sugar and starch are broken down particularly quickly; fat, wax, cellulose and lignin much more slowly and sometimes incompletely. With the progress of self-purification, the microbial population changes. Thus, where the pollution is due to domestic sewage, the proportion of proteolytic bacteria gradually decreases and that of organisms which decompose cellulose increases. This has been observed over and over again in the Elbe, in sections of the river where self-purification can proceed undisturbed (Rheinheimer, 1960). As shown in Figure 101, the saprophyte numbers usually decreased considerably between Dömitz (km 505) and Zollenspieker (km 598), whilst the numbers of cellulose-decomposing organisms rose. With progressive self-purification of the river, not only the concentration of pollutants but also the bacterial content diminished, whilst decreasing oxygen consumption reduced the O_2 deficit (Figure 102). Investigations of the Elbe (Rheinheimer, 1965a) showed that, in the section between Schnackenburg (km 475) and Bunthaus (km 610), not only saprophyte numbers but also the numbers of putrefying bacteria, thiobacilli and coliform bacteria went down progressively (Figure 103), and so did the yeast counts. On the other hand, during the warm season of the year, nitrite bacteria increased markedly, in addition to the cellulose-decomposing bacteria. The final step of remineralization of organic nitrogen compounds, nitrification, reaches its peak only when the proteinaceous pollutants have been decomposed to a large degree.

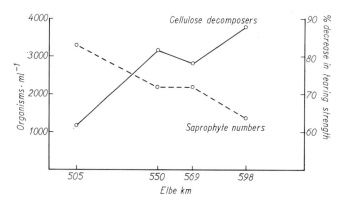

Figure 101   Saprophyte numbers and decrease in tearing strength (per cent after 7 days) of a cotton thread, as an expression of the numbers of cellulose decomposers in the Elbe water between Dömitz and Zollenspieker. Mean values from the middle of May to the end of September, 1958. With the progressive self-purification of the river, the saprophyte numbers decrease and the cellulose decomposers increase

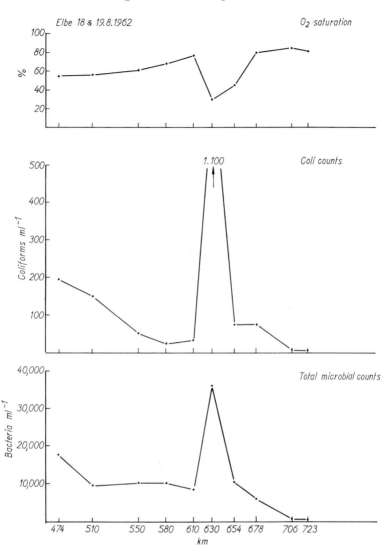

Figure 102   Oxygen saturation, coli counts and saprophyte numbers of the Elbe water between
Schnackenburg and Cuxhaven on August 18 and 19, 1962

Vigorous nitrification causes a corresponding oxygen consumption and in polluted
waters can contribute to oxygen depletion. For the conversion of 1 mg ammonium-N
to nitrate, 4.57 mg oxygen are required.

After the introduction of unpurified communal sewage into a river, the total
numbers of eubacteria usually decline from the beginning — whilst the sheathed
bacteria at first often continue to increase — then follow peaks of the protozoa and
finally of the algae (Figure 104).

A measure of the oxygen consumption by the living processes of the micro-organisms
present in the water is the so-called biochemical oxygen demand. It is determined,
as a rule, using water samples incubated at 20 °C in the dark for 2 (BOD_2) or 5 (BOD_5)
days (see: *Standard Methods for Examination of Water and Waste Water*, 14th, edition,
1976. American Public Health Association, Washington, DC). The BOD_5 is used to

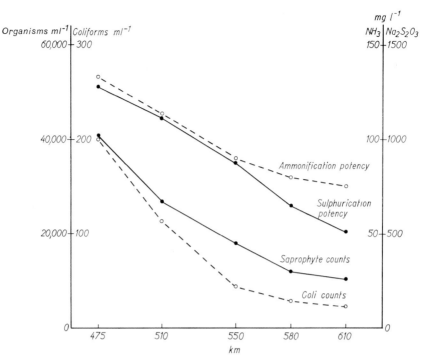

Figure 103   Saprophyte numbers, coli counts, ammonification potency (as an expression of the numbers of proteolytic bacteria) and potency of sulphur oxidation (as an expression of the numbers of thiobacilli) at five locations along the Elbe between Schnackenburg and Bunthaus. The falling curves show the progress of self-purification

assess the loading of the water with aerobically degradable substances and accordingly permits the effectiveness of the self-purification of waters and also of the purification plant to be determined.

Waters differ a great deal in their power of self-purification. It is greatest where lively movement of the water causes a rapid distribution of the sewage and a brisk exchange of gases with the atmosphere; a really extensive decomposition of pollutants is possible only in the presence of oxygen, which must be replenished all the time. These conditions are provided in most streams and rivers and in coastal waters with pronounced tidal movements or vigorous currents due to wind. In waters with little movement, the sewage may get dammed up and an insufficient oxygen supply leads to an early oxygen deficit and thereby to a collapse of self-purification.

The power of self-purification also varies with the time of year and is, in our latitudes, greater during the summer months than in winter. There are two reasons for this. First, bacterial activity is aided by the higher water temperatures in summer; second, in brighter light the phytoplankton supplies more oxygen. The greater metabolic activity of most micro-organisms during the warm season causes the nutrients, supplied with the sewage, to be used up much more quickly. This, in turn, leads to a decrease in bacterial numbers. In the fresh water of the Elbe as in the brackish water of the Schlei Fjord, a decrease of saprophyte numbers could always be observed during the warm season (see Figure 44 and Rheinheimer, 1965a). The same is true for polluted sea water, for example in the areas of large harbours. However, the frequently quite large seasonal variations in saprophyte numbers of sewage-loaded waters are not due solely to changes in the nutritional conditions. Other

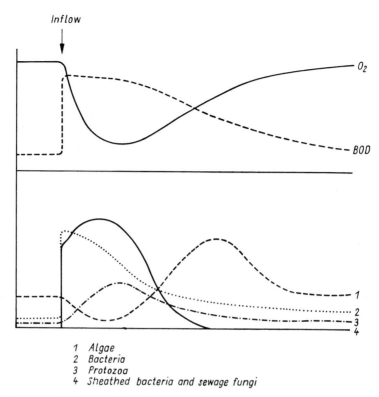

Inflow

O_2

BOD

1 Algae
2 Bacteria
3 Protozoa
4 Sheathed bacteria and sewage fungi

Figure 104   Influence of a sewage inflow and the subsequent self-purification on the development
of different groups of micro-organisms in a river. (Based on Hynes, 1971. modified)

factors are also involved: for example, the autolysis of bacteria and fungi is accelerated
at higher temperature; protozoa feeding on bacteria are more active; and, in some
places, bright daylight may perhaps exercise a bactericidal effect (see Section 8.1).
Correspondingly, the warm waters of the tropics show more rapid self-purification
than the cold ones of the arctic. The so-called thermal pollution can also have similar
effects; due to the constantly increasing demands on our rivers for the supply of
cooling water both for conventional as well as for nuclear power stations this gains
greater and greater importance. The temperature rise caused in this way will only
result in an accelerated self-purification if an adequate supply of oxygen is ensured.
However, this might often present considerable difficulties, since with increasing
temperature oxygen consumption increases and the oxygen saturation of the water
decreases.

   Even under favourable conditions of temperature and of oxygen supply, the breakdown of organic pollutants in the sea progresses more slowly than in comparable
inland waters. According to Müller (1953) and Mann (1956), the self-purification of
sea water polluted with domestic sewage requires about double the time of that for
corresponding fresh water. In this case the 'bacterial self-purification' proceeds more
rapidly than the 'biochemical' — i.e. the number of sewage microbes declines more
quickly in the sea than the content of pollutants. That is the consequence, above
all, of the inhibitory activity of sea water on many non-marine bacteria. As the
sewage water flows into the sea, most of the bacteria it carries with it die very soon,
and a new, predominantly marine, flora develops (Figure 105), which now causes
remineralization of the organic pollutants (Rheinheimer, 1966). This process, of

14   Rheinheimer, Microbiology, 3. A.

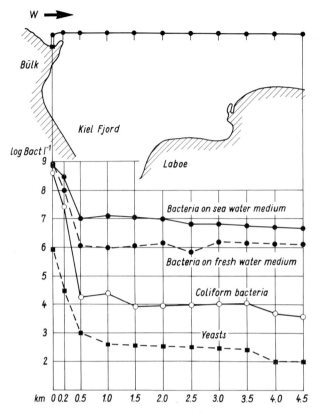

Figure 105   Numbers of bacteria and yeasts at a sewage outlet in the western Baltic Sea. The curves show that, with increasing distance from the mouth of the outlet, the microbial population changes progressively. The number of marine halophilic bacteria declines much more slowly than that of the fresh water and sewage bacteria, coliforms and yeasts, so that the proportion of the former in the microflora is greatly increased

course needs time. In fresh water, on the other hand, the microbial breakdown of pollutants can proceed, as a rule, without interruption; at the worst, there may be a temporary slowing down. In this connection, the fact that many marine bacteria have a lower metabolic activity than corresponding fresh water bacteria may be of significance. Fair and Geyer (1961) state that the velocity of breakdown in each case greatly depends on the dilution. At high sewage water concentrations it should be greater in sea water, and at low concentrations, on the other hand, it should be greater in fresh water. Since on introduction of sewage into the flowing tide a very strong dilution usually occurs, this means under natural conditions a more rapid self-purification of sewage-loaded fresh water.

Natural self-purification functions, however, only under conditions where composition and quantity of pollutants do not overtax the power of self-purification of the receiving body of water. Frequently, much more refuse and sewage is led into waters than they can process even under the most favourable conditions. Then disturbances in the self-purification process ensue with, as a rule, very disagreeable consequences. The vigorous oxygen consumption often results in completely anaerobic zones where putrefying processes and bacterial sulphate reduction lead to the production of hydrogen sulphide. This, in turn, causes the death not only of almost all

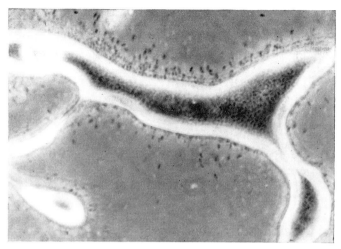

Figure 106   Colonization of an oil drop in water by oil-degrading bacteria. The degradation is limited to the oil-water interface ($\times 1\,000$). (W. Gunkel)

higher organisms but also of many micro-organisms, and a microbial population consisting of only few species develops which can carry out only partial breakdown of the organic pollutants. The formation of iron sulphide leads to evil-smelling black sapropel which causes the death of the original fauna at the bottom. It is particularly dangerous if, in lakes and coastal waters, the H_2S-containing water from the deeper parts suddenly reaches the surface, for instance after severe storms. This may cause the extensive dying of fish and may be detrimental to fisheries in the affected waters.

Disturbances in the self-purification process occur also through the direct introduction of poisonous substances, which mainly get into waters with the sewage and refuse from industrial plants and cause the death of the organisms involved in the remineralization processes. This happens particularly with compounds of heavy metals, cyanides and organic poisons. Metallic mercury and mercury ions have a lower toxic effect than organic mercury compounds, which, for example, can be formed by bacterial methylation. These are fat soluble and therefore can penetrate into the human nervous system. Accordingly, the activities of mercury-resistant bacteria which can degrade mercury methyl compounds become of much significance. In sediments they counteract an enrichment of the dangerous organic mercury compounds (Spangler et al., 1973).

In the purification of oil-polluted waters micro-organisms play an important part. Hydrocarbon-decomposing bacteria and, occasionally, fungi accumulate in them (see Section 11.3). Whilst the oil components which are soluble in water are broken down very quickly by micro-organisms, lawns and flat slicks of oil can only be attacked from the surface (Figure 106). Bacteria may penetrate into the oil but apparently are not capable of multiplication there (Heyer, 1966). Only thin oil films thus disappear very quickly. In oil-polluted waters (particularly near ports) the oil-decomposing organisms are particularly numerous in summer; moreover, they are very active at high water temperatures. Their activity, though, does not depend solely on the water temperature, but also on the concentration of inorganic nutrients — particularly phosphorus and nitrogen compounds (Gunkel, 1967a). In oligotrophic waters, therefore, the microbial breakdown of oil may be helped by the supply of such compounds. Thicker layers of oil, above all those which are the result of tanker catastrophes, are broken down only very slowly by microbes from the surface. Hence their importance in cases of acute oil-pollutions on a large scale is small.

The bacterioneuston plays a significant role in microbial degradation. According to Tsiban *et al.* (1980), the degradation rate in the Baltic Sea ranges from 30 to 120 μg of oil per m² per day. The total degradation would be estimated at 1200–5000 m³ oil.

In areas with submarine oil production, the proportion of oil-degrading bacteria in the microflora has clearly increased, as demonstrated by an investigation of Oppenheimer *et al.* (1977) in the North Sea. On the other hand, the hydrocarbon concentrations in the sediment have changed only little. From that it can be deduced that here the oil-degrading micro-organisms play an important role in the self-purification.

Saltzmann (1982), with the aid of ¹⁴C-labelled substances, investigated the decomposition of naphthalene and benzpyrene in the neighbourhood of oil fields in the North Sea. The rate of consumption of naphthalene was greatest within a circle of 0.9 km around the bore platform and then decreased with increasing distance. The rate varied from 424 to 8000 μg of naphthalene per m² per day in the topmost centimetre of the sediment. The consumption rates of benzpyrene, in contrast, was some three orders of magnitude lower.

Crude oil alters rapidly in the sea, mainly due to physical processes. Within 8–14 days the lighter components volatilize, so that almost the only components which remain are those which are not easily volatilized, and these mix with the sea water to form a brown mushy suspension. This contains around 23 per cent heavy oil, 4 per cent solids and 73 per cent sea water, and after several weeks it sinks to the bottom. Forming a suspension improves the chances of bacterial attack. Thus, Gunkel (personal communication) found 50 million bacteria per ml in oil globules from the North Sea, which were composed of a water-in-oil emulsion with about 50 per cent water content, whilst the surrounding sea water contained only about 50 per ml. For that reason artifical emulsifiers are frequently used in fighting oil pollution, but these are mostly poisonous to many aquatic organisms and, therefore, often cause more damage than the oil itself. Some of these substances inhibit bacterial growth and, as a result, instead of aiding the microbial decomposition of oil, they counteract it (Hellmann *et al.*, 1966). At present, the chemical industry is endeavouring to develop non-toxic emulsifiers.

After the oil slicks have sunk to the bottom, they often turn resinous, which makes colonization of the material virtually impossible, so that no further breakdown by micro-organisms can take place. Thus at the mouth of the Hudson River near New York, a completely dead asphalt-like layer about 50 cm thick covers the bottom. Similar findings — though on a much smaller scale — have been reported for the Elbe downstream of Hamburg (Gunkel, 1967 b).

Frequently oil, in varying quantities, reaches beaches and banks. Here it is often overlaid with sand (occasionally also ploughed in), and oil-decomposing bacteria and fungi quickly accumulate. Where the oil is well mixed with sand, the breakdown proceeds very rapidly, at least during the warm season; but larger cakes and clumps, particularly if the oil is already converted into resin, often remain unchanged even for years (Wallhäuser, 1967). This is true also for plant and animal material that is soaked with oil. Thus, even one year after an oil pollution, oil-soaked birds which were properly mummified and showed only small traces of putrefaction, were found in the sand of the North Sea coast (Figure 107).

Considerable importance is attached to oil-decomposing micro-organisms in the purification of sewage from the oil industry. Here the addition of river water has proved of value as it contains a rich population of bacteria and fungi which break down hydrocarbons.

Under natural conditions, several kinds of bacteria and sometimes of fungi are associated with the microbial breakdown of oil. Some micro-organisms are specialized for certain fractions, others live on intermediate products. Nevertheless, only rarely

Figure 107 An oil-covered
sea bird, which was washed
free from the sand one year
after an oil spill

is crude oil completely remineralized. In the laboratory, a floccular debris remains which does not change even after a long time.

Maciejowska and Rakowska (1973) established that the microbial degradation of oil proceeds most effectively at very low concentrations. In water of the Baltic Sea, 32–61 per cent of 0.5 g of oil per litre were degraded after 4 weeks in contrast to only 28–40 per cent of 2.5 g of oil per litre; after 8 weeks, 44–73 per cent and 41–70 per cent respectively were degraded.

Under anaerobic conditions the breakdown of oil is due, above all, to *Desulphovibrio desulphuricans*, provided there is sufficient sulphate. In the process, sapropel is formed (see above).

Also, in the purification of phenol-containing sewage from the chemical industry and from hospitals, micro-organisms like, for example, *Nocardia rubra*, play a similarly important role and, during the warm season, cause a rapid breakdown of the phenols present in the water. If suitable technical procedures are applied, not only univalent and polyvalent phenols, but also condensed phenols, cresols, cyanides, rhodanides and formaldehyde may be broken down (Bringmann, 1970).

From areas used for agriculture and forestry, due to rainfall flowing off the surface and to drainage, plant protection chemicals and soil improvement agents as well as inorganic plant nutrients get into waters. Among these are insecticides such as DDT (dichlorodiphenyltrichloroethane) and herbicides like 2,4-D (2,4-dichlorophenoxyacetic acid) which is used to kill dicotyledonous weeds especially in cereal crops and on grassland. Such substances can be taken up by plankton algae in inland and coastal waters and accumulate so strongly in the food chain that they cause damage to the end members.

Some of these substances, however, proved to be toxic also to certain micro-organisms. Thus, Filipkowska et al. (1983) observed an inhibition of nitrification which was accompanied by stimulation of ammonification and denitrification. As a consequence the ammonium concentration in the affected waters increased so much that it became toxic to higher water organisms. Consequently the breakdown of insecticides by bacteria and fungi acquires great importance. DDT is relatively stable and is only very slowly broken down by micro-organisms. Organophosphorus insecticides like parathion and similar compounds are attacked by various micro-organisms and are less persistent than DDT (Walker, 1975). That is also true of the herbicide, 2,4-D, which is quickly broken down by bacteria and fungi.

A function similar to the natural self-purification of waters must be attributed to micro-organisms involved in the biological purification of sewage in sewage disposal plants, whether they work by the activated sludge method or that of biological filters (Uhlmann, 1975).

The activated sludge communal sewage disposal plant has a characteristic population of micro-organisms. This consists chiefly of bacteria and protozoa — whereas fungi and metazoa play only a subordinate role. The bacteria are represented especially by *Pseudomonas, Zoogloea, Achromobacter, Arthrobacter, Sphaerotilus* and *Nocardia*. These likewise are found in biological filters but here fungi also are of greater significance and, in addition to protozoa, metazoa occur in greater quantity — mainly worms and insect larvae. In sewage plants, the intestinal bacteria decline sharply and scarcely take part at all in the breakdown processes. The protozoa are primarily responsible for this decline and, to a lesser extent, bacteriophages and *Bdellovibrio* also (Pike and Curds, 1971). The following table by Bonde (1977) shows the marked decline of *Escherichia coli* due to biological purification:

	E. coli MPN (ml^{-1})	E. coli (%)
Untreated sewage	30 000	100
Mechanically clarified sewage	27 000	90
Biologically clarified sewage	1 600	5

The final rotting down of fresh sludge taken from sewage, which contains about 95 per cent water and 3–3.5 per cent organic matter, is almost entirely due to bacteria. This anaerobic process is carried out in special sludge digestion plants. In the process a gas mixture develops which contains (Husmann, 1964):

65–90%	methane	0–10%	nitrogen
10–35%	carbon dioxide	0–0.5%	hydrogen sulphide
0–5%	hydrogen		

As can be seen, there is essentially a bacterial methane fermentation (see Section 11.3). The evolved gases are frequently used in sewage plants for heat production.

In the so-called third step of purification, during which the inorganic plant nutrients, particularly nitrate and phosphate, are withdrawn from the sewage which has already been purified mechanically and biologically, bacteria again play an important role. Thus denitrifiers (see Section 11.5), under suitable (anaerobic) conditions, can completely reduce nitrate and nitrite to free nitrogen, after ammonia has previously (aerobically) been oxidized to nitrate by nitrifying bacteria.

The preparation of drinking water from surface water is also carried out with the help of micro-organisms. The sand filters of waterworks contain a characteristic microflora which consists, apart from algae and protozoa, mainly of bacteria. In the top layer they free the water from foreign matter by oxidation and reduction processes. When this has been achieved most of the bacteria are removed, partly by filtration, partly by ingestion through protozoa, so that a largely purified water leaves the filter. Drinking water to be supplied to the consumer should contain neither ciloforms nor streptococci in 100 ml, and the colony count should be below 100 per ml.

Difficulties in drinking water preparation are occasionally encountered due to actinomycetes, particularly of the genus *Streptomyces*, which produce an earthy smell and an unpleasant taste (Mucha and Horska, 1981).

In recent times, the purification of stale air by a biological air wash has been gaining importance (see Claus, 1979). This exploits the ability of micro-organisms to break down odorous substances. With suitable bacteria, enriched wash water is used which can be circulated through aeration tanks. The method is useful in agriculture, *inter alia*, for de-odorizing large piggeries. In south Germany, a large-scale installation has been introduced in a foundry for purifying 120000 m² of air per hour containing, in particular, phenols, formaldehyde and tertiary amines. Biological air cleansing is also important for the chemical industry.

15.4. Microbiological Parameters for Water Control

Besides physical and chemical parameters, biological ones have also been utilized for a long time for the evaluation of water quality in inland and coastal waters. Kolkwitz and Marsson (1908) already used bacteria, besides other vegetable and animal creatures, as indicator organisms for the characterization of water condition and also suggested various bacteria for the individual saprobic stages — for example, *Zoogloea ramigera* and *Sphaerotilus natans* for the polysaprobic region.

Bacteriological parameters are of still greater importance in the hygienic evaluation of waters. For this purpose, besides the saprophyte numbers or colony counts (see Chapter 7), bacteria from the human intestine are determined as faecal indicators (Bonde, 1977; Daubner and Peter, 1974; Kohl, 1975). The coli and coliform numbers as well as the coli titre are of first importance. Members of the Enterobacteriaceae which can ferment lactose to carbonic acid with gas formation (H_2, CO_2) are designated coliform bacteria. They belong to the genera *Escherichia, Citrobacter, Enterobacter* and *Klebsiella, Escherichia coli* develops in the intestine of humans and warm blooded animals. Its concentration in human faeces amounts to more than 100 million cells in 1 g wet weight. The other coliform bacteria likewise are often of faecal origin, but they multiply also in sewage and in surface waters. For the determination of coliform numbers, Endo's lactose-sulphite-fuchsin-agar is usually used as a selective medium. The total coliform number is estimated after incubation for 24–48 h at 37 °C and the faecal coliform number at 44 °C. If definitive statements have to be made as to the presence of *Escherichia coli* organisms, then differentiation by means of suitable biochemical tests is necessary (colour series, IMViC test — Daubner and Peter, 1974; Reichardt, 1978).

In most countries permissible limiting values for drinking water, bathing water and water for other purposes were established. Drinking water must not contain any pathogens. That is considered to be adequately safeguarded if no coliforms (and correspondingly also *E. coli*) are present in 100 ml water. In a number of countries (European community) water for bathing should conform to the following quality requirements: the total coliform numbers if possible should be below 500 and in any case less than 10000 per 100 ml — the faecal coliform numbers should correspondingly be below 100 or 2000 respectively per 100 ml of water. The former value is denoted as a guide and the latter as a compulsory limit. If this is exceeded with at least 14 days' sampling in more than 20 per cent of the investigations then a bathing prohibition notice must be issued for the water in question.

As a supplement to the coliform test, the number of faecal streptococci (enterococci) is occasionally determined (Ritter, 1974). As a rule, this is essentially less than that of the coliforms. A higher proportion of faecal streptococci is an indication of fresh faecal pollution. Not infrequently a qualitative test for salmonellae is made to which, as the causative agents of typhoid and paratyphoid fever and enteritis, great hygienic significance is attached. They must not be present in either drinking water or water for bathing.

In addition, the number of bacterial spores can be determined. With increasing self-purification the proportion of vegetative bacterial cells in a water as a rule declines and that of the spores increases. The determination of the number of *Clostridium perfringens* spores, usual in some countries, is of hygienic interest (Bonde, 1977). It is a possible indication of earlier faecal contaminations, if coliforms and streptococci have already disappeared.

The bacteriological investigation of a water may only reflect the present condition of a water. This can alter very rapidly especially in small rivers and coastal waters affected by the tide. However, if one wishes to make more far-reaching statements on the sewage load, then the sediment must also be included in the investigation. That often makes possible some indication of the extent of previous sewage loading and hence a better appraisal of the actual water quality.

According to Kohl (1975), investigation of the surface growth of water plants is also to be recommended. He found in adequate amounts, many times, distinctly higher numbers of different intestinal bacteria and, in 14 out of 18 plant samples from the rivers Gurk and Glan, he was able to detect salmonellae, whereas the parallel water samples always gave negative results. The following comparison of data obtained in November, 1973 (15.30 h) from the Erlabach in Austria also shows comparable findings:

	Saprophytes (1 ml^{-1})	Coliforms (100 ml^{-1})	Streptococci (100 ml^{-1})	Salmonellae (225 ml^{-1})
Water	3 000	2 600	2 100	Not detectable
Water plants	400 000	220 000	120 000	*S. stanleyville*

The filtering action of water plants certainly plays a role in this enrichment. Living conditions on the water plants are also more favourable so that bacteria have a longer survival time.

Since fish may contain the bacteria present in the water on the skin, the gills and in the gut (see Section 10.2), their investigation may likewise give information on the bacteriological hygienic conditions in waters. Thus, Kohl (1975) found salmonella in 32 out of 50 fish in the Vienna Danube Canal.

The efficiency of sewage purification plants can likewise be tested with the help of the bacteriological parameters mentioned (see Section 15.3). As long as the presence of pathogens in the outlets, particularly in the vicinity of bathing areas, remains a possibility, the effluent, before being led into the flowing tide, must be submitted to disinfection (e.g. by chlorination). But the self-purification processes in the tide can be impaired by this, and there exists the danger that poisonous chlorinated organic compounds may be formed. Consequently the disinfection of sewage with chemical agents should only be considered as an emergency.

In the case of industrial sewage, toxicity testing is frequently necessary. In addition to various animals and algae, bacteria are also used for this. *Pseudomonas fluorescens* (Bringmann and Kühn, 1975) or, more rarely, *P. putida*, serves as a test organism. These, however, possess a relatively low sensitivity towards many poisons. Therefore the results from the usual bacterial tests can be related only very conditionally to the behaviour of the bacterial population of the tide. Part of the aquatic bacteria show a distinctly greater sensitivity and already react to much lower concentrations, so that even where the test is negative a diminution in activity of the micro-organisms participating in the self-purification processes in the flowing tide is possible. Therefore supplementation of the usual test procedure by involving the microbial population at the site should be attempted.

A very promising toxicity test with luminescent bacteria has been recently recommended (see Krebs, 1983). In this, the light intensity is assessed with the help of a bioluminescence photometer. Owing to the great sensitivity to poisons, a reaction time of about 30 min is already adequate and so the results become available very quickly.

16. The Economic Significance of Aquatic Micro-organisms

Bacteria and fungi participate in the production of some algal and fish products, but on the other hand, also in the destruction of economically important goods. They not only cause spoilage of food obtained from the water, but also the rotting of nets and ropes, of cork, rubber and wood, and the destruction of hydraulic buildings of metal and concrete. From year to year they cause losses which in some coastal regions run into millions of pounds.

16.1. Fermentation of Algae and Fish

In south and east Asia there are several foods prepared from various algae and fish which, before consumption, are subjected to certain fermentation processes. Micro-organisms certainly play an important role here (Wood, 1967). Thus in Japan a product called 'shoyu' is made from the brown algae *Ecklonia radiata* and *Eisenia* species. The algae are first acidified, then boiled for several hours, neutralized with soda and salted a little. The liquid is boiled again for a short time and then made to undergo a process of maturation the details of which are not known. Before being sold the product is pasteurized for 15 min at 80 °C.

In the preparation of other food products made from *Laminaria* and *Ecklonia* species, micro-organisms are also involved. The algae are dried for about 2 weeks during which time various fungi like *Penicillium*, *Aspergillus* and, occasionally, *Alternaria* grow on them. Possibly they break down mannitol and mannonic acid which makes the product more easily digestible for man. Micro-organisms are probably involved in the production of 'nori' (Levring *et al.*, 1969), which is obtained from *Porphyra* species.

Besides these algal preparations — particularly popular in Japan — in various Asiatic countries fish products are made in which — as in cheese production — micro-organisms are involved. The fish, salted with sea salt, are subjected to fermentation of longer or shorter duration during which bacteria, particularly clostridia, are said to be active. In the production of 'katsuobushi', on the other hand, mould fungi are involved. Very popular in Japan are sauces obtained from fish by means of microbial fermentation processes.

Some of these products belong to the staple diet of the population in some parts of Asia, and their production has risen within recent years.

16.2. Spoilage of Fish and other Aquatic Products

Great economic importance must be attributed to the microbial spoilage of food obtained from water, such as fish, shellfish, crabs, etc. Such spoilage is mainly due to bacteria; fungi — with the exception of some yeasts — are not involved. This concerns both the unprocessed hauls as well as the commerical products. Because of their high protein content they all make excellent media for putrefying bacteria.

The organisms are already present on the skin and in the alimentary tract of the living animals (see Chapter 10); while the animals are being caught their organs can be damaged and this leads to infection of the flesh. Most frequent amongst the organisms involved are members of the genera *Pseudomonas, Achromobacter, Flavobacterium, Micrococcus, Bacillus* and *Corynebacterium*. Putrefying bacteria mutiply quickly and can spoil the food within a short time by proteolysis. Apart from their poor keeping quality, fish with a high bacterial content have usually a disagreeable smell and taste and, therefore, are unsuitable for consumption. It must be admitted, though, that the quality demanded differs a good deal in different countries. A product which, for instance, is rejected in Europe or America, may still be perfectly marketable in the Far East (Wood, 1967). The disagreeable smell of fish is caused mainly by ammonia and volatile amines which develop during the breakdown process. Whilst in bonefish mainly trimethylamine is produced, in cartilaginous fish (sharks and rays), with their high urea content, ammonia predominates. In every case mono- and diamines are also present. The peak for the development of odour comes, as a rule, earlier than that for proteolysis. As some of the putrefying bacteria are psychro-phils, they can grow even in fish kept on ice. Therefore, in modern sea fishery the tendency to deep-freeze the catch on board immediately or after gutting and filletting is gaining ground. Besides careful methods of fishing, this is the surest way to protect sea fish against bacterial decomposition. Coastal and inland fisheries, on the other hand, are dependent on quick landing and marketing of the catch.

In canned fish which has been prepared in an inexpert way, *Clostridium botulinum* may grow. The spores of this organism are widely distributed and relatively heat-resistant. If they have not been killed during the canning process they can germinate in the absence of oxygen (as for instance in tins of fish) and the cells may multiply rapidly. They produce a toxin which causes in man the most dangerous form of food poisoning, so-called botulism. This involves damage to the nerves which by paralysis of the breathing leads to death. The toxin is destroyed by normal cooking procedures but the spores are not destroyed. Freshly cooked fish dishes are, therefore, harmless, but not the leftovers which have been kept and either eaten cold or warmed up, as fresh toxin may have been produced in them. There are different types of *C. buto-linum* whose toxins differ in their potency for man. They are also toxic for animals, so that infected fodder may lead to severe, often fatal, poisoning in farm animals. During the growth of *C. botulinum*, gas is sometimes produced which blows out the tins.

The meat of lobster and crabs is sometimes stained black due to the activity of putrefying bacteria; it is then no longer suitable for consumption. Reed and MacLeod (1924) state that members of the genera *Pseudomonas, Flavobacterium* and *Bacillus* in particular produce, in addition to ammonia, H_2S which then reacts with iron to produce black iron sulphide.

16.3. Rotting of Nets, Ropes and Woven Textiles

Nets, ropes and fabrics made from natural fibres are destroyed in water by bacteria and fungi with varying speed. This, even nowadays, causes considerable economic damage to the shipping and fishing industries, but it was very much more serious before the introduction of nylon and other fully synthetic fibres which are very little affected by micro-organisms. However, both the shipping and the fishing industries still use natural fibres to some extent for a variety of purposes, and protecting them against microbial destruction is of considerable importance. Two groups of fibres must be distinguished — on the one hand, those which consist essentially of cellulose, such as cotton, hemp and linen, on the other the proteinaceous fibres like silk, wool and gut. Both groups serve as food for numerous bacteria and fungi. For instance, the much-used cotton nets are attacked by cellulose-decomposing bacteria and fungi and are destroyed — particularly in warm waters — in a relatively short time (see Section 11.3). Contamination of nets and rigging may take place from dry land or from the water. Rotting proceeds particularly quickly if the nets cannot be dried at once, on board. Thus nets and rigging stored away in a damp condition, at summer temperatures, may sometimes fall to pieces after only a few days. While in rivers and lakes destruction of cellulose fibres in due mainly to bacteria (see Section 11.3), on ships and on land it is due predominantly to fungi.

As a protection against rotting, the yarns or textiles are impregnated with substances which inhibit the growth of bacteria and fungi. Soaking the fabrics in hot tannin liquor (i.e. infusion of barks containing much tannin, for example oak, birch, spruce, quebracho, catechu, etc.) and, after drying, immersing in copper sulphate solution is a process of proved value. This treatment preserves the suppleness of the nets and does not stain. Often tar or mineral oil products are also used.

16.4. Destruction of Wood, Cork and Rubber

In all waters, natural materials like wood, cork, rubber, etc. are attacked and eventually destroyed by micro-organisms. The greatest economic damages are due to destruction of wooden buildings in water; even today, many landing-stages, piers, dolphins and quay walls, as well as boats and ships, are made of wood, which has only a limited life span in water, in spite of effective impregnation. In the destruction of wood, fungi which decompose cellulose and lignin play a prominent part in inland waters as well as in the sea. The fungi in question are predominantly ascomycetes and their close relations, Fungi Imperfecti, which have been isolated in large numbers from submerged wood (Höhnk, 1955; Kohlmeyer, 1960; Meyers and Reynolds, 1960; Schaumann, 1969). Cellulose-decomposing bacteria also take part in the destruction of wood particularly *Cytophaga* and *Sporocytophaga* species), but are much less prominent than fungi.

Micro-organisms also seem to aid the activity of marine wood-borers like *Teredo navalis* and *Limnoria lignorum*, which destroy the stability of wood by their tunnelling. Although the animals are able to digest wood directly, this is aided by lignin-decomposing bacteria and fungi (Meyers and Reynolds, 1959).

Cork and rubber are gradually destroyed in water by micro-organisms, again with a certain amount of damage to the shipping and fishing industries. Cork consists mainly of a lignin-cellulose-subarin complex and is attacked by some of the wood-destroying micro-organisms. They break down the cell walls so that the cork crumbles, which makes it useless (ZoBell, 1946a).

Rubber is a nutrient for some bacteria and is broken down particularly by organisms of the genera *Pseudomonas*, *Micrococcus*, *Bacillus*, *Nocardia* and *Micromonospora* (ZoBell and Beckwith, 1944). Rubber consists of unsaturated hydrocarbons, and the rubber-decomposing bacteria are mostly able to break down other hydrocarbons also (see Section 11.3). Synthetic rubber, too, is attacked by bacteria. Some kinds, indeed, are subject to a more rapid bacterial destruction than natural rubber.

16.5. Corrosion

Bacteria participate prominently in the corrosion of iron and steel in water. Hydraulic plants, as well as ships and naval installations, are affected. Corrosion may proceed under aerobic or under anaerobic conditions. Anaerobic corrosion is due frequently to sulphate reducers, particularly *Desulphovibrio desulphuricans*. The hydrogen sulphide which these organisms produce can react directly with iron. Corrosion cells form between iron sulphide and metallic iron. Reducing zones of metallic iron act as anodes towards other areas which are in contact with oxygen-containing water, so that dissolution of iron ensues. Iron structures in mud present particularly favourable conditions for such corrosion phenomena, because they are situated in an anaerobic environment rich in nutrients and, moreover, containing abundant sulphate (see Section 11.6). The same processes, however, can also take place in water. The necessary lowering of the redox potential is provided by the so-called fouling, microbial growth which quickly covers all objects, especially in sea water. Besides algae and numerous lower animals, various heterotrophic organisms are involved which form a dense slimy layer on the surface of the structures in the water.

As a rule, fouling begins with the formation of a primary film of biopolymers which very quickly spread over the surface. The film consists of proteins, polysaccharides and lipids that are produced by organisms living in the sea. The surface tension of the film is lower than that of the metal surface and so favours the attachment of micro-organisms. Bacteria freely swimming in the water are evidently attracted by the higher nutrient concentrations and attach themselves, at first, reversibly. Soon afterwards, the attachment becomes irreversible and the excretion of acidic polysaccharides ensues. In addition to bacteria, after some time, fungi, protozoa and small algae come in and, later, higher algae and metazoa also, so that the coating becomes thicker and thicker. Differing populations develop on different surfaces and, in this regard, the chemical composition of the metal plays an important part, but not the physical condition of the surface. With aluminium, there is a correlation between the development of a large bacterial population and rapid corrosion. Heat transfer from metal to water (e.g. in condensers of power stations) is also hindered by the biofilm (Berk *et al.*, 1981). Owing to degradation of proteins in the fouling layer, which covers the surface in structures existing in the water, ammonia and hydrogen sulphide are liberated causing a marked consumption of oxygen.

The heterotrophic aufwuchs flora, however, also causes severe corrosion directly through the release of CO_2 and partly also of H_2. It can exceed considerably that caused by the sulphate reducers (Frenzel, 1966). Other metals, particularly aluminium, are corroded in a similar way.

To prevent fouling and corrosion, hydraulic structures and ships are painted with protective coats which contain poisonous substances like copper, mercury, arsenic, etc.; for some time, organic compounds like PCB (polychlorinated biphenols) have also been applied. Some bacteria, however, are relatively resistant to these poisons, so that fouling can not be prevented permanently. Moreover, since with all these substances it is a question of environmental poisons some of which may accumulate in food chains, new ways for controlling fouling need to be found.

Even concrete structures can be destroyed by bacteria. This happens particularly through the activity of the acidophilic thiobacilli, which oxidize various sulphur compounds to sulphuric acid (see Section 11.6). *Thiobacillus concretivorus* grows optimally at pH 2–4, and can even grow at pH 1. By its vigorous acid production this organism may destroy sewage pipes, sewage purification plants, port buildings, bridge piers, etc., constructed of concrete, and causes enormous losses. A similar role may be played by other acidophilic thiobacilli, like *T. thiooxydans* and *T. ferrooxydans*. They can all cause the corrosion of metals and other acid-sensitive materials. In spite of their economic importance, the factors which influence the bacterial destruction of hydraulic constructions — positively or negatively — have not yet been the object of much research.

16.6. Leaching

Leaching is the term used to describe a hydrometallurgical-biotechnical procedure for obtaining heavy metals, in which the metals are washed out through the activity of micro-organisms (Schwartz, 1973). This is of particular importance in the exploitation of low-grade ores and wastes where a mining operation would not be justified

Hitherto, the chemoautotrophic acidophilic bacteria *Thiobacillus ferrooxydans* and *T. thiooxydans*, in particular, have been used for the processing of sulphide ores and uranium ores. These oxidize reduced sulphur compounds to sulphuric acid (see Section 11.6), by which the metals are leached out and suitably enriched. Thereby the reaction falls to pH 2 or even lower. The copper from copper pyrites goes into solution as sulphate. The liquor contains 1–3 g of copper per litre. Uranium is dissolved from uranite as uranium sulphate. The metals can then be separated from the sulphuric acid solutions by hydrometallurgical methods. With the exception of silver, the participating bacteria can withstand heavy metal concentrations of 0.05–1.8 M. By the activity of thermophilic thiobacilli and *Sulpholobus* species, self-heating of waste ore dumps (e.g. marcasite) can be caused. Halophilic thiobacilli play a role in the processing of salt-containing ores.

Recently, attempts at leaching have also been made with the use of heterotrophic bacteria and with fungi (Schwartz, 1978; Schwartz and Näveke, 1980). In this case the formation of water-soluble organic metal complexes is intended from which the heavy metals may then be separated. Suitable complex-forming organic compounds include citric acid, lactic acid and other fermentation products and also peptides. Up to the present experiments with mixed populations or individual *Bacillus* and *Pseudomonas* species as well as with acid-forming bacteria and fungi have been made. As organic nutrients, waste products such as spent sulphite lyes and similar materials can be employed. The main interest here lies in the extraction of noble metals.

17. Outlook

A survey of our present knowledge of aquatic microbiology, such as this book tries to present, shows the multifarious roles which bacteria and fungi play in the life of rivers, lakes and seas, and what importance they have for economics and human welfare. It also makes clear that we are, so far, very inadequately informed on certain branches of aquatic microbiology. Research in this subject will, in the future, be required on a larger scale than up to now. Work on methods must on no account be neglected. Their importance starts with the collecting of samples, and the application of modern methods of investigation must be made possible in this field. It should be our aim to study the activity of micro-organisms in waters directly, under the conditions which prevail at their natural locations. For such research, small models of waters might be useful which would allow continuous control of the various factors involved.

Special attention should be devoted from now on to the role of bacteria and fungi in energy flow in aquatic ecosystems. In recent years some very promising new methods have been devised for the estimation of microbial production and these are worth developing further. The scanning electron microscope gives us the possibility of studying surface growth on dead and living substrates and thereby obtaining a better basic knowledge of bacterial colonization of floating materials and of aquatic organisms. The mutual relationships between the microbial populations of water and sediment likewise represent an area of aquatic microbiology that is still little explored. Also of great importance are taxonomic and autoecological investigations, because knowledge of the bacteria and fungi living in the water and their physiological and biochemical performance is a prerequisite for any further ecological research. In this respect, serological methods will acquire greater importance, especially immuno-fluorescence techniques.

Applied aquatic microbiology merits more attention than it has received in the past. Although extensive research has been done for many years on the self-purification of polluted waters, it is not yet clear, for example, what is the nature of the interrelationships between sewage load, primary production and bacterial development or why sea water is bactericidal for limnic and terrestrial organisms; and we know all too little about the behaviour of human pathogenic organisms in different waters and sediments. Very valuable also would be more research on the diseases of aquatic plants and animals due to micro-organisms and how to combat them. The same holds for the role of micro-organisms in fouling, corrosion and leaching.

For the solution of some of these problems of applied aquatic microbiology, modern genetics, maybe in the near future, might well provide a contribution. This is especially true in the case of insertion of suitable plasmids for influencing various processes.

Of great interest would be further clarification of the effect of bacteria and fungi on the formation and transformation of sediments and the mineral resources of the earth.

This short summary of the tasks for research makes one realize that in the future aquatic microbiology, while growing in extent, must become more and more specialized. In years to come, moreover, the emphasis in our knowledge of waters will shift from zoology and production biology more and more towards microbiology and biochemistry.

References

Adair, E. J., and Vishniac, H. S. (1958): Marine fungus requiring vitamin B $_2$. *Science* **127**, 147–148.

Ahne, W., and Wolf, K. (1980): Viruserkrankungen der Fische. In Reichenbach-Klinke: *Krankheiten und Schädigungen der Fische*. Fischer, Stuttgart, 56–105.

Ahrens, R. (1968): Taxonomische Untersuchungen an sternbildenden *Agrobacterium*-Arten aus der westlichen Ostsee. *Kieler Meeresforsch.* **24**, 147–173.

— (1969): Ökologische Untersuchungen an sternbildenden *Agrobacterium*-Arten aus der Ostsee. *Kieler Meeresforsch.* **25**, 190–204.

— (1971): Untersuchungen zur Verbreitung von Phagen der Gattung *Agrobacterium* in der Ostsee. *Kieler Meeresforsch.* **27**, 102–112.

— and Rheinheimer, G. (1967): Über einige sternbildende Bakterien aus der Ostsee. *Kieler Meeresforsch.* **23**, 127–136.

— Moll, G., and Rheinheimer, G. (1968): Die Rolle der Fimbrien bei der eigenartigen Sternbildung von *Agrobacterium luteum*. *Arch. Mikrobiol.* **63**, 321–330.

Ainsworth, C. G. (1971): Dictionary of the fungi. Commonwealth Mycological Inst. Kew. 6. Ed. 663 pp.

Aldermann, D. J. (1976): Fungal diseases in marine animals. In: Jones, E. B. G. (ed.), *Recent Advances in Aquatic Mycology*, 223–260. Elek Science, London.

Aliversieva-Gamidova, L. A. (1969): Mikrobiologische Prozesse im Mektheb See (russ.). *Mikrobiologija* **38**, 1096–1100.

Amlacher, E. (1981): *Taschenbuch der Fischkrankheiten*. 4. Aufl. Fischer Jena and Stuttgart. 474 pp.

Ammerman, J. W., and Azam, F. (1982): Uptake of cyclic AMP by natural populations of marine bacteria. *Appl. Environ. Microbiol.*, **43**, 869–876.

Amy, P. S., and Morita, R. Y. (1983): Starvation-survival patterns of sixteen freshly isolated open-ocean bacteria. *Appl. Environ. Microbiol.*, **45**, 1109–1115.

Anagnostidis, K., and Overbeck, J. (1966): Methanoxydierer und hypolimnische Schwefelbakterien. Studien zur ökologischen Biocönotik der Gewässermikroorganismen. *Ber. Dtsch. Bot. Ges.* **79**, 163–174.

— and Schwabe, G. H. (1966): Über artenreiche Bestände von Cyanophyten und Bacteriophyten in einem Farbstreifensandwatt sowie über das Auftreten von *Gomontiella*-deformierter *Oscillatoria*-Trichome. *Nova Hedwigia* **9**, 417–441.

Anastasiou, C. J. (1963): Fungi from salt lakes. II. Ascomycetes and fungi imperfecti from the Salton Sea. *Nova Hedwigia* **6**, 243–276.

Anderson, J. I. W. (1962): Studies on micrococci isolated from the North Sea. *J. appl. Bact.* **25**, 362–368.

Apostol, S. (1977): Die Zersetzung der Zellulose in den Gewässern. In Daubner, I. 2. *Internationales hydromikrobiologisches Symposium*. 151–161. Bratislava, Veda.

Azam, F., Fenchel, T., Field, J. G., Gray, J. S., Meyer-Reil, L. A., and Thingstad, F. (1983): The ecological role of water-column microbes in the sea. *Mar. Ecol. Prog. Ser.* 10, 257–263.

Baars, J. K. (1930): Over sulfaatreductie door bacterien. Dissertation, Delft. (after Thimann, 1964).

Baas Becking, L. M. G. (1925): Studies on the sulfur bacteria. *Ann. Botany* **39**, 613–650.

— and Wood, E. J. F. (1955): Biological processes in the estuarine environment. *Kon. Ned. Akad. Weten. Proc.* B, **58**, 160—181.

Babenzien, H. D. (1966): Untersuchungen zur Mikrobiologie des Neustons. *Verh. int. Ver. Limnol.* **16**, 1503–1511.

Bailey, C. A., Neihof, R. A., and Tabor, P. S. (1983): Inhibitory effect of solar radiation on amino acid uptake in Chesapeake Bay bacteria. *Appl. Environ. Microbiol.* **46**, 44–49.

Baker, J. H., and Bradham, L. A. (1976): The role of bacteria in the nutrition of aquatic detriti-vores. *Oecologia* **24**, 95–104.

— and Farr, I. S. (1977): Origins, characterization and dynamics of suspended bacteria in two chalk streams. *Arch. Hydrobiol.* **80**, 308–326.

Bansemir, K. (1969): Bakteriologische Untersuchungen von Wasser und Sedimenten aus dem Gebiet der Island-Färöer-Schwelle. *Ber. Dtsch. wiss. Komm. Meeresforsch.* **20**, 282–287.

— (1970): Mikrobiologische Untersuchungen zur Schwefelwasserstoffbildung in einer Vertiefung der Kieler Förde. Dissertation, Universität Kiel.

Bärlocher, F., and Kendrick, B. (1976): Hyphomycetes as intermediaries of energy flow in streams. In: Jones, E. B. G. (ed.), *Recent Advances in Aquatic Mycology*, Elek Science, London, 435–446.

Barth, R. (1967): Observacoes sobre occorencia em massa de cyanophycea. *Publ. Inst. Perg. Mar.* **6**, 1–8 (after Carr and Whitton, 1973).

Bauerfeind, S. (1982): Versuche zum Abbau von Plankton- und Detritusmaterial durch natür-liche Bakterienpopulationen der Schlei und Ostsee. Dissertation, University of Kiel.

Baumann, P., and Baumann, L. (1981): The marine gram-negative eubacteria: genera *Photobacterium, Beneckea, Alteromonas, Pseudomonas* and *Alcaligenes*. In: Starr *et al. The Procaryotes* 1302–1331. Springer, Berlin, Heidelberg and New York.

Bavendamm, W. (1932): Die mikrobiologische Kalkfällung in der tropischen See. *Arch. Microbiol.* **3**, 205–276.

Becking, J. H. (1981): The family Azotobacteraceae. In: Starr *et al., The Procaryotes* 795–817. Springer, Berlin, Heidelberg, New York.

Beerstecher, E. (1954): *Petroleum microbiology*. Elsevier, Houston Texas, 375 pp.

Bell, C. R., and Albright, L. J. (1981): Attached and free-floating bacteria in the Fraser River estuary, British Columbia, Canada. *Mar. Ecol. Prog.* Ser. **6**, 317–327.

Bell, R. T., Ahlgren, G. M., and Ahlgren, I. (1983): Estimating bacterioplankton production by measuring ^{3}H-thymidine incorporation in an eutrophic Swedish lake. *Appl. Environ. Microbiol.* **45**, 1709–1721.

Bell, W. H., Lang, J. M., and Mitchell, R. (1974): Selective stimulation of marine bacteria by algal extracellular products. *Limnol. Oceanogr.* **20**, 833–839.

Berk, S. G., Mitchell, R., Bobbie, R. J., Nickels, J. S., and White, D. C. (1981): Microfouling on metal surfaces exposed to seawater. *Int. Biodeterioration Bull.* **17**, 29–37.

Bernard, F. (1963): Density of flagellates and Myxophyceae in the heterotrophic layers related to environment. In Oppenheimer, C. H. *Symposium on marine microbiology.* 215–228. Springfield Ill., Thomas.

— and Lecal, J. (1960): Plancton unicellulaire recolté dans l'Ocean Indien par le Charcot (1950) et le Norsel (1955–56). *Bull. Inst. Oceanogr., Monaco* **1166**, 1–59.

Bertrand, J. C., and Vacelet, J. (1971): L'association entre Esponges cornées et bacteries. *C. R. Acad. Sci. (Paris) Sér. D* **272**, 638–641.

Bianchi, A. J., and Bianchi, M. A. (1982): Evolution a court terme des effectifs et de l'activité des communautes bactériennes dans des ecosystemes d'aquaculture. *Publ.* CNEXO **13**, 61–71.

Billen, G. (1978): A budget of nitrogen recycling in North Sea sediments off the Belgian coast. *Estuar.* and *Coast. Mar. Sci.* **7**, 127–146.

Birge, E. A., and Juday, C. (1934): Particulate and dissolved organic matter in inland lakes. *Rcol. monographs* **4**.

Blackburn, T. H., and Henriksen, K. (1983): Nitrogen cycling in different types of sediments from Danish waters. *Limnol. Oceanogr.* **28**, 477–493.

Blanchard, D. C., and Syzdek, L. (1970): Mechanism for the water-to-air transfer and concentration of bacteria. *Science* **170**, 626–628.

Bock, E. (1965): Vergleichende Untersuchungen über die Wirkung sichtbaren Lichtes auf *Nitrosomonas europaea* und *Nitrobacter winogradskyi. Arch. Mikrobiol.* **51**, 18–41.

Bölter, M. (1981): DOC-turnover and microbial production. *Kieler Meeresforsch., Sonderh.* **5**, 304–310.

— Meyer-Reil, L.-A., and Probst, B. (1977): Comparative analysis of data measured in the brakishwater of the Kiel Fjord and the Kiel Bay. In: Rheinheimer, G., *Microbial ecology of a brackish water environment. Ecological Studies* **25**, 249–280. Springer, Heidelberg.

Bonde, G. J. (1967): Pollution of a marine environment. *J. Water Poll. Contr. Fed. Washington* **2**, 45–63.

— (1977): Bacterial indication of water pollution. In Droop, M. R., and Jannasch, H. W., *Advances in aquatic microbiology.* 273–364. London and New York, Academic Press.

Boylen, C. W., Shick, M. O., Roberts, D. A., and Finger, R. (1983): Microbiological survey of Adirondack Lakes with various pH values. *Appl. Environ. Microbiol.* **45**, 1538–1544.

Brandt, K. (1902): Über den Stoffwechsel im Meere. *Wiss. Meeresunters. (Abt. Kiel)* **6**, 23–79.

Brandt, R. P. (1923): Potash from kelp. Early development and growth of the giant kelp, *Macrocystis pyrifera*. *U. S. Dep. Agr. Bull.* **1191**, 1–40 (after Wood, 1965).

Brandt, A. von (1953): Zelluloseabbau in Fließgewässern. *Ber. Limnol. Flußst. Freudenthal* **4**, 17–19.

— and Klust, G. (1950): Zelluloseabbau im Wasser. *Arch. Hydrobiol.* **43**, 223–233.

Breed, R. S., Murray, E. D. G., and Smith, N. R. (1957): *Bergey's manual of determinative bacteriology.* 7. Ed. Baltimore, Williams and Wilkins, 1094 pp.

Bringmann, G. (1970): Mikrobiologie des Wassers. In: Schormüller, J., *Handbuch der Lebensmittelchemie.* Heidelberg, Springer **8**, 119–162.

— and Kühn, R. (1975): Die beginnende Hemmung der Zellvermehrung von *Pseudomonas* als wassertoxikologischer Test. *Vom Wasser* **44**, 119–129.

Brinton, C. C. (1965): The structure, function, synthesis and genetic control of bacterial pili and a molecular model for DNA and RNA transport in gram negative bacteria. *Transact. N. York Acad. Sc. II*, **27**, 1003–1054.

Brisou, J. (1955): *Microbiologie du milieu marin.* Paris, Flammarion. 271 pp.

Brock, T. D. (1967a): Life at high temperatures. *Science* **158**, 1012–1019.

— (1967b): Relationship between standing crop and primary productivity along a hot spring thermal gradient. *Ecology* **48**, 566–571.

— (1976): *Principles of microbial ecology.* Prentice–Hall inc., Englewood cliffs, N. Jersey USA, 306 pp.

Brown, R. M. (1972): Algal viruses. *Advan. Virus Res.* **17**, 243–277.

Buchanan, R. E., and Gibbons, N. E. (1974): *Bergey's manual of determinative bacteriology.* 8. Ed. Baltimore, Williams and Wilkins, 1246 pp.

Buchner, P. (1953): *Endosymbiose der Tiere mit pflanzlichen Mikroorganismen.* Stuttgart, 771.

— (1960): *Tiere als Mikrobenzüchter.* Heidelberg, Springer, 160 pp.

Bullock, G. L., Conroy, D. A., and Snieszko, S. F. (1971): *Bacterial diseases of fishes.* T. F. H. Public. Neptune City, N. J. 151 pp.

Butkewitsch, V. S. (1928): Die Bildung der Eisenmanganablagerungen am Meeresboden und die daran beteiligten Mikroorganismen. *Trudy Morsk. Nauchn. In-ta* **3**, 5–81 (after Kusnezow et al., 1963).

Butlin, K. R., and Postgate, J. R. (1954): The microbiological formation of sulphur in Cyrenaican lakes. In *Biology of Deserts*, Inst. Biol. London, 112.

Camman, L. M., and Walker, J. A. (1982): Distribution and activity of attached and free-living bacteria in the Bay of Fundy. *Canad. J. Fish Aquat. Sci.* **39**, 1655–1663.

Campell, L. L., and Williams, O. B. (1951): A study of chitin decomposing microorganisms of marine origin. *J. gen. Microbiol.* **5**, 894–905.

Campbell, R. (1981): *Mikrobielle Ökologie*, Weinheim, Verlag Chemie, 243 pp.

Cantacuzene, A. (1930): Contributions à l'étude des tumeurs bacterienne chez les algues marines. Dissertation Paris (after Wood, 1965).

Canter, H. M., and Lund, J. W. G. (1948): Studies on plankton parasites. I. Fluctuations in the numbers of *Asterionella formosa* Hoss. in relation of fungal epidemics. *New Phytol.* **47**, 238–261.

— and Lund, J. W. G. (1951): Studies on plankton parasites III. Examples of the interaction between parasitism and other factors determining the growth of diatoms. *Ann. Bot.* (London) **15**, 359–371.

— and Lund, J. W. G. (1953): Studies on plankton parasites II. The parasitism of diatoms with special reference to lakes in the English lake district. *Trans. Brit. Mycol. Soc.* **36**, 13–37.

Cappenberg, Th. E., Hordijk, C. A., and Hagenaars, C. P. M. M. (1984): A comparison of bacterial sulfate reduction and methanogenesis in the anaerobic sediments of a stratified lake ecosystem. *Arch. Hydrobiol. Beih.* **19** *Ergeb. Limnol.* 191—199.

Caron, D. A., Davis, P. G., Madin, L. P., and Sieburth, J. McN. (1982): Heterotrophic bacteria and bacteriovorous protozoa in oceanic macroaggregates. *Science* **218**, 795–797.

Carr, N. G., and Whitton, B. A. (Edit.) (1973): The biology of bluegreen algae. *Botanical Monogr.* 9. Oxford, Blackwell, 676 pp.

Caspers, H. (1959): Die Einteilung der Brackwasserregionen in einem Ästuar. *Arch. Oceanogr. Limnol.* **11**, Suppl. 133–169.

Castenholz, R. W. (1973): Ecology of blue-green algae. In: Carr, N. G., and Whitton, B. A. The biology of blue-green algae. *Botanical Monogr.* **9**, 379–414, Oxford, Blackwell.

Cavanaugh, C. M. (1983): Symbiotic chemoautotrophic bacteria in marine invertebrates from sulphide-rich habitats. *Nature* **302**, 58–61.

Cavanaugh, C. M., Gardiner, S. L., Jones, M. L., Jannasch, H. W., and Waterbury, J. B. (1981): Procaryotic cells in the hydrothermal vent tube worm *Riftia pachyptila* Jones: Possible chemoautotrophic symbionts. *Science* **213**, 340–342.

Christophersen, J. (1955): Bakterien. In: Precht, H., Christophersen, J., and Hensel, H., Temperatur und Leben. Heidelberg, Springer.

Chrost, R. J. (1975a): Inhibitors produced by algae as an ecological factor affecting bacteria in water ecosystems. I. Dependence between phytoplankton and bacteria development. *Acta Microbiol. Pol. Ser. B$_1$* **7**, 125–133.

— (1975b): Inhibitors produced by algae as an ecological factor affecting bacteria in water. II. Antibacterial activity of algae during blooms. *Acta Microbiol. Pol. Ser. B$_1$* **7**, 167–176.

Claus, G. (1979): Bakterien bauen Geruchsstoffe ab. *Forum Mikrobiologie*, **2**, 85.

Cobet, A. B., Jones, G. E., Albright, J., Simon, H., and Wirsen, C. (1971): The effect of nickel on a marine bacterium: fine structure of Arthrobacter marinus. *J. gen. Microbiol.* **66**, 185–196.

Cohen, Y., Krumbein, E., and Shilo, M. (1977): Solar lake (Sinai) 3. Bacterial distribution and production. *Limnol. Oceanogr.* **22**, 621–634.

Collins, V. G., and Willoughby, L. G. (1962): The distribution of bacteria and fungal spores in Blelham Tarn with particular reference to an experimental overturn. *Arch. Mikrobiol.* **43**, 294–307.

Colwell, R. R. (1979): Human pathogens in the marine environment. In: Colwell and Foster: *Aquatic microbial ecology*. 337–344. College Park, Maryland Sea Grant Publ.

— Deming, J. W., and Tabor, B. S. (1983): Bacteria associated with the digestive tract of deep-sea animals. CNRS.

Cooper, L. (1937): Oxidation-reduction potential in sea water. *J. marin. biol. Ass.* **22**, 167–176.

Cowley, G. T., and Chrzanowski, T. H. (1980): Yeasts in the habitat and nutrition of *Uca pugilator. Bot. Mar.* **23**, 397–403.

Cviic, V. (1960): Contribution à la connaissance du rôle des bacteries dans l'alimentation des larve de langouste (*Palinurus vulgaris* Lart.). *Rapp et Proc. Verb. Comm. Int. Expl. Sci. Med.* **25**, 45–47.

Czaya, E. (1981): *Ströme der Erde*. Aulis, Köln. 248 pp.

Czeczuga, B., and Gradski, F. (1973): Relationship between extracellular and cellular production in the sulphuric green bacterium *Chlorobium limicola* Nads. (Chlorobacteriaceae) as compared to primary production of phytoplankton. *Hydrobiologia* **42**, 85–95.

Daniltschenko, P., and Tschigirin, N. (1926): Zur Frage nach dem Ursprung des Schwefelwasserstoffes im Schwarzen Meer (russ.). *Tr. Osoboj. zool. labor. i. Sewastopolsk. biol. stanzii AN SSSR, ser, II,* **10**, 141 (after Kriss, 1961).

Daubner, I. (1972): *Mikrobiologie des Wassers*. Berlin, Akademie-Verlag. 440 pp.

— and Peter, H. (1974): *Membranfilter in der Mikrobiologie des Wassers*. Berlin, New York, de Gruyter, 216 pp.

Dawson, R., and Pritchard, R. G. (1978): The determination of α-amino acids in seawater using a fluorimetric analyser. *Marine Chemistry* **6**, 27–40.

Denk, V. (1950): Zur Frage der Ammonentstehung im Stoffkreislauf der Natur. *Arch. Mikrobiol.* **15**, 308–314.

Desikachary, R. V. (1959): *Cyanophyta*. New Delhi, Indian Counc. Agric. Res. 686 pp.

Deufel, J. (1965): Plötzliche Zunahme von *Azotobacter* im Bodensee. *Naturwissensch.* **52**, 192–193.

Dietrich, G., Düing, W., Grasshoff, K., and Koske, P. H. (1966): Physikalische und chemische Daten nach Beobachtungen der F. S. „Meteor" im Indischen Ozean, 1964–1965. *Meteor Forschungsergebn.* A 2.

— Kalle, K., Krauss, W., and Siedler, G. (1975): *Allgemeine Meereskunde; eine Einführung in die Ozeanographie*. 3. Aufl. Gebrüder Bornträger, Berlin, Stuttgart, 593 pp.

Dimitroff, V. T. (1926): Spircheates in Baltimore market oysters. *J. Bact.* **12**, 135–177.

Dodds, A. (1979): Viruses of marine algae. In: Andrews, J. H., Pathology of seaweeds: Current status and future prospects. *Experientia* **35**, 440–442.

Dorier, A., and Degrange, C. (1961): L'évolution de l'*Ichthyosporidium (Ichthyophonus) hoferi* (Plehn et Mulsow) chez le salmonides d'élevage. *Trav. Labor. Hydrobiol.* Grenoble **52**, 7–44.

Drake, C. H. (1965): Occurrence of *Siderocapsa treubii* in certain waters of the Niederrhein. *Gewässer u. Abw.* **39/40**, 41–63.

Dubinina, G., and Derjugina, Z. (1972): Vergleichende elektronenmikroskopische Untersuchungen der Mikroflora von Moorerz und Seewasser. *Arch. Hydrobiol.* **71**, 90–102.

Ehrhardt, M. (1969): The particulate organic carbon and nitrogen, and the dissolved organic carbon in the Gotland Deep in May 1968. *Kieler Meeresforsch.* **25**, 71–80.

Ehrlich, H. L. (1972): The role of microbes in manganese nodule genesis and degradation. In: Horn, D., *Ferromanganese deposits on the ocean floor.* Nat. Science Found. Washington, 63–70.

Ellenberg, H. (1973): *Ökosystemforschung.* Berlin, Heidelberg, New York, Springer, 280 pp.

Engel, H. (1958): Nitrifikation. In: Ruhland, *Handbuch der Pflanzenphysiologie.* Heidelberg, Springer **8**, 1107–1127.

— (1960): Die Nitrifikanten. In: Ruhland, *Handbuch der Pflanzenphysiologie.* Heidelberg, Springer **5**, 664–681.

Enoksson, V. (1980): Nitrification in marine ecosystems. In: Rosswall, *Processes in the nitrogen cycle.* The Swedish Environmental Protection Board. PM 1213, 157–166. (in Swedish).

Esterly, C. O. (1916): The feeding habits and food of pelagic copepods and the question of nutrition by organic substances in solution in water. *Univ. Calif. Publ. in Zool.* **16**, 171–184.

Exerzew, W. A. (1948): Bestimmung der Mächtigkeit der mikrobiologisch wirksamen Schicht der Schlammablagerungen einiger Seen. *Mikrobiologija* **16**, 6 (russ.) (after Kusnezow, 1959).

Fair, G. W., and Geyer, J. C. (1961): Wasserversorgung und Abwasserbeseitigung. *Grundlagen, Technik und Wirtschaft.* München, Verl. R. Oldenbuorg. 969 pp.

Fell, J. W. (1967): Distribution of yeasts in the Indian Ocean. *Bull. marin. Sci. Gulf Caribb.* **17**, 454–470.

— and Uden, N. van (1963): Yeasts in marine environments. In: Oppenheimer, C. H., *Symposium on marine microbiology.* C. C. Thomas, Springfield Ill., 329–340.

Filipkowska, Z., Misetic, S., and Niewolak, S. (1983): Effect of some herbicides on nitrogen transformation in lake water. II. Nitrification. *Zesz. nauk. ART Olszt.* **12**, 33–43.

Fletcher, M., and Marshall, K. C. (1983): Are solid surfaces of ecological significance to aquatic bacteria? *Adv. Microbiol. Ecol.* **6**, 199–236.

Fogg, G. E. (1973): Physiology and ecology of marine blue-green algae. In: Carr, N. G., and Whitton, B. A., The biology of blue-green algae. *Botanical Monogr.* **9**, 368–378. Oxford, Blackwell.

— Stewart, W. D. P., Fay, P., and Walsby, A. E. (1973): *The Blue-green Algae.* London and New York, Academic Press, 459 pp.

Fott, B. (1959): *Algenkunde*, 2. Ed. Jena, Fischer, 482 pp.

Frenzel, H. J. (1966): Wechselbeziehungen zweier an der Korrosion von Eisen beteiligter Bakteriengruppen. *Material und Organismen* **1**, 275–286.

Fuhrman, J. A., and Azam, F. (1982): Thymidine incorporation as a measure of heterotrophic bacterioplankton production in marine surface waters: evaluation and field results. *Mar. Biol.* **66**, 109–120.

Fuhs, G. W. (1961): Der mikrobielle Abbau von Kohlenwasserstoffen. *Arch. Mikrobiol.* **39**, 374–422.

Gaertner, A. (1964): Elektronenmikroskopische Untersuchungen zur Struktur der Geißeln von *Thraustochytrium* spec. Veröff. *Inst. Meeresforsch. Bremerhaven* **9**, 25–30.

— (1967): Niedere mit Pollen köderbare Pilze in der südlichen Nordsee. *Veröff. Inst. Meeresforsch. Bremerhaven* **10**, 159–165.

— (1968a): Niedere, mit Pollen köderbare marine Pilze diesseits und jenseits des Island-Färöer-Rückens im Oberflächenwasser und im Sediment. *Veröff. Inst. Meeresforsch. Bremerhaven* **11**, 65–82.

— (1968b): Die Fluktuationen niederer Pilze in der Deutschen Bucht 1965 und 1966. *Veröff. Inst. Meeresforsch. Bremerhaven Sonderb.* **3**, 105–120.

— (1968c): Eine Methode des quantitativen Nachweises niederer, mit Pollen köderbarer Pilze in Meerwasser und Sediment. *Veröff. Inst. Meeresforsch. Bremerhaven*, Sonderb. **3**, 75–91.

Gak, D. Z. (1959): Fizjologičeskaja aktivnost' i sistematičeskoe polozenie mobilizujusčich fosfor mikroorganizmov, videlennych iz vodoemov pribaltiki. *Mikrobiol.* **28**, 551–556.

Gast, V. (1983): Untersuchungen über die Bedeutung der Bakterien als Nahrungsquelle für das Mikrozooplankton der Schlei und der Ostsee unter besonderer Berücksichtigung der Ciliaten. Dissertation, Universität Kiel.

Gäumann, E. (1964): *Die Pilze.* Basel u. Stuttgart, Birkhäuser. 2. Ed., 451 pp.

Geesey, G. G., Mutch, R., and Costerton, J. W. (1978): Sessile bacteria: an important component of the microbial population in small mountain streams. *Limnol. Oceanogr.* **23**, 1214–1223.

Geitler, L. (1932): Cyanophyceae. In: Rabenhorst, L., *Kryptogamen-Flora von Deutschland, Öster-reich und der Schweiz*, **14**. Leipzig, Akademische Verlagsges.

Geller, A. (1983): Degradability of dissolved organic lake water compounds in cultures of natural bacterial communities. *Arch. Hydrobiol.* **99**, 69–79.

Gentile, J. H., and Maloney, T. E. (1969): Toxicity and environmental requirements of a strain of *Aphanizomenon flos-aquae* (L.) Ralfs. *Can. J. Microbiol.* **15**, 165–173.

Gessner, F. (1955): *Hydrobotanik I*. Berlin, Deutscher Verlag d. Wissenschaften, 517 pp.

— (1957): *Meer und Strand*. 2. Ed. Berlin, Deutscher Verlag d. Wissenschaften, 426 pp.

Gessner, R. V. (1977): Seasonal occurrence and distribution of fungi associated with *Spartina alterniflora* from a Rhode Island estuary. *Mycologia* **69**, 477–491.

Ghiorse, W. C., and Hirsch, P. (1982): Isolation and properties of ferromanganese-depositing budding bacteria from Baltic Sea ferromanganese concretions. *Appl. Environ. Microbiol.* **43**, 1464–1472.

Giesy, J. P. (Ed.) (1980): *Microcosms in ecological research*. Publ. Technical Information Center, US Dept. Energy, Washington. 1010 pp.

Glombitza, K. W. (1969): Antibakterielle Inhaltsstoffe in Algen. *Helgoländer wiss. Meeresunters.* **19**, 376–384.

Gocke, K. (1969): Untersuchungen über Abgabe und Aufnahme von Aminosäuren und Polypep-tiden durch Planktonorganismen. Dissertation Universität Kiel.

— (1975a): Studies on short-term variations of heterotrophic activity in the Kiel Fjord. *Marine Biology* **33**, 49–55.

— (1975b): Untersuchungen über die Aufnahme von gelöster Glukose unter natürlichen Verhält-nissen durch größenfraktioniertes Nano- und Ultrananoplankton. *Kieler Meeresforsch.* **31**, 87–94.

— (1977): Heterotrophic activity. In: Rheinheimer, G., Microbial ecology of a brackish water environment. *Ecological Studies* **25**, 198–222. Berlin, Heidelberg, New York, Springer.

Gode, P. (1970): Untersuchungen über nitrifizierende Bakterien in einem geschichteten eutrophen See. Dissertation Universität Kiel.

Godlewska-Lipowa, W. A. (1974a): Heterotrophic activity of bacterial microflora in Mazurian Lakes of various trophy. *Polskie Arch. Hydrobiol.* **21**, 51–58.

— (1974b): Uptake of organic matter by natural bacteria groups in the Mazurian Lakes. *Polskie Arch. Hydrobiol.* **21**, 59–67.

— (1974c): Organic matter decomposition in aquatic ecosystems of different trophic types. *Bull. Academic Polon. Sciences*, Cl. II. **22**, 41–45.

Golobic, S. (1976): Taxonomy of extant stromatolite-building cyanophytes. In: Walter, Stromato-lites. *Developments in Sedimentology* **20**, 127–140. Amsterdam, Elsevier.

Goreau, T. J., Kaplan, W. A., Wofsy, S. C., McElroy, M. B., Valois, F. C., and Watson, S. W. (1980): Production of NO_2^- and N_2O by nitrifying bacteria at reduced concentrations of oxygen. *Appl. Environ. Microbiol.* **40**, 526–532.

Gorkom, H. J. van, and Donze, M. (1971): Localization of nitrogen fixation in Anabaena. *Nature* **234**, 231–232.

Gorlenko, W. M. (1977): Die phototrophen Bakterien in den stratifizierten Seen und ihre Ökologie In: Daubner, I., *2. Internationales hydromikrobiologisches Symposium*. 91–112. Bratislava, Veda

— and Kusnezow, S. I. (1972): Über die photosynthetisierenden Bakterien des Kononjer-Sees *Arch. Hydrobiol.* **70**, 1–13.

— Dubinina, G. A., and Kusnezow, S. I. (1977): Edolognja Vodn'ich Mikroorganizmov. *Moskau izdatel'stvo Nauka*. 288 pp.

— Dubenina, G. A., and Kusnezow, S. I. (1983): *The ecology of aquatic micro-organisms*. Schweit-zerbart, Stuttgart, 252 pp.

Götting, K. J., Kilian, E. F., and Schnetter, R. (1982): *Einführung in die Meeresbiologie*. Band 1. *Marine Organismen*. Vieweg, Braunschweig, 179 pp.

Graevenitz, A. von, and Carrington, G. O. (1973): Halophilic vibrios from extraintestinal lesions in man. *Infection* **1**, 54–58.

Granhall, U., and Berg, B. (1972): Antimicrobial effects of *Cellvibrio* on blue-green algae. *Arch. Mikrobiol.* **84**, 234–242.

— and Lundgren, A. (1971): Nitrogen fixation in Lake Erken. *Limnol. Oceanogr.* **16**, 711–719.

Greenfield, L. H. (1963): Metabolism and concentration of calcium and magnesium and precipi-tation of calcium carbonate by a marine bacterium. *Ann. N. Y. Acad. Sci.* **109**, 23–45.

Guillard, R. (1963): Organic sources of nitrogen for marine centric diatoms. In: Oppenheimer, C. H., *Symposium on marine microbiology*. Springfield, Ill., Thomas, 93–104.

Gundersen, K. (1966): The growth and respiration of *Nitrosocystis oceanus* at different partial pressures of oxygen. *J. gen. Microbiol.* **42**, 387–396.

Gunkel, W. (1966): Bakteriologische Untersuchungen im Indischen Ozean. *Veröff. Inst. Meeresforsch. Bremerhaven* Sonderb. **2**, 255–264.

— (1967a): Experimentell-ökologische Untersuchungen über die limitierenden Faktoren des mikrobiellen Ölabbaues im marinen Milieu. *Helgoländer wiss. Meeresunters.* **15**, 210–225.

— (1967b): Arbeitssitzung über Gewässerverölung. Ölbekämpfung und Ölabbau. *Helgoländer wiss. Meeresunters.* **16**, 285–384.

— and Oppenheimer, C. H. (1963): Experiments regarding the sulfide formation in sediment of the Texas Gulf Coast. In: Oppenheimer, C. H., *Symposium on marine microbiology*. Springfield Ill., Thomas, 674–684.

— Gassmann, G., Oppenheimer, C. H., and Dundas, I. (1980): Preliminary results of baseline studies of hydrocarbons and bacteria in the North Sea: 1975, 1976 and 1977. In: Ponencias del simposio internacional en: Resistencia a los antibioticos y microbiologia marina. Santiago de Compostela, España, 223–247.

— Crow, S., and Klings, K. W. (1983): Yeast population increases during degradation of *Desmerestia viridis* (Phaeophyceae) in sea water model micro-ecosystems. *Mar. Biol.* **75**, 327–332.

Hagedorn, H. (1971): Die ökologische Bedeutung des Thiamins. *Ber. Dtsch. Bot. Ges.* **84**, 479–482.

Hagström, A., Larsson, U., Hörstedt, P., and Normark, S. (1979): Frequency of dividing cells, a new approach to the determination of bacterial growth rates in aquatic environments. *Appl. Environ. Microbiol.* **37**, 5, 805–812.

Hamilton, R. D., and Holm-Hansen, O. (1967): Adenosine triphosphate content of marine bacteria. *Limnol. Oceanogr.* **12**, 319–324.

Hanaoka, H. (1973): Cultivation of three species of pelagic microcrustacean plankton. *Bull. Plankton Soc. Japan* **20**, 19–29.

Hanert, H. (1974): In situ Untersuchungen zur Analyse und Intensität der Eisen(III)-Fällung in Dränungen. *Z. Kulturtechnik u. Flurbereinigung* **15**, 80–90.

— (1981): The genus *Gallionella*. In: Starr et al., *The Procaryotes* 509–515. Springer, Berlin, Heidelberg, New York.

Hanson, R. B., Shafer, D., Ryan, T., Pope, D. H., and Lowery, H. K. (1983): Bacterioplankton in Antarctic Ocean waters during late Austral winter: abundance, frequency of dividing cells and estimates of production. *Appl. Environ. Microbiol.* **45**, 1622–1632.

Harms, H., and Engel, H. (1965): Der Einfluß sichtbaren Lichts auf das Cytochromsystem ruhender Zellen von *Micrococcus denitrificans* Beijerinck. *Arch. Mikrobiol.* **53**, 224–230.

Harrison, J. L. (1972): The salinity tolerance of freshwater and marine zoosporic fungi, including some aspects of the ecology and ultrastructure of the Thraustochytriaceae. Ph. D. thesis, University of London.

Hart, T. J. (1966): Some observations on the relative abundance of marine phytoplankton populations in nature. In: Barnes, H., Some contemporary studies in marine science, 375–393. London, Allen and Unwin.

Haskins, R. H., and Weston, W. H. Jr. (1950): Studies in the lower Chytridiales. I. Factors affecting pigmentation, growth, and metabolism of a strain of *Karlingia (Rhizophlyctis) rosea*. *Ann. J. Botany* **37**, 739–750.

Hastings, J. W., and Mitchell, G. (1971): Endosymbiotic bioluminescent bacteria from the light organ of pony fish. *Biol. Bull.* **141**, 261–268.

— and Nealson, K. H. (1981): The symbiotic luminous bacteria. In: Starr et al., *The Procaryotes*, 1332–1345. Springer, Berlin, Heidelberg, New York.

Hawker, L. E. (1966): Environmental influences on reproduction. In: Ainsworth, C. G., and Sussman, A. S., *The Fungi* **2**, 435–469. New York and London, Academic Press.

Hedén, C. (1964): Effects of hydrostatic pressure on microbial systems. *Bacteriol. Rev.* **28**, 14–29.

Heinen, W. (1962): Siliciumstoffwechsel bei Mikroorganismen. II. Beziehungen zwischen dem Silicat- und Phosphat-Stoffwechsel bei Bakterien. *Arch. Mikrobiol.* **41**, 229–246.

— (1965): Siliciumstoffwechsel bei Mikroorganismen. VI. Enzymatische Veränderung des Stoffwechsels bei der Umstellung von Phosphat auf Silicat bei Proteus mirabilis. *Arch. Mikrobiol.* **52**, 49–68.

Heitzer, R. D., and Ottow, J. C. G. (1976): New denitrifying bacteria isolated from Red Sea sediments. *Marine Biol.* **37**, 1–10.

Hellmann, H., Klein, K., and Knöpp, H. (1966): Untersuchungen über die Eignung von Emulgatoren für die Beseitigung von Öl auf Gewässern. *Dtsch. gewässerk. Mitt.* **11**, 91–95.

Hendrie, M., Hodgkinss, W., and Shewan, J. M. (1970): The identification, taxonomy and classification of luminous bacteria. *J. Gen. Microbiology* **64**, 151–159.

Henningson, M. (1978): Physiology of aquatic lignicolous fungi from Swedish coastal waters. *Material and Organisms* **13**, 129–168.

Henrici, A. T., and Johnson, D. E. (1935): Studies of freshwater bacteria. II. Stalked bacteria, a new order or Schizomycetes. *J. Bact.* **30**, 61–93.

Henriksen, K., Rasmussen, M. B., and Jensen, A. (1983): Effect of bioturbation on microbial nitrogen transformations in the sediment and fluxus of ammonium and nitrate to the overlying water. *Ecol. Bull.* **35**, 193–205.

Hentzschel, G. (1977): Charakterisierung mariner Bdellovibrionen. Dissertation, Universität Hamburg.

— (1980): Wechselwirkungen bakteriolytischer und saprophytischer Bakterien aus der Nordsee. *Mitt. Inst. Allg. Bot. Hamburg*, **17**, 113–124.

Herbert, R. A., Brown, C. M., and Stanley, S. O. (1977): Nitrogen assimilation in marine environments. In: Skinner, F. A., and Shewan, J. M., *Aquatic Microbiology*. 161–177. New York and London, Acedemic Press.

Hess, E. (1937): A shell disease in lobsters *(Homarus americanus)* caused by chitinovorous bacteria. *J. biol. Board Canada* **3**, 358–362.

Heyer, J. (1966): Beobachtungen über das Wachstum von Bakterien in nichtwäßrigen Medien. In: Malek, I., and Schwartz, W., *Erdölmikrobiologie*. Berlin, Akademie-Verlag, 227–229.

Hirsch, P. (1958): Stoffwechselphysiologische Untersuchungen an *Nocardia petroleophila*. *Arch. Mikrobiol.* **29**, 368–393.

— (1960): Einige weitere, von Luftverunreinigungen lebende Actinomyceten und ihre Klassifizierung. *Arch. Mikrobiol.* **35**, 391–414.

— and Engel, H. (1956): Über oligocarbophile Actinomyceten. *Ber. Dtsch. Bot. Ges.* **69**, 441–454.

— and Rheinheimer, G. (1968): Biology of Budding Bacteria. V. Budding Bacteria in aquatic habitats: Occurrence, enrichment and isolation. *Arch. Mikrobiol.* **62**, 289–306.

— and Rades-Rohkohl, E. (1983): Microbial diversity in a groundwater aquifer in northern Germany. *Developments in Industrial Microbiology* **24**, 183–200.

Hitchner, E. R., and Snieszko, S. F. (1947): A study of a microorganism causing a bacterial disease of lobsters. *J. Bact.* **54**, 48.

Hoffmann, B. (1966): Wachstum und Vermehrung von *Escherichia coli* bei niederen Temperaturen. Dissertation Universität Kiel.

Hoffmann, C. (1942): Beiträge zur Vegetation des Farbstreifensandwattes. *Kieler Meeresforsch.* **4**, 85–108.

Höhnk, W. (1953): Studien zur Brack- und Seewassermykologie III. Oomycetes: *Zweiter Teil. Inst. Meeresforsch. Bremerh.* **2**, 52–108.

— (1955): Studien zur Brack- und Seewassermykologie. V. Höhere Pilze des submersen Holzes. *Veröff. Inst. Meeresforsch. Bremerhaven* **3**, 199–227.

— (1969a): Zur Entfaltung der marinen Mykologie. *Ber. Dtsch. Bot. Ges.* **81**, 380–390.

— (1969b): Über den pilzlichen Befall kalkiger Hartteile von Meerestieren. *Ber. Dtsch. wiss. Kommission Meeresforsch.* **20**, 129–140.

Hoppe, H. G. (1970): Ökologische Untersuchungen an Hefen aus dem Bereich der westlichen Ostsee. Dissertation Universität Kiel.

— (1976): Determination and properties of actively metabolizing heterotrophic bacteria in the sea investigated by means of microautoradiography. *Marine Biol.* **36**, 291–302.

— (1977): Analysis of actively metabolizing bacterial populations with the autoradiographic method. In: Rheinheimer, G., Microbial ecology of a brackish water environment. *Ecological Studies* **25**, 179–197. Berlin, Heidelberg, New York, Springer.

— (1978): Relations between active bacteria and heterotrophic potential in the sea. *Netherl. J. Sea Res.* (in Press).

— (1981): Blue-green algae agglomeration in surface water: a micro-biotope of high bacterial activity. *Kieler Meeresforsch. Sonderh.* **5**, 291–303.

— (1983): Significance of exoenzymatic activities in the ecology of brackish water: measurements by means of methylumbelliferyl-substrates. *Mar. Ecol. Prog. Ser.* **11**, 299–308.

Horne, A. J., and Goldman, C. R. (1972): Nitrogen fixation in Clear Lake, California. I. Seasonal variation and the role of heterocysts *Limnol. Oceanogr.* **17**, 678–692.

Horowitz, A., Krichevsky, M. I., and Atlas, R. M. (1983): Characteristics and diversity of sub-arctic marine oligotrophic, stenoheterotrophic and euryheterotrophic bacterial populations. *Can. J. Microbiol.* **29**, 527–535.

Horstmann, U. (1975): Eutrophication and mass production of blue-green algae in the Baltic. *Havsforskningsinst. Skr.* **239**, 83–90.

Hübel, H., and Hübel, M. (1974a): In situ-Messungen der diurnalen Stickstoff-Fixierung an Mikro-benthos der Ostseeküste. *Arch. Hydrobiol. Suppl.* **46**, 39–54.

— — (1974b): Stickstoff-Fixierung in Küstengewässern der mittleren Ostsee. *Z. Allg. Mikrobiol.* **14**, 617–619.

Hughes, G. C. (1974): Geographical distribution of the higher marine fungi. *Veröff. Inst. Meeres-forsch. Bremerhaven. Suppl.* **5**, 419–441.

Husmann, W. (1964): *Praxis der Abwasserreinigung.* Heidelberg, Springer, 191 pp.

Hynes, H. B. N. (1971): *The Biology of Polluted Waters.* Liverpool, University Press, 202 pp.

Ingold, C. T. (1953): *Dispersal in fungi.* Oxford, Clarendon Press. 197 pp.

— (1976): The morphology and biology of freshwater fungi excluding phycomycetes. In: Jones, E. B. G. (ed.), *Recent Advances in Aquatic Mycology,* 335–358. Elek Science, London.

Ingraham, J. L. (1962): Temperature relationships. In: Gunsalus, I. C., and Stanier, R. Y., The bacteria. IV, 265–296. New York and London, Academic Press.

Ingvorsen, K., Zeikus, J. G., and Brock, T. D. (1981): Dynamics of bacterial sulfate reduction in a eutrophic lake. *Appl. Environ. Microbiol.* **42**, 1029–1036.

Iturriaga, R. (1979): Bacterial activity related to sedimenting particulate matter. *Mar. Biol.* **55**, 157–169.

— and Hoppe, H.-G. (1977): Observations of heterotrophic activity on photoassimilated organic matter. *Marine Biol.* **40**, 101–108.

— and Rheinheimer, G. (1972): Untersuchungen über das Vorkommen von phenolabbauenden Mikroorganismen in Gewässern und Sedimenten. *Kieler Meeresforsch.* **28**, 213–218.

— — (1975): Eine einfache Methode zur Auszählung von Bakterien mit aktivem Elektronen-transportsystem in Wasser- und Sedimentproben. *Kieler Meeresforsch.* **31**, 83–86.

Iversen, N., and Blackburn, T. H. (1981): Seasonal rates of methane oxidation in anoxic marine sediments. *Appl. Environ. Microbiol.* **41**, 1295–1300.

Jakob, H. E. (1970): Redox Potential. In: Norris, J. R., and Ribbons, D. W., *Methods in micro-biology II,* 91–123. London and New York, Academic Press.

Jankowski, G. J., and ZoBell, C. E. (1944): Hydrocarbon production by sulfate-reducing bacteria. *J. Bact.* **47**, 447.

Jannasch, H. W. (1955): Zur Ökologie der zymogenen planktischen Bakterienflora natürlicher Gewässer. *Arch. Mikrobiol.* **23**, 146–180.

— (1960): Versuche über die Denitrifikation und die Verfügbarkeit des Sauerstoffs in Wasser und Schlamm. *Arch. Hydrobiol.* **56**, 355–369.

— (1967): Growth of marine bacteria at limiting concentrations of organic carbon in sea water. *Limnol. Oceanogr.* **12**, 264–271.

— (1970): Threshold concentrations of carbon sources limiting bacterial growth in sea water. In: Hood, H. W., Organic matter in natural waters. *Inst. Mar. Sci. U. of Alaska Pub. No.* **1**, 231–328.

— and Wirsen, C. O. (1977): Retrieval of concentrated and undecompressed microbial populations from the deep sea. *Appl. Environm. Microb.* **33**, 664–666.

— Eimhjellen, K., Wirsen, C. O., and Farmanfarmaian, A. (1971): Microbial degradation of organic matter in the Deep Sea. *Science* **171**, 672–675.

— — (1981): Morphological survey of microbial mats near deep-sea thermal vents. *Appl. Environ. Microbiol.* **41**, 528–538.

— — (1982): Microbial activities in undecompressed and decompressed deep-water samples. *Appl. Environ. Microbiol.* **43**, 1116–1124.

Jansson, O. J. (1972): Ecosystem approach to the Baltic problem. *Bull. Ecolog. Res. Comm. Stock-holm* **16**, 82 pp.

Jassby, A. D. (1975): The ecological significance of sinking to planktonic bacteria. *Can. J. Mikro-biol.* **21**, 270–274.

Jawetz, E., Melnick, L. J., and Adelberg, E. A. (1963): *Medizinische Mikrobiologie.* Heidelberg, Springer, 600 pp.

Jegorowa, A. A., Derjugina, S. P., and Kusnezow, S. I. (1952): Die Charakterisierung der saprophytischen Mikroflora des Wassers einer Anzahl von Seen verschiedenen Trophiegrades. *Tr. in — ta. mikrobiologii AN SSSR*, t 2 (after Kusnezow, 1959).

Johannes, R. E. (1968): Nutrient regeneration in lakes and oceans. In: Droop, M. R., and Ferguson Wood, E. J., *Advances in microbiology of the sea* 1, 203–213. London and New York, Academic Press.

Johnson, F. H., Eyring, H., and Polissar, M. J. (1954): *The kinetic basis of molecular biology*. John Wiley and Sons, New York, 874 pp.

Johnson, T. W. (1968): Saprobic marine fungi. In: Ainsworth, C. G., and Sussman, A. S., *The fungi II*. 95–104. London and New York, Academic Press.

— and Sparrow, F. K. (1961): *Fungi in oceans and estuaries*. Weinheim, J. Cramer, 668 pp.

Jones, E. B. G., and Byrne, P. J. (1976): Physiology of the higher marine fungi. In: Jones, E. B. G. (ed.), *Recent Aadvances in Aquatic Mycology*, 135–176. Elek Science, London.

— and Harrison, J. L. (1976): Physiology of marine phycomycetes. In: Jones, E. B. G. (ed.), *Recent Advances in Aquatic Mycology*, 261–277. Elek Science, London.

Jones, J. G. (1980): Some differences in the microbiology of profundal and littoral lake sediments. *J. Gen. Microbiol.*, 117, 285–292.

Jordan, M. J., and Likens, G. E. (1980): Measurement of planktonic bacterial production in an oligotrophic lake. *Limnol. Oceanogr.* 25, 719–732.

Jörgensen, B. B. (1977): The sulfur cycle of a coastal marine sediment (Limfjorden, Denmark). *Limnol. Oceanogr.* 22, 814–832.

— and Revsbech, N. P. (1983): Colorless sulfur bacteria, *Beggiatoa* sp. and *Thiovulum* sp., in O_2 and H_2S micro-gradients. *Appl. Environ. Microbiol.* 45, 1261–1270.

Kadota, H. (1951): Microbiological studies on the weakening of netting cords. 1. On the aerobic cellulose-decomposing bacteria in seawater. *Bull. Jap. Soc. sci. Fish.* 16, 63–70.

— and Miyoshi, H. (1963): Organic factors responsible for the stimulation of growth of *Desulfovibrio desulfuricans*. In: Oppenheimer, C. H., *Symposium on marine microbiology*. Springfield, Ill., C. C. Thomas, 442–452.

Kaestner, A. (from 1955): *Lehrbuch der speziellen Zoologie. Band I : Wirbellose*. Jena, Fischer.

Kandler, O. (1981): Archaebakterien und Phylogenie der Organismen. *Naturwissenschaften* 68, 183–192.

— Baumgartner, H., and Biologh, P. (1978): Zur Bakterienflora des Ammersees. Verh. Ges. Ökol. Kiel 1977, 151–152.

Kang, H., and Seki, H. (1983): The gram stain characteristics of the bacterial community as a function of the dynamics of organic debris in a mesotrophic irrigation pond. *Arch. Hydrobiol.* 98, 39–58.

Kaplan, I. R., and Rittenberg, S. C. (1964): Microbiological fractionation of sulphur isotopes. *J. gen. Microbiol.* 34, 195–212.

Karl, D. M., Wirsen, C. O., and Jannasch, H. W. (1980): Deep sea primary production at the Galapagos hydrothermal vents. *Science* 207, 1345–1347.

Karlson, P. (1974): *Kurzes Lehrbuch der Biochemie für Mediziner und Naturwissenschaftler*, 9. Ed. Stuttgart, Thieme, 412 pp.

Kepkay, P. E., and Novitsky, J. A. (1980): Microbial control of organic carbon in marine sediments: coupled chemoautotrophy and heterotrophy. *Mar. Biol.* 55, 261–266.

Khrutskaya, Z. T. (1970): Clogging of drains with iron compounds. M. "Kolos", Moscow.

Kim, S. J. (1983): Vergleichende Untersuchung über die Bakterienpopulation des Neustons und des darunter liegenden Wassers in einem Meeresgebiet. Dissertation, Universität Kiel.

King, G. M., Klug, M. J., and Lovley, D. R. (1983): Metabolism of acetate, methanol and methylated amines in intertidal sediments of Lowes Cove, Maine. *Appl. Environ. Microbiol.* 45, 1848 to 1853.

Kirchman, D. L., Mazella, L., Alberte, R. S., and Mitchell, R. (1984): Epiphytic bacterial production on *Zostera marina*. *Mar. Ecol. Prog. Ser.* 15, 117–123.

Kjelleberg, S., Humphrey, B. A., and Marshall, K. C. (1982): Effect of interfaces on small, starved marine bacteria. *Appl. Environ. Microbiol.* 43, 1166–1172.

— Stenström, T. A., and Odham, G. (1979): Comparative study of different hydrophobic devices for sampling lipid surface films and adherent micro-organisms. *Mar. Biol.* 53, 21–25.

Klie, H., and Schwartz, W. (1963): Untersuchungen über Lebensweise und Kultur von *Labyrinthula*. *Z. Allg. Mikrobiol.* 3, 15–24.

Knorr, M., and Sonnabend, W. (1964): Beobachtungen über den Keimtransport durch das Phyto-plankton des Bodensees. *Aus Bodensee-Projekt der Deutschen Forschungsgemeinschaft, Erster Bericht, 1. Juli 1963.*

Kohl, W. (1975): Über die Bedeutung bakteriologischer Untersuchungen für die Beurteilung von Fließgewässern, dargestellt am Beispiel der österreichischen Donau. *Arch. Hydrobiol. Suppl.* **44**, 392–461.

Kohlmeyer, J. (1960): Wood-inhabiting marine fungi from the Pacific Northwest and California. *Nova Hedwigia* **2**, 293–343.

— (1969): The role of marine fungi in the penetration of calcareous substances. *American Zoologist* **9**, 741–746.

— (1977): New genera and species of higher fungi from the deep sea (1615–5315 m) (1). *Rev. Mycologie* **41**, 189–206.

— and Kohlmeyer, E. (1964–1968): *Icones fungorum maris. Fasc.* 1–6. Weinheim, J. Cramer.

— — (1979): *Marine mycology – the higher fungi.* Academic Press, New York, London, 690 pp.

Koike, I., and Hattori, A. (1978): Denitrification and ammonia formation in anaerobic coastal sediments. *Appl. Environ. Microbiol.* **35**, 278–282.

Kolkwitz, R., and Marsson, M. (1908): Ökologie der pflanzlichen Saprobien. *Ber. Deutsche Bot. Ges.* **26**a, 505–519.

Koop, K., Carter, R. A., and Newell, R. C. (1982): Mannitol-fermenting bacteria as evidence for export from kelp beds. *Limnol. Oceanogr.* **27**, 950–954.

Koske, P. H., Krumm, H., Rheinheimer, G., and Szekielda, K.-H. (1966): Untersuchungen über die Einwirkung der Tide auf Salzgehalt, Schwebstoffgehalt, Sedimentation und Bakteriengehalt in der Unterelbe. *Kieler Meeresforsch.* **22**, 47–63.

Krassilnikow, N. A. (1959): *Diagnostik der Bakterien und Actinomyceten.* Jena, Fischer, 813 pp.

Krause, R. (1960): Bildung der Oogonien bei *Saprolegnia ferax* (Gruith.) Thuret. *Arch. Mikrobiol.* **36**, 373–386.

Krebs, F. (1983): Toxizitätstest mit gefriergetrockneten Leuchtbakterien. *Gewässerschutz, Wasser, Abwasser,* **63**, 173—230.

Kriss, A. E. (1961): *Meeresmikrobiologie.* Jena, Fischer, 570 S.

— Mishustina, I. E., Mitskevich, N., and Zemtsova, E. V. (1967): *Microbial population of oceans and seas.* London, Edward Arnold Ltd., 287 pp.

Krumbein, W. (1979): Cyanobakterien – Bakterien oder Algen? Universität Oldenburg, 130 pp.

— Cohen, Y., and Shilo, M. (1977): Solar Lake (Sinai) 4. Stromatolitic, cyanobacterial mats. *Limnol. Oceanogr.* **22**, 635–656.

Kühl, H., and Rheinheimer, G. (1968): Veränderungen der Bakterienflora, des Planktons und einiger chemischer Faktoren während einer Tide in der Elbmündung bei Cuxhaven. *Kieler Meeresforsch.* **24**, 27–37.

Kuparinen, J., Leppänen, J. M., Sarvala, J., Sundberg, A., and Virtanen, A. (1983): Production and utilization of organic matter in a Baltic ecosystem off Tvärminne, south west of Finland. *Rapp. P. v. Reun, Cons. int. Explor. Mer.,* 177–189.

Kurbatowa-Belikowa, N. M., (1954): The results of a study of the microbiological activity in natural peat bog deposits. *Trudy in-ta Torfa AN SSSR,* No. 3. Russ. (after Kusnezow et al., 1963).

Kusnezow, S. I. (1959): *Die Rolle der Mikroorganismen im Stoffkreislauf der Seen.* Berlin, Deutscher Verlag d. Wissenschaften, 301 pp.

— (1966): Die Rolle der Mikroorganismen bei der Bildung von Calcitkristallen im Schlamm des Sewan-Sees. *Z. Allg. Mikrobiol.* **6**, 289–295.

— (1970): *Mikroflora der Seen und ihre geochemische Aktivität (russisch).* Isdatelstwo nauka Leningradskoje otdelenie, Leningrad. 440 pp..

— Iwanow, M. V., and Lyalikowa, N. N. (1963): *Introduction to geological microbiology.* New York, McGraw-Hill Company, 252 pp.

— and Romanenko, V. J. (1966): Produktion der Biomasse heterotropher Bakterien und die Ge-schwindigkeit ihrer Vermehrung im Rybinsk-Stausee. *Verh. internat. Verein Limnol.* **16**, 1493–1500.

Ladd, T. I., Ventullo, R. M., Wallis, P. M., and Costerton, J. W. (1982): Heterotrophic activity and biodegradation of labile and refractory compounds by groundwater and stream microbial populations. *Appl. Environ. Microbiol.* **44**, 321–329.

Lampert, W. (1982): Further studies on the inhibitory effect of the toxic blue-green *Microcystis aeruginosa* on the filtering rate of zooplankton. *Arch. Hydrobiol.* **95**, 207–220.

Lange, W. (1970): Cyanophyta-bacteria systems: effects of added carbon compounds or phosphate on algal growth at low nutrient concentrations. *J. Phycol.* **6**, 230–234.

Larsen, H. (1962): Halophilism. In: Gunsalus, I. C., and Stanier, R. Y., *The Bacteria IV*, 297–342. New York and London, Academic Press.

— (1981): The family Halobacteriaceae. In: Starr et al., *The Procaryotes* 985–994. Springer, Berlin, Heidelberg, New York.

Larsson, U., and Hagström, A. (1979): Phytoplankton exudate release as an energy source for the growth of pelagic bacteria. *Mar. Biol.* **52**, 199–206.

Lauckner, G. (1983): Diseases of mollusca: Bivalia. In: O. Kinne, *Diseases of marine animals* II, 477—962. Hamburg, Biol. Anst. Helgol.

Lecianova, L. (1981): Die Anwesenheit von Mycobakterien im Wasserbiotop. In: Daubner, I., *3. Internationales hydromikrobiologisches Symposium* 99–105. VEDA, Bratislava.

Lehnberg, B. (1972): Ökologische Untersuchungen an aeroben agar- und zellulosezersetzenden Bakterien in Nord- und Ostsee. Dissertation Universität Kiel.

Lein, A. Y., Namsaraev, G. B., Trotsyuk, V. Y., and Ivanov, M. V. (1981): Bacterial methanogenesis in holocene sediments of the Baltic Sea. *Geomicrobiol. J.* **2**, 4, 299–315.

Levring, T., Hoppe, H. A., and Schmid, O. J. (1969): *Marine algae a survey of research and utilisation.* Hamburg, Cram, De Cruyter u. Co., 421 pp.

Liebert, F. (1915): Über mikrobiologische Nitrit- und Nitratbildung im Meere. *Rapp. Verh. Rijksinst. Vissch. Onderz.* **1**, 3 (after ZoBell, 1946).

Liebezeit, G., and Dawson, R. (1982): The analysis of natural organic compounds in sea water. *Kontakte* (Merck), **2**, 19—28.

Linley, E. A. S., Newell, R. C., and Bosma, S. A. (1981): Heterotrophic utilization of mucilage released during fragmentation of kelp (*Ecklonia maxima* and *Laminaria pallida*). I. Development of microbial communities associated with the degradation of kelp mucilage. *Mar. Ecol. Prog. Ser.* **4**, 31–41.

Liston, J., and Colwell, R. R. (1963): Host and habital relationships of marine commensal bacteria. In: Oppenheimer, C. H., *Symposium on marine microbiology*. Springfield, Ill., Thomas, 611–624.

Litchfield, C. D., and Hood, D. W. (1965): Microbiological assay for organic compounds in seawater. I. Quantitative assay procedures and biotin distribution. *Appl. Microbiol.* **13**, 886–894.

Lovley, D. R., and Klug, M. J. (1983): Methanogenesis from methanol and methylamines and acetogenesis from hydrogen and carbon dioxide in the sediments of a eutrophic lake. *Appl. Environ. Microbiol.* **45**, 1310–1315.

Lucas, M. I., Newell, R. C., and Velimirov, B. (1981): Heterotrophic utilization of mucilage released during fragmentation of kelp (*Ecklonia maxima* and *Laminaria pallida*) II. Differential utilization of dissolved organic components from kelp mucilage. *Mar. Ecol. Prog. Ser.* **4**, 43–55.

Lucht, F. (1964): Hydrographie des Elbeaestuars. *Arch. Hydrobiol.* 29, Suppl. Elbe-Aestuar **2**, 1–96.

Luck, J. M., Sheets, G., and Thomas, J. O. (1931): The role of bacteria in the nutrition of protozoa. *Quart. Rev. Biol.* **6**, 46–58.

Macdonell, M. T., and Hood, M. A. (1982): Isolation and characterization of ultramicro bacteria from a gulf coast estuary. *Appl. Environ. Microbiol.* **43**, 566–571.

Maciejowska, M. (1968): Wplyw czynnikow fizykochemicznych srodowiska na rozklad blonnika w wodach slanowych. *Prace Morskiego Instytutu Rybackiego* **15**, A, 33–54.

— and Rakowska, E. (1973): Badania nad rozkladem palier cieklych przez mikroorganizmy morskie. *Prace Morskiego Instytutu Rybackiego* **17**, A, 181–203.

MacLeod, R. A. (1965): The question of the existence of specific marine bacteria. *Bact. Rev.* **29**, 9–23.

— (1968): On the role of inorganic ions in the physiology of marine bacteria. In: Droop and Wood *Advances in microbiology of the sea* 1. London and New York, Academic Press, 95–126.

— and Onofrey, E. (1956): Nutrition and metabolism of marine bacteria. II. Observations on the relation of seawater to the growth of marine bacteria. *J. Bact.* **71**, 661–667.

Madri, P. P., Hermel, M., and Claus, G. (1971): The microbial flora of the spônge *Microcionia prolifera* Verrill and its ecological implications. *Bot. mar. (Berlin)* **14**, 1–5.

Maeda, M., and Taga, N. (1980): Alkalitolerant and alkaliphilic bacteria in sea water. *Mar. Ecol. Prog. Ser.* **2**, 105–108.

Mah, R. A., and Smith, M. R. (1981): The methanogenic bacteria. In: Starr et al., *The Procaryotes*, 948–977. Springer, Berlin, Heidelberg, New York.

Mann, H. (1956): Aufarbeitung von Abwasser in Tideflüssen — dargestellt am Beispiel der Elbe. *Arb. dtsch. Fisch. Verb.* **7**, 8–16.

— (1970): Über den Befall der Plattfische der Nordsee mit *Lymphocystis*. *Ber. Deutsche Komm. Meeresforsch.* **21**, 219–223.

Marinucci, A. C., Hobbie, J. E., and Helfrich, J. V. K. (1983): Effect of litter nitrogen on decomposition and microbial biomass in *Spartina alternifolia*. *Microb. Ecol.* **9**, 27–40.

Marshall, K. C. (1976): *Interfaces in microbial ecology*. Harvard University Press, Cambridge, Mass.

Mason, J. (1972): The cultivation of the European mussel, *Mytilus edulis* Linnaeus. *Oceanogr. Mar. Biol. Ann. Rev.* **10**, 437–460.

Masters, M. J. (1976): Freshwater phycomycetes on algae. In: Jones, E. B. G. (ed.), *Recent Advances in Aquatic Mycology*, 489–512. Elek Science, London.

Mattheis, T. (1964): Ökologie der Bakterien im Darm von Süßwassernutzfischen. *Z. Fischerei* **12**, 507–600.

Matulewich, V. A., and Finstein, M. S. (1978): Distribution of autotrophic nitrifying bacteria in a polluted river (the Passaic). *Appl. Environ. Microbiol.* **35**, 67–71.

McElroy, W. (1961): Bacterial luminescence. In: Gunsalus, I. C., and Stanier, R. Y., The bacteria II. New York and London, Academic Press, 479–508.

Meffert, M. E., and Overbeck, J. (1981): Interactions between *Oscillatoria redekei* (Cyanophyta) and bacteria. *Verh. Internat. Verein. Limnol.* **21**, 1432–1435.

Melchiorri-Santolini, U., and Hopton, J. W. (1972): Detritus and its role in aquatic ecosystems. *Mem. 1st. Ital. Idrobiol.* **29**, Suppl., 540 pp.

Meyer-Reil, L. A. (1972): Untersuchungen über die Salzansprüche von Ostseebakterien. Dissertation Universität Kiel.

— (1977): Bacterial growth rates and biomass production. In: Rheinheimer, G., Microbial ecology of a brackish water environment. *Ecological Studies* **25**, 223–243. Berlin, Heidelberg, New York, Springer.

— (1983): Benthic response to sedimentation events during autumn to spring in a shallow water station in the western Kiel Bay: II. Analysis of benthic bacteria populations. *Mar. Biol.* **77**, 247–256.

— Bölter, M., Dawson, R., Liebezeit, G., Szwerinski, H., and Wolter, K. (1980): Interrelationships between microbiological and chemical parameters of sandy beach sediments, a summer aspect. *Appl. Environ. Microbiol.* **39**, 797–802.

— Dawson, R., Liebezeit, G., and Tiedge, H. (1978): Fluctuations and interactions of bacterial activity in sandy beach sediments and overlying waters. *Marine Biol.* **48** (2), 161–172.

— Faubel, A. (1980): Uptake of organic matter by meiofauna organisms and interrelationships with bacteria. *Mar. Ecol.* **3**, 251–256.

Meyers, S. P. (1968): Observations on the physiological ecology of marine fungi: In: Kadota, H., and Taga, N., Proceedings of the U.-S.-Japan Seminar on marine micro-biology. *Bull. Misaki Marine Biolog. Inst. Kyoto University* **12**.

— and Reynolds, E. S. (1959): Marine fungi and wood borer attack, *Science* **130**, 46.

— (1960): Occurrence of lignicolous fungi in northern Atlantic and Pacific marine localities. *Can. J. Bot.* **38**, 217–226.

— Ahearn, D. G., and Roth, F. J. (1967): Mycological investigations of the Black Sea. *Bull. marine Sci.* **17**, 576–596.

Michajlenko, L. E. (1981): Die vieljährige Bakterienplanktondynamik in eutrophen Dnepr-Stauseen und ihre Rolle in der Ernährung von Wassertieren. In: Daubner, I., 3. *Internationales hydromikrobiologisches Symposium*, 139–142. VEDA Bratislava.

Miklosovicova, L. (1981): Die Anwesenheit von Myxobakterien im Wasserbiotop. In: Daubner, I.; 3. *Internationales hydromikrobiologisches Symposium*, 107–113. VEDA Bratislava.

Mitchell, R. (1968): The effect of water movement on lysis of non-marine microorganisms by marine bacteria. *Sarsia* **34**, 236–266.

— (1972): Ecological control of microbial imbalances. In: Mitchell, R., *Water pollution microbiology*. New York, Wiley Interscience, 273–288.

Moebus, K. (1972): Seasonal changes in antibacterial activity of North Sea water. *Mar. Biol.* **13**, 1–13.

Moewus, L. (1963): Studies on a marine parasitic ciliate as a potential virus vector. In: Oppenheimer, C. H., *Symposion on marine microbiology*. Springfield, Ill., Thomas, 366–379.

Moll, G., and Ahrens, R. (1970): Ein neuer Fimbrientyp. *Arch. Mikrobiol.* **70**, 361–368.

Moriarty, D. J. W., and Hayward, A. C. (1982): Ultrastructure of bacteria and the proportion of gram negative bacteria in marine sediments. *Microbiol. Ecol.* **8**, 1–14.

— and Pollard, P. C. (1982): Diel variation of bacterial productivity in seagrass *(Zostera capricorni)* beds measured by rate of thymidine incorporation into DNA. *Mar. Biol.* **72**, 165–173.

Morita, R. Y. (1966): Marine psychrophilic bacteria. *Oceanogr. marin. biol. Ann. Rev.* **4**, 105–121.

— (1967): Effects of hydrostatic pressure on marine microorganisms. *Oceanogr. marin. biol. Ann. Rev.* **5**, 187–203.

Morita, R. Y. (1968): The basic nature of marine psychrophilic bacteria. *Bull. Misaki Marine Biolog. Inst. Kyoto University* **12**, 163–177.

— (1975 a): Psychrophilic bacteria. *Bacteriol. Rev.* **39**, 144–167.

— (1975 b): Microbial contribution to the evolution of the 'steady state' carbon dioxide system. *Origins of Life* **6**, 37–44.

— (1979 a): Deep sea microbial energetics, *Sarsia* **64**, 9–12.

— (1979 b): Current status of the microbiology of the deep sea. Ambio special report **6**, 33–36.

— (1980 a): Low temperature, energy, survival and time in microbial ecology. *Microbiology* 323–324.

— (1980 b): Calcite precipitation by marine bacteria. *Geomicrobiol. J.* **2**, 63–82.

— and Albright, L. J. (1965): Cell yields of *Vibrio marinus*, an obligate psychrophile, at low temperature. *Canad. J. Microbiol.* **11**, 221–227.

— and Haight, R. D. (1964): Temperature effects on the growth of an obligate psychrophilic marine bacterium. *Limnol. Oceanogr.* **9**, 103–106.

— Iturriaga, R., and Gallardo, V. A. (1981): *Thioploca;* methylotroph and significance in the food chain. *Kieler Meeresforsch. Sonderh.* **5**, 384–389.

— and Zobell, C. E. (1956): Effect of hydrostatic pressure on the succinic dehydrogenase system in *Escherichia coli. J. Bacteriol.* **71**, 668–672.

Mucha, V., and Horska, E. (1981): Die limnologische und hygienische Bedeutung von Bakterien der Gattung *Streptomyces.* In: Daubner, I., *3. Internationales hydromikrobiologisches Symposium*, 91–97. VEDA Bratislava.

Müller, W. (1953): Die Einleitung von Abwässern ins Meer. *Rohrleger u. Gesundheits-Ing.* **74**, 286–288.

Müller-Haeckel, A., and Rheinheimer, G. (1983): Studies on the annual cycle of bacteria and fungi in the Ängeran, a coastal stream in northern Sweden. *Aquilo Ser. Zool.* **22**, 51–56.

Müller-Neuglück, M., and Engel, H. (1961): Photoinaktivierung von *Nitrobacter winogradskyi. Buch. Arch. Mikrobiol.* **39**, 130–138.

Murray, C. L., and Lovett, J. S. (1966): Nutritional requirements of the chytrid *Karlingia asteriocysta*, an obligate chitinophile. *Ann. J. Bot.* **53**, 469–476.

Mütze, B. (1963): Der Einfluß des Lichtes auf *Micrococcus denitrificans* Beijerinck. *Arch. Mikrobiol.* **46**, 402–408.

— and Engel, H. (1960): Untersuchungen über bakterielle Schwefeloxydation in der Elbe. *Arch. Mikrobiol.* **35**, 303–309.

Naguib, M. (1982): Methanogenese im Sediment der Binnengewässer. 1. Methanol als dominanter Methan-„Precursor" im Sediment eines eutrophen Sees. *Arch. Hydrobiol.* **95**, 317–329.

Nedwell, D. B., and Floodgate, G. D. (1972): Temperature-induced changes in the formation of sulphide in a marine sediment. *Marine Biology* **14**, 18–24.

Nellen, W. (1967): Ökologie und Fauna (Makroevertebraten) der brackigen und hypotrophen Ostseeförde Schlei. *Arch. Hydrobiol.* **63**, 273–309.

Neofitowa, V. K. (1953): The Mycoflora of the upper, undrained peat layers and its role in the process of peat formation. *Vestn. Leningr. Ges. Un-ta*, No. **10** (russ.) (after Kusnezow et al., 1963).

Newell, R. C., and Field, J. G. (1983): Relative flux of carbon and nitrogen in a kelp-dominated system. *Mar. Biol. Letters* **4**, 249–257.

— Lucas, M. I., and Linley, E. A. S. (1981): Rate of degradation and efficiency of conversion of phytoplankton debris by marine micro-organisms. *Mar. Ecol. Prog. Ser.* **6**, 123–136.

Newell, S. Y. (1976): Mangrove fungi: The succession in the mycoflora of red mangrove *(Rhizophora mangle* L.) seedlings. In: Jones, E. B. G. (ed.), *Recent Advances in Aquatic Mycology*, 51–91. Elek Science, London.

Niewolak, S. (1965): Badanie intensywnosci niektorych przemian azotu w jeziorach ilawskich w latach 1962–1963. *Zeszyty Naukowe Wyszszej Szkoly Rolniczej w Olsztynie* **20**, No. 425, 69–84.

— (1971 a): A microbiological study on the hyponeuston of the Ilawa lakes in the summer season. *Acta Hydrobiol.* **13**, 295–311.

— (1971b): The microbiological decomposition of tribasic calcium phosphate in the Ilawa Lakes. *Acta Hydrobiol.* **13**, 131–145.

— (1972): Fixation of atmospheric nitrogen by *Azotobacter* sp. and other heterotrophic oligonitrophilous bacteria in the Ilawa Lakes. *Acta Hydrobiol.* **14**, 287–305.

— (1973): Oligonitrophilic strains of heterotrophic bacteria in the lakes of Ilawa. *Roczniki Nauk Rolniczych* **95**, 55–65.

— (1974a): Production of bacterial biomass in the water of Ilawa Lakes. *Acta Hydrobiol.* **16**, 101–112.

— (1974b): Vertical distribution of the bacterioplankton and the thermal-oxygen relations in the water of the Ilawa Lakes. *Acta Hydrobiol.* **16**, 173–187.

— (1977): The occurrence of yeasts in some of the Masurian Lakes. *Acta Mycol.*, **12**, 241—256.

— and Korycka, A. (1972): Wiazanie azotu atmosferycznego przez drobnoustroje w wodzie jezior ilawskich. *Roczniki Nauk Rolniczych* **94**, 107–126.

— and Sobierajska, M. (1971): The participation of some bacteria in the synthesis of vitamin B_{12} in the water of the Ilawa lakes. *Acta Hydrobiol.* **13**, 147–158.

Nishio, T., Koike, I., and Hattori, A. (1983): Estimates of denitrification and nitrification in coastal and estuarine sediments. *Appl. Environ. Microbiol.* **45**, 444–450.

Norkrans, B. (1967): Cellulose and Cellulolysis. *Adv. appl. Microbiol.* **9**, 91–130.

Novitsky, J. A. (1983): Heterotrophic activity throughout a vertical profile of seawater and sediment in Halifax Harbor, Canada. *Appl. Environ. Microbiol.* **45**, 1753–1760.

Odum, E. P. (1980): *Grundlagen der Ökologie.* Thieme, Stuttgart, 836 pp.

Odum, H. T. (1957): Trophic structure and productivity of Silver Springs, Florida. *Ecol. Monogr.* **27**, 55–112.

— (1969): An energy circuit language for ecological and social systems: its physical basis. *Progr. Rep. U.S. Atom Energy Com.* **1** (40–1)–3666, 3–90.

Ogura, N. (1970): The relation between dissolved organic carbon and apparent oxygen utilization in the western North Pacific. *Deep-Sea Res.* **17**, 221–231.

Ohle, W. (1958): Die Stoffwechseldynamik der Seen in Abhängigkeit von der Gasausscheidung ihres Schlammes. *Vom Wasser* **25**, 127–149.

— (1965): Nährstoffanreicherung der Gewässer durch Düngemittel und Meliorationen. *Münchner Beiträge* **12**, 54–83.

Ohwada, K., Tabor, P. S., and Colwell, R. R. (1980): Species composition and barotolerance of gut microflora of deep sea benthic macrofauna collected at various depths in the Atantic Ocean. *App. Environ. Microbiol.* **40**, 746–755.

Olah, J. (1969a): The quantity, vertical and horizontal distribution of the total bacterioplankton of Lake Balaton in 1966/67. *Anal. Biol. Tihany* **36**, 185–195.

— (1969b): A quantitative study of the saprophytic and total bacterioplankton in the open water and the littoral zone of Lake Balaton in 1968. *Anal. Biol. Tihany* **26**, 197–212.

— (1972): Leaching, colonization and stabilization during detritus formation. *Mem. Ist. Ital. Idrobiol.* **29**, Suppl. 105–127.

Olrik, K. (1976): Giftige alger. *Vand* **4**, 1–7.

Olson, R. J. (1981): Differential photoinhibition of marine nitrifying bacteria: a possible mechanism for the formation of the primary nitrite maximum. *J. Mar. Res.* **39**, 227–238.

Oppenheimer, C. H., Gunkel, W., and Gassmann, G. (1977): Microorganisms and hydrocarbons in the North Sea during July–August 1975. *Proceed. 1977 Oil Spill Confr. New Orleans, USA.* 593–609.

Oremland, R. S., and Polcin, S. (1982): Methanogenesis and sulfate reduction: Competitive and non-competitive substrates in estuarine sediments. *Appl. Environ. Microbiol.* **44**, 1270–1276.

Oren, A. (1983): Population dynamics of halobacteria in the Dead Sea water column. *Limnol. Oceanogr.* **28**, 1094–1103.

Ostertag, H. (1950): Abbau der Cellulose im städtischen Abwasser und im Vorfluter. *Städtehygiene* **1**, 206–210.

Ottow, J. C. G. (1983): Ökologische Folgen der Manganknollengewinnung. *Naturwiss. Rundschau* **36**, 48–59.

Overbeck, J. (1968): Prinzipielles zum Vorkommen der Bakterien im See. *Mitt. Int. Ver. Limnol.* **14**, 134–144.

— (1972): Experimentelle Untersuchungen zur Bestimmung der bakteriellen Produktion im See. *Verh. internat. Verein Limnol.* **18**, 176–187.

— (1973): Über die Kompartimentierung der stehenden Gewässer — ein Beitrag zur Struktur und Funktion des limnischen Ökosystems. *Verh. Ges. Ökol.*, 211–223.

— (1974): Microbiology and biochemistry. *Mitt. Internat. Ver. Limnol.* **20**, 198–228.

— (1979a): Dark CO_2 uptake, biochemical background and its relevance to *in situ* bacterial production. *Arch. Hydrobiol. Beih. Ergebn. Limnol.* **12**, 38–47.

— (1979b): Studies on heterotrophic functions and glucose metabolism of microplankton in Lake Pluss. *Arch. Hydrobiol. Beih. Ergebn. Limnol.* **13**, 56–76.

Overbeck, J. (1981): A new approach for estimating the overall heterotrophic activity in aquatic ecosystems. *Verh. internat. Verein. Limnol.* **21**, 1355–1358.

— (1982): Bakterien und Kohlenstoffumsatz im Plusssee; Probleme und Möglichkeiten der Mikrobenökologie. *Forum Mikrobiol.* **5**, 292–303.

— and Babenzien, H. D. (1964a): Bakterien und Phytoplankton eines Kleingewässers im Jahreszyklus. *Z. allg. Mikrobiol.* **4**, 59–76.

— — (1964b): Über den Nachweis von freien Enzymen im Gewässer. *Arch. Hydrobiol.* **60**, 107–114.

— Höfle, M. G., Krambeck, C., and Witzel, K.-P. (eds.) (1984): Proceedings of the 2 Workshop on measurement of microbial activity in the carbon cycle of aquatic ecosystems. *Arch. Hydrobiol. Beih. Ergebn. Limnol.*, **19**, 279 pp.

Padan, E., and Shilo, M. (1969): Distribution of cyanophages in natural habitats. *Verh. Int. Verein. Limnol.* **17**, 747–751.

— — (1973): Cyanophages-viruses attacking Blue-Green Algae. *Bacteriological Reviews* **37**, 343–370.

Paluch, J., and Szulicka, J. (1967): Mikrobiologiczny rozklad zwiazkow fosforowych w wodach powierzchniowych. *XVI Zjazd Pol. Tow. Mikrobiol., Streszcz. nadesl. doniesień, Lublin*, 352.

Pankow, H. (1976): *Algenflora der Ostsee II Plankton*. Jena, Fischer, 493 pp.

Parkin, T. B., and Brock, T. D. (1981): The role of phototrophic bacteria in the sulfur cycle of a meromictic lake. *Limnol. Oceanogr.* **26**, 880–890.

Parsons, T. R., and Strickland, J. D. H. (1962): On the production of particulate organic carbon by heterotrophic processes in the sea. *Deep Sea Res.* **8**, 211–222.

Pawlaczyk, M. (1965): Effect of glucose and urea on the rate of phenol degradation by *Pseudomonas fluorescens*. *Acta Mikrobiologica Polonica* **14**, 207–214.

— and Solski, A. (1967): Distribution of nitrogen bacteria in water of the Olawa river in relation to its chemical character. *Polskie Arch. Hydrobiol.* **14**, 17–38.

Perfiljew, B. V. (1926): New data on the role of microbes in ore formation. *Izvestiya Geol. Komiteta 45*, No. 7 (russ.) (after Kusnezow et al., 1963).

— and Gabe, D. R. (1969): *Capillary methods of investigating microorganisms*. Edinburgh, Oliver and Boyd, 627 pp.

Perry, J. J. (1979): Microbial co-oxidations involving hydrocarbons. *Microbiol. Rev.* **43**, 59–72.

Petterson, H., Höglund, H., and Landberg, S. (1934): Submarine daylight and the photosynthesis of phytoplankton. *Medd. Göteborg Högskolas Oceanogr. Inst.* **9**.

Pike, E. B., and Curds, C. R. (1971): The microbial ecology of the activated sludge process. In: Sykes, G., and Skinner, F. A., *Microbial aspects of pollution*. London and New York, Academic Press, 123–137.

Pope, D. H., and Berger, L. R. (1973): Inhibition of metabolism by hydrostatic pressure: What limits microbial growth? *Arch. Mikrobiol.* **93**, 367–370.

Precht, H., Christophersen, J., Hensel, H., and Larcher, W. (1973): *Temperature and life*. Heidelberg, Springer.

Prieur, D. (1981): Experimental studies of trophic relationships between marine bacteria and bivalve molluscs. *Kieler Meeresforsch. Sonderh.* **5**, 376–383.

— (1982): Étude expérimental de l'installation d'une microflore associée au tractus digestif de la moule, *Mytilus edulis* (L.) *Publ. CNEXO (Actes Colloq.)* **13**, 97–104.

Primavesi, C. A. (1970): Der Einfluß virusbelasteten Abwassers auf den Vorfluter und die Frage der Viruseliminierung. *Gewässerschutz, Wasser, Abwasser* **3**, 1–8.

Pschenin, L. N. (1963): Distribution and ecology of Azotobacter in the Black Sea. In: Oppenheimer, C. H., *Symposium on marine Microbiology*. Springfield, Ill., Thomas, 383–391.

Rasumov, A. S. (1932): Eine direkte Methode der Zählung von Wasserbakterien. Ihr Vergleich mit der Kochschen Methode. *Mikrobiologija* **1**, 2 (after Kusnezow, 1959).

Reed, G. B., and MacLeod, D. J. (1924): A bacteriological and chemical study of certain problems in lobster canning. *Contr. Can. Biol. Fish. N. s.* **2**, 1–30·

Reichardt, W. (1978): *Einführung in die Methoden der Gewässermikrobiologie.* Stuttgart, Fischer, 250 pp.

Reichenbach-Klinke, H. H. (1980): *Krankheiten und Schädigungen der Fische.* 2. Aufl. Fischer, Stuttgart, 472 pp.

Renn, C. E. (1936): The wasting disease of *Zostera marina.* Biol. Bull. **70**, 148–158.

Rheinheimer, G. (1960): Untersuchungen über den mikrobiellen Celluloseabbau in der Elbe. *Arch. Mikrobiol.* **36**, 124—130.

— (1964): Beobachtungen über den Einfluß des strengen Winters 1962/63 auf das Bakterienleben eines Flusses. *Kieler Meeresforsch.* **20**, 218–226.

— (1965a): Mikrobiologische Untersuchungen in der Elbe zwischen Schnackenburg und Cuxhaven. *Arch. Hydrobiol.* **29**, Suppl. Elbe-Aestuar **2**, 181–251.

— (1965b): Beobachtungen über die Bakterienverteilung im Elbe-Aestuar. *Botanica Gothoburgensia* **3**, 185–193.

— (1965c): Der Jahresgang der Nitrit- und Nitratbakterienzahl im Wasser der Elbe bei Bleckede. *Kieler Meeresforsch.* **21**, 122–123.

— (1966): Einige Beobachtungen über den Einfluß von Ostseewasser auf limnische Bakterienpopulationen. *Veröff. Inst. Meeresforsch. Bremerhaven. Sonderbd.* **2**, 237–244.

— (1967): Ökologische Untersuchungen zur Nitrifikation in Nord- und Ostsee. *Helgoländer wiss. Meeresunters.* **15**, 243–252.

— (1968a): Beobachtungen über den Einfluß von Salzgehaltsschwankungen auf die Bakterienflora der westlichen Ostsee. *Sarsia* **34**, 253–262.

— (1968b): Ergebnisse und Probleme einer mikrobiologischen Aestuaruntersuchung. *Mitt. Internat. Verein. Limnol.* **14**, 155–163.

— (1970a): Mikrobiologische und chemische Untersuchungen in der Flensburger Förde. *Ber. Dt. Wiss. Kommission Meeresforsch.* **21**, 420–429.

— (1970b): Turbidity — bacteria, fungi and blue-green algae. In: Kinne, O., *Marine ecology* I, **2**, 1167–1175.

— (1971): Über das Vorkommen von Brackwasserbakterien in der Ostsee. *Vie et Milieu,* Suppl. **22**, 281—291.

— (1977a): Mikrobiologische Untersuchungen in Flüssen. I. Fluoreszenzmikroskopische Analyse der Bakterienflora einiger norddeutscher Flüsse. *Arch. Hydrobiol.* **81**, 106–118.

— (1977b): Mikrobiologische Untersuchungen in Flüssen. II. Die Bakterienbiomasse in einigen norddeutschen Flüssen. *Arch. Hydrobiol.* **81**, 259–267.

— (Edit.) (1977c): Microbial ecology of a brackish water environment. *Ecological Studies* 25. Heidelberg, Berlin, New York, Springer, 291 pp.

— (1981): Investigations on the role of bacteria in the food web of the western Baltic. *Kieler Meeresforsch. Sonderh.* **5**, 284–290.

— and Gunkel, W. (1974): Bakteriologische Untersuchungen im Persischen Golf. *"Meteor"-Forsch.-Ergebnisse D* **17**, 1–16.

— and Schmaljohann, R. (1983): Investigations on the influence of coastal upwelling and polluted rivers on the microflora of the north eastern Atlantic off Portugal. I. Size and composition of the bacterial population. *Bot. Mar.* **26**, 137–152.

Rhode, W., Hobbie, J. E., and Wright, T. R. (1966): Phototrophy and heterotrophy in high mountain lakes. *Verh. Intern. Ver. Limnol.* **16**, 302–313.

Rieper, M. (1976): Investigation on the relationships between algal blooms and bacterial populations in the Schlei Fjord (western Baltic Sea). *Helgol. Wiss. Meeresunters.* **28**, 1–18.

— (1978): Bacteria as food for marine harpacticoid copepods. *Marine Biol.* **45**, 337–345.

Rinne, I., Melvasalo, T., Niemi, A., and Niemestö, L. (1978): Nitrogen fixation by blue-green algae in the Baltic Sea. *Kieler Meeresforsch, Sonderh.* **4**, 178–187.

Rippel-Baldes, A. (1955): *Grundriß der Mikrobiologie.* Heidelberg, Springer. 404 pp.

Rittenberg, S. C. (1963): Marine bacteriology and the problem of mineralisation. In: Oppenheimer, C. H., *Symposion on marine microbiology.* Springfield, Ill., Thomas, 48–60.

Ritter, R. (1974): Zur Methodik der hydrobakteriologischen Untersuchung II. *Gewässerschutz, Wasser, Abwasser* **115**, 448–455.

— (1977): Ökologische und physiologische Bakteriengruppen: Ergebnisse hydrobakteriologischer Untersuchungen. In: Daubner, I., *2. Internationales hydromikrobiologisches Symposium.* 113 bis 136. Bratislava, VEDA.

Rodina, A. G. (1972): *Methods in aquatic microbiology.* London, Butterworths, 461 pp.

Romanenko, V. I. (1959): Anteil der methanoxydierenden Bakterien im Wasser, ermittelt mit der Methode der Radioautographie der Kolonien auf Membranfiltern. *Bjull.* Inst. Biol. *Vodochranilišč* **R**, 40–42 (after Fuhs, 1961).

Rönner, V. (1983): Biological nitrogen transformations in marine ecosystems with emphasis on denitrification. Dissertation, University Goteborg.

Roswall, T. (1972): Modern methods in the study of microbial ecology. *Bull. Ecolog. Res. Comm. Stockholm* **17**, 511 pp.

Roth, F. J., Ahearn, D. G., Fell, J. W., Meyers, S. P., and Meyer, S. A. (1962): Ecology and taxonomy of yeasts isolated from various marine substrates. *Limnol. Oceanogr.* **7**, 178–185.

Round, F. E. (1968): *Biologie der Algen*. Stuttgart, Thieme, 315 pp.

Ruby, E. G., Wirsen, C. O., and Jannasch, H. W. (1981): Chemolithotrophic sulfur-oxidizing bacteria from the Galapagos Rift hydrothermal vents. *Appl. Environ. Microbiol.* **42**, 317–324.

Rüger, H. J. (1975): Bakteriensporen in marinen Sedimenten (Nordatlantik, Skagerrak, Biskaya und Auftriebsgebiet von Nordwestafrika); quantitative Untersuchungen. *Veröff. Inst. Meeresforsch. Bremerh.* **15**, 227–236.

— (1982): Psychrophilic sediment bacteria in the upwelling area off N.W. Africa. *Naturwissensch.* **69**, 448–449.

Ruschke, R., and Rath, M. (1966): *Sporocytophaga cauliformis* Knorr und Gräf, eine Mycobakterien-Art mit großer Bedeutung für den Abbau organischen Materials. *Arch. Hydrobiol./ Suppl.* **28**, 377–402.

Ruttner, F. (1962): *Grundriß der Limnologie*. 3. Ed. Berlin, W. de Gruyter.

Saffermann, R. S. (1973): Phycoviruses. In: Carr, N. G., and Whitton, B. A., The biology of blue-green algae. *Botanical Monogr.* **9**, 214–237. Oxford, Blackwell.

Saltzmann, H. A. (1982): Biodegradation of aromatic hydrocarbons in marine sediments of three North Sea oil fields. *Mar. Biol.* **72**, 17–26.

Schäperclaus, W. (1979): *Fischkrankheiten*, 4. Ed. Berlin, Akademie-Verlag, 1025 pp.

Schaumann, K. (1969): Über marine höhere Pilze von Holzsubstraten der Nordseeinsel Helgoland. *Ber. Dtsch. Bot. Ges.* **82**, 307–327.

— (1975): Ökologische Untersuchungen über höhere Pilze im Meer- und Brackwasser der Deutschen Bucht unter besonderer Berücksichtigung der holzbesiedelnden Arten. *Veröff. Inst. Meeresforsch. Bremerh.* **15**, 79–182.

Scheuring, L., and Höhnl, G. (1956): Sphaerotilus natans, seine Ökologie und Physiologie. *Schriften Ver. Zellstoff- und Papier-Chemiker und Ingenieure* **26**, 1–152.

— and Zehender (1962): Untersuchungen zur Stoffwechselphysiologie des "Abwasserpilzes" *Fusarium aquaeductuum* Lagh. Verwertung von Kohlenhydraten. *Schweiz. Z. Hydrologie* **24**, 152–171.

Schlegel, H. G. (1981): *Allgemeine Mikrobiologie*, 5. Ed. Stuttgart, Thieme, 559 pp.

Schlieper, C. (ed.) (1968): *Methoden der meeresbiologischen Forschung*. Jena, Fischer, 322 pp.

Schmaljohann, R. (1984): Morphological investigations on bacterioplankton of the Baltic Sea, Kattegat and Skagerrak. *Bot. Mar.* **27**, 425–436.

Schmidt, G. W. (1970): Numbers of bacteria and algae and their interrelations in some Amazonian waters. *Amazonia II* **4**, 393–400.

Schmidt, W. D. (1976): Zur Biologie der Eisenbakterien — *Siderocapsa* — *Leptothrix echinata* — des Plusssees. Diplomarbeit, Universität Kiel.

— (1979): Morphologie und Physiologie manganoxydierender Mikroorganismen — Kultur und *in situ* Untersuchungen zur ökologischen Charakterisierung von *Metallogenium* sp. und *Siderocapsa geminata* im Plusssee. Dissertation, Universität Kiel.

Schmoller, H. (1960): Kultur und Entwicklung von *Labyrinthula coenocystis* n. sp. *Arch. Mikrobiol.* **36**, 365–372.

Schneider, J. (1969): Über Niedere Pilze der westlichen Ostsee. *Ber. Deutsch. Bot. Ges.* **81**, 369–374.

— (1976): Lignicole marine Pilze (Ascomyceten und Deuteromyceten) aus zwei Ostseeförden. *Bot. Mar.* **19**, 295–307.

— (1979): Stromatolithische Milieus in Salinen der Nord Adria. In: Krumbein, *Cyanobakterien — Bakterien oder Algen*? Schrift der Universität Oldenburg, 93–106.

Schöberl, P. (1967): Stoffwechselphysiologische Untersuchungen an einem kohlenwasserstoff-oxydierenden *Pseudomonas*-Stamm aus der Elbe. *Arch. Mikrobiol.* **56**, 354–370.

— and Engel, H. (1964): Das Verhalten der nitrifizierenden Bakterien gegenüber gelöstem Sauerstoff. *Arch. Mikrobiol.* **48**, 393–400.

Scholes, R. B., and Shewan, J. M. (1964): The present status of some aspects of marine microbiology. *Adv. mar. Biol.* **2**, 133–169.

Scholz, E. (1958): Über niedere Phycomyceten aus Salzböden und ihr Verhalten in Salzlösungen.
Arch. Mikrobiol. **30**, 119–146.

Schön, G. (1964): Untersuchungen über den Nutzeffekt von *Nitrobacter winogradskyi* Buch. Dissertation Universität Hamburg.

Schulz, E. (1937): Das Farbstreifen-Sandwatt und seine Fauna, eine ökologisch-biozönotische
Untersuchung an der Nordsee. *Kieler Meeresforsch.* **1**, 359–378.

Schulz, H. (1961): Qualitative und quantitative Planktonuntersuchungen im Elbe-Aestuar. *Arch.
Hydrobiol.* 26, Suppl. Elbe-Aestuar **1**, 5–105.

Schwartz, W. (1973): Verwendung von Mikroorganismen zu Leaching-Prozessen. *Metall* **27**,
1202–1206.

— (1978): Hydrometallurgie und Biotechnik. *Metall* **32**, 337 339.

— and Näveke, R. (1980): Biotechnische Laugung armer Erze mit heterotrophen Mikroorganismen. *Metall.* **34**, 847–850.

— and Schwartz, A. (1961): *Grundriß der allgemeinen Mikrobiologie II. Slg. Göschen 1157.* Berlin,
De Gruyter, 142 pp.

Schwarz, J. R., Yayanos, A. A., and Colwell, R. R. (1976): Metabolic activities of the intestinal
microflora of a deep-sea invertebrate. *Appl. Environ. Microbiol.* **31**, 46–48.

Schweisfurth, R. (1969): Manganoxidierende Pilze. *Zentralbl. Bakteriol. Parasitenk., Infektions-
krankh. Hygiene I*, **212**, 486–491.

Seki, H. (1966): Role of bacteria as food for plankton. *Bull. Planktol. Japan* **13**, 54–62.

— (1982): Organic materials in aquatic systems. CRC Press, Boca Raton, Florida, 201 pp.

— Nakai, T., and Otobe, H. (1972): Regional differences on turnover rate of dissolved materials
in the Pacific Ocean at summer 1971. *Arch. Hydrobiol.* **71**, 79–89.

— and Taga, N. (1963): Microbiological studies on the decomposition of chitin in marine environment. I. Occurrence of chitinoclastic bacteria in the neritic region. *J. Oceanograph. Soc. Japan*
19, 101–108.

— and Taga, N. (1965): Microbiological studies on the decomposition of chitin in marine
environment. VIII. Distribution of chitinoclastic bacteria in the pelagic and neritic waters.
J. Oceanograph. Soc. Japan **21**, 174–187.

— Yamaguchi, Y., and Ichimura, S. (1975): Turnover rate of dissolved organic materials in a
coastal region of Japan at summer stagnation period of 1974. *Arch. Hydrobi l.* **75**, 297–305.

Selenka, F., and Oppelt, O. (1965): Über Fundorte und Verdriftung von Salmonellen durch
Plankton und Watten im Bodensee. *Arch. Hyg. Bakt.* **149**, 288–303.

Sguros, P. L., Rodrigues, J., and Simms, J. (1973): Role of marine fungi in the biochemistry of
the oceans. V. Patterns of constitutive nutritional growth responses. *Mycologia* **65**, 161–174.

Sherr, B. F., Sherr, E. B., and Berman, T. (1983): Grazing, growth, and ammonium excretion
rates of a heterotrophic microflagellate fed with four species of bacteria. *Appl. Environ.
Microbiol.* **45**, 1196–1201.

Shiba, T., and Taga, N. (1980): Heterotrophic bacteria attached to seaweeds. *J. Exp. Mar. Biol.
Ecol.* **47**, 251–258.

— (1981): Effects of the extracellular products of *Enteromorpha linza* on its epiphytic bacterial
flora. *Bull. Japan Soc. Sci. Fish.* **47**, 1193–1197.

Sieburth, J. McN. (1965): Organic aggregation in seawater by alkaline precipitation of inorganic
nuclei during the formation of ammonia by bacteria. *J. gen. Microbiol.* **41**.

— (1968a): Observations on bacteria planctonic in Narragansett Bay. Rhode Island; a résumé.
Bull. Misaki Marine Biological Inst. Kyoto University **12**, 49–64.

— (1968b): The influence of algae antibiosis on the ecology of marine microorganisms. In: Droop,
M. R., and Ferguson Wood, E. J., *Advances in microbiology of the sea* **1**, 63–94. London and New
York, Academic Press.

— (1971): Distribution and activity of oceanic bacteria. *Deep-Sea Res.* **18**, 1111–1121.

— (1979): *Sea microbes.* New York, Oxford University Press, pp. 491.

Skuja, H. (1956): Taxonomische und biologische Studien über das Phytoplankton schwedischer
Binnengewässer. *Nova Acta Reg. Soc. Sci. Uppsala Ser. IV*, **16**, No. 3, 1–404.

Smith, R. L., and Oremland, R. S. (1983): Anaerobic oxalate degradation: widespread natural
occurrence in aquatic sediments. *Appl. Environ. Microbiol.* **46**, 106–113.

Sommerstorff, H. (1911): Ein Tiere-fangender Pilz (*Zoophagus insidians*, nov. gen. nov. sp.).
Österr. bot. Z. **61**, 361–373.

Sörensen, J. (1978a): Denitrification rates in a marine sediment as measured by the acetylene inhibition technique. *Appl. Environ. Microbiol.* **36**, 139–143.

— (1978b): Capacity for denitrification and reduction of nitrate to ammonia in a coastal marine sediment. *Appl. Environ. Microbiol.* **35**, 301–305.

— (1982): Reduction of ferric ion in anaerobic, marine sediment and interaction with reduction of nitrate and sulfate. *Appl. Environ. Microbiol.* **43**, 319–324.

Sorokin, J. I. (1964): A quantitative study of the microflora in the Central Pacific Ocean. *J. Cons. Charlottenlund* **29**, 25–40.

— (1965): On the trophic role of chemosynthesis and bacterial biosynthesis in water bodies. *Mem. 1st. Ital. Idrobiol.* **18**, Suppl., 187–205.

Sorokin, J. I. (1973): On the feeding of some scleractinian coralls with bacteria and dissolved organic matter. *Limnol. Oceanogr.* **18**, 380–385.

Sorokina, W. A. (1938): Die Bildung eines Bakterienhäutchens an der Oberfläche von Seeschlamm und ihr Einfluß auf den Stoffkreislauf zwischen Schlamm und Wasser. *Mikrobiologija* **7**, 7 (russ.) (after Kusnezow, 1959).

Spangler, W. J., Spigarelli, J. L., Rose, J. M., Flippin, R. S., and Miller, H. H. (1973): Degradation of methylmercury by bacteria isolated from environmental samples. *Appl. Microbiol.* **25**, 488–493.

Sparrow, F. K. (1960): *Aquatic Phyceomycetes*, 2. Ed. Ann. Arbor, The University of Michigan Press, 1187 pp.

— (1968): Ecology of fresh water fungi. In *The fungi* **3**, ed. by G. C. Ainsworth and A. S. Sussmann, 41–93. New York and London, Academic Press.

Spencer, R. (1955): The taxonomy of certain luminous bacteria. *J. gen. Microbiol.* **13**, 111–118.

— (1963): Bacterial viruses in the sea. In: Oppenheimer, C. H., *Symposium on marine microbiology.* Springfield, Ill., Thomas. 350–365.

Staley, J. T., Lehmicke, L. G., Palmer, F. E., Peet, R. W., and Wissmar, R. C. (1982): Impact of Mount St. Helen's eruption on bacteriology of lakes in the blast zone. *Appl. Environ. Microbiol.* **43**, 664–670.

Starkey, R. L. (1945): Transformation of iron by bacteria in water. *J. Americ. Water Works Ass.* **37**, 963–984.

Starr, M. P., Stolp, H., Trüper, H. G., Balows, A., and Schlegel, H. G. (1981): *The Procaryotes.* A handbook of habitats, isolation and identification of bacteria. Springer, Berlin, Heidelberg, New York, 2284 pp.

Steinberg, C. (1977): Vergleich der gelösten organischen Stoffe verschiedener Holsteinischer Seen. *Arch. Hydrobiol.* **80**, 297–307.

Steinmann, J. (1974): Ökologische Untersuchungen zum bakteriellen Abbau von Harnstoff und Harnsäure in Gewässern. Dissertation Universität Kiel.

— (1976): Untersuchungen über den bakteriellen Abbau von Harnstoff und Harnsäure in der westlichen Ostsee. *Bot. Mar.* **19**, 47–58.

Stewart, W. D. P., Fitzgerald, G. P., and Burris, R. H. (1967): In situ studies on N_2 fixation using the acetylene reduction technique. *Proc. Nat. Acad. Sci. USA* **58**, 2071–2078.

Stokes, J. L. (1962): In: *Recent progress in microbiology. Symposia, VIII Intern. Congr. Microbiol.*, Toronto, Canada, Univ. Toronto Press, 187–192.

Stolp, H., and Starr, M. P. (1963): *Bdellovibrio bacteriovorus* gen. et. sp. n., predatory, ecoparasitic, and bacteriolytic microorganism. *A. v. Leeuwenhoek* **29**, 217–248.

Straskrabova, V., and Desortova, B. (1981): Bakterielle Ausnutzung der extrazellulären Algenprodukte. In: Daubner, I., *3. Internationales hydromikrobiologisches Symposium*, 423–434. VEDA, Bratislava.

— and Legner, M. (1966): Interrelations between bacteria and protozoa during glucose oxidation in water. *Int. Revue ges. Hydrobiol.* **51**, 279–293.

Sturm, L. D. (1948): Die Sulfatreduktion durch fakultativ aerobe Bakterien (russ.). *Mikrobiologija* **17**, 6 (after Kusnezow, 1959).

Stüven, K. (1960): Beiträge zur Physiologie und Systematik sulfatreduzierender Bakterien. *Arch. Mikrobiol.* **35**, 152–180.

Suberkropp, K., Arsuffi, T. L., and Anderson, J. P. (1983): Comparison of degradative ability, enzymatic activity and palatability of aquatic hyphomycetes grown on leaf litter. *Appl. Environ. Microbiol.* **46**, 237–244.

Suckow, R. (1966): Schwefelmikrobengesellschaften der See- und Boddengewässer von Hiddensee. *Z. allg. Mikrobiol.* **6**, 309–315.

Suehiro, S. (1963): Studies on the marine yeasts. III. Yeasts isolated from the mud of tide land. *Sci. Bull. Agric. Fac. Kyuschu Univ.* **12**, 223–227.

Suzuki, S. (1960a): Distribution of aquatic Phycomycetes in a river polluted by municipal wastes *J. Waterworks Sewerage Assoc. (Tokyo)* **35**, 51–54 (japan.).

— (1960b): Microbiological studies on the lakes of Volcano Bandai. I. Ecological studies on aquatic Phycomycetes in the Goshikinuma Lake group. *Japan. J. Ecol.* **10**, 215–218.

— (1961): Distribution of aquatic Phycomycetes in some inorganic acidotrophic lakes of Japan. *Botan. Mag. (Tokyo)* **74**, 317–320.

Szwerinski, H. (1981): Investigations on nitrification in the water and the sediment of the Kiel Bay (Baltic Sea). *Kieler Meeresforsch. Sonderh.* **5**, 396–407.

Taylor, C. B. (1942): Bacteriology of freshwater, III. The types of bacteria present in lakes and streams and their relationship to the bacterial flora of soil. *Hygiene* **42**, 284–296.

— and Lochhead, A. G. (1938): Qualitative studies of soil micro-organisms. II. A survey of the bacterial flora of soils differing in fertility. *Can. J. Res., Sec. C* **16**, 162–173.

Testrake, D. (1959): Estuarine distribution and saline tolerance of some Saprolegniaceae. Master Thesis, Duke University, Durham, N. Carolina.

Thomas, J., and David, K. A. V. (1972): Site of nitrogenase activity in the blue-green alga *Anabaena* spec. *Nature* **238**, 219–221.

Thomsen, P. (1910): Über das Vorkommen von Nitrobakterien im Meere. *Wiss. Meeresunters. (Abt. Kiel)* **11**, 1–27.

Torella, F., and Morita, R. Y. (1979): Evidence by electron micrographs for a high incidence of bacteriophage particles in the waters of Yaquina Bay, Oregon: ecological and taxonomical implications. *Appl. Environ. Microbiol.* **37**, 774–778.

Trüper, H. G., and Genovese, S. (1968): Characterization of photosynthetic sulfur bacteria causing red water in Lake Faro (Messina, Sicily). *Limnol. Oceanogr.* **13**, 225–232.

— and Imhoff, J. F. (1981): The genus *Ectothiorhodospira*. In: Starr, et al., *The Procaryotes*, 274–278. Springer, Berlin, Heidelberg, New York.

Tsiban, A. V. (1971): Marine Bacterioneuston. *J. Oceanogr. Soc. Japan*, **27**, No. 2, 56–66.

— Panov, G. V., Daksh, L. V., and Jurskowskaya, V. A. (1980): Bacterial population of the open waters of the Baltic Sea. In: Investigations of the ecosystem of the Baltic Sea. I. Results of the Soviet-Swedish expedition in the Baltic Sea. *Hydromet.*

Tyler, P. A., and Marshall, K. C. (1967): Form and function in manganese oxidizing bacteria. *Arch. Mikrobiol.* **56**, 344–353.

Uden, N. van, and Castelo Branco, R. (1961): *Metschnikowiella zobellii* sp. nov. and *M. krissii* sp. nov. two yeasts from the Pacific Ocean pathogenic for *Daphnia magna*. *J. gen. Microbiol.* **26**, 141–148.

— and Fell, J. W. (1968): Marine yeasts. In: Droop, M. R., and Ferguson Wood, E. J., *Advances in microbiology of the sea* **1**, 167–201. London and New York, Academic Press.

Uhlmann, D. (1975): *Hydrobiologie. Ein Grundriß für Ingenieure und Naturwissenschaftler*. Jena, Fischer, 345 pp.

Ulken, A. (1963): Die Herkunft des Nitrits in der Elbe. *Arch. Hydrobiol.* **59**, 486–501.

— (1964): Über einige Thraustochytrien des polyhalinen Brackwassers. *Veröff. Inst. Meeresforsch. Bremerhaven* **9**, 31–42.

— (1965): Zwei neue Thraustochytrien aus der Außenweser. *Veröff. Inst. Meeresforsch. Bremerhaven* **9**, 289–295.

— (1969): Über das Vorkommen niederer saprophytischer Phycomyceten (Chytridiales) im Bassin d'Arcachon (Frankreich). *Veröff. Inst. Meeresforsch. Bremerhaven* **11**, 303–308.

— (1980): On some chytrids found in estuarine habitats. *Bot. Mar.* **23**, 343–352.

— (1981): On the role of phycomycetes in the food web of different mangrove swamps with brackish waters and waters of higher salinity. *Kieler Meeresforsch. Sonderh.* **5**, 425–428.

Vargues, H., and Brisou, J. (1963): Researches on nitrifying bacteria in ocean depth on the coast of Algeria. In: Oppenheimer, C. H., *Symposium on marine Microbiology*. Springfield, Ill., Thomas, 415- 425.

Waksman, S. A. (1933): On the distribution of organic matter in the sea bottom and the chemical nature of marine humus. *Soil Science* **36**, 125–147.

— Carey, C. L., and Allen, M. C. (1934): Bacteria decomposing alginic acid. *J. Bact.* **28**, 213–220.

Walker, N. (1975): Microbial degradation of plant protection chemicals. In: Walker, N., *Soil microbiology*, 181–192. London and Boston, Butterworths.

Wallhäuser, K. H. (1967): Ölabbauende Mikroorganismen in Natur und Technik. *Helgoländer wiss. Meeresunters.* **16**, 328–335.

Walter, R., Diener, W., and Namaschk, A. (1981): Zur Problematik der Virusinaktivierung in Wässern. In: Daubner, I., *3. Internationales hydromikrobiologisches Symposium*, 251–262. VEDA, Bratislava.

Ward, B. B. (1982): Oceanic distribution of ammonium-oxidizing bacteria determined by immunofluorescent assays. *J. Mar. Res.* **40**, 1155–1172.

Ward, T. G., and Frea, J. I. (1980): Sediment distribution of methanogenic bacteria in Lake Erie and Cleveland Harbor. *Appl. Environ. Microbiol.* **39**, 597–603.

Watson, S. W. (1963): Autotrophic nitrification in the ocean. In: Oppenheimer, C. H., *Symposium on marine microbiology*. Springfield, Ill., Thomas, 73–84.

Watson, S. W. (1965): Characteristics of a marine nitrifying bacterium, *Nitrosocystis oceanus* sp. n. *Limnol. Oceanogr.* **10** (Suppl.) 274–289.

— and Waterbury, J. B. (1971): Characteristics of two marine nitrite oxidizing bacteria, *Nitrospina gracilis* nov. gen. nov. spec. and *Nitrococcus mobilis* nov. gen. nov. spec. *Arch. Mikrobiol.* **77**, 203–230.

Weise, W., and Rheinheimer, G. (1978): Scanning electron microscopy and epifluorescence investigation of bacterial colonization of marine sand sediments. *Microb. Ecol.* **4** (3), 175–188.

— — (1979): Fluoreszenzmikroskopische Untersuchungen über die Bakterienbesiedlung mariner Sandsedimente. *Bot. Mar.* **22**, 99–106.

Weiss, R. L. (1973): Survival of bacteria at low pH and high temperature. *Limnol. Oceanogr.* **18**, 877–883.

Westheide, W. (1968): Zur quantitativen Verteilung von Bakterien und Hefen in einem Gezeitenstrand der Nordseeküste. *Marine Biology* **1**, 336–347.

Wetzel, R. G. (1968): Dissolved organic matter and phytoplanktonic productivity in mare lakes. *Mitt. Int. Ver. Limnol.* **14**, 261–270.

— (1975): *Limnology*. Philadelphia, Saunders, 743 pp.

— Rich, P. H., Miller, M. C., and Allen, H. L. (1972): Metabolism of dissolved and particulate detrital carbon in a temperate hard water lake. *Mem. Ist. Ital. Idrobiol.* **29**, 185–243.

Weyland, H. (1969): Actinomycetes in North Sea and Atlantic Ocean sediments. *Nature* **223**, 858.

— (1981): Distribution of actinomycetes on the sea floor. *Zb. Bakt. Suppl.* **11**, 185–193.

Wilde, V. (1975): Untersuchungen zum Symbioseverhältnis zwischen *Hirudo officinalis* und Bakterien. *Zool. Anz.* **195**, 289–306.

Williams, P. J. LeB. (1981): Incorporation of micro-heterotrophic processes into the classical paradigm of the planktonic food web. *Kieler Meeresforsch. Sonderh.* **5**, 1–28.

— and Askew, C. (1968): A method of measuring the mineralization by microorganisms of organic compounds in sea-water. *Deep-Sea Res.* **15**, 365–375.

Willoughby, L. G. (1962): The occurrence and distribution of reproductive spores of Saprolegniales in fresh water. *J. Ecol.* **50**, 733–759.

— (1969): Salmon disease in Windermere and the river Leven; the fungal aspect. *Salmon* and *Trout Mag.* **186**, 124–130.

— Smith, S. M., and Bradshaw, R. M. (1972): Actinomycete virus in fresh water. *Freshwat. Biol.* **2**, 19–26.

Winfrey, M. R., and Ward, D. M. (1983): Substrates for sulfate reduction and methane production in intertidal sediments. *Appl. Environ. Microbiol.* **45**, 193–199.

Winogradsky, S. (1890): Recherches sur les organismes de la nitrification. *Ann. Inst. Pasteur* **4**, 213–231, 257–275, 760–771.

— (1891): Recherches sur les organismes de la nitrification. *Ann. Inst. Pasteur* **5**, 92–100.

— (1925): Études sur la microbiologie du sol. I. Sur la methode. *Ann. Inst. Pasteur* **39**, 299–354.

Witzel, K.-P., and Overbeck, H. J. (1979): Heterotrophic nitrification by *Arthrobacter* sp. (strain 9006) as influenced by different cultural conditions, growth state and acetate metabolism. *Arch. Microbiol.* **122**, 137–143.

— — and Moaledj, K. (1982): Microbial communities in Lake Pluss; an analysis with numerical taxonomy of isolates. *Arch. Hydrobiol.* **94**, 38–52.

Wolter, K. (1982): Bacterial incorporation of organic substances released by natural phytoplankton populations. *Mar. Ecol. Prog. Ser.* **7**, 287–295.

Wolters, N., and Schwartz, W. (1956): Untersuchungen über Vorkommen und Verhalten von Mikroorganismen in reinen Grundwässern. *Arch. Hydrobiol.* **51**, 500–541.

Wood, E. J. F. (1967): *Microbiology of oceans and estuaries*. Elsevier Oceanography Series Vol. 3. Amsterdam, London, New York, Elsevier Publ. Co., 319 pp.

Wood, H. G., and Stjernholm, R. L. (1962): Assimilation of carbon dioxide by heterotrophic organisms. In: Gunsalus, I. C., and Stanier, R. Y., *The bacteria* III, 41–117. New York and London, Academic Press.

Wright, R. T., and Coffin, R. B. (1983): Planktonic bacteria in estuaries and coastal waters of northern Massachusetts: spatial and temporal distribution. *Mar. Ecol. Prog. Ser.* **11**, 205–216.

— and Hobbie, J. E. (1966): Use of glucose and acetate by bacteria and algae in aquatic systems. *Ecology* **47**, 447–464.

Wunderlich, M. (1973): Freie gelöste Zellulasen im natürlichen Gewässer und ihre ökologische Bedeutung. Ein Beitrag zum Umsatz der Zellulose im Gewässer. Dissertation Universität Kiel.

Yingst, J. Y., and Rhoads, D. C. (1980): The role of bioturbation in the enhancement of bacterial growth rates in marine sediments. In: Tenore, K. R., and Coull, B. C., *Marine Benthic Dynamics*. The Belle W. Baruch Library in Marine Science, No. 11, University S. Carolina Press.

Zaiss, U., Blass, M., and Kaltwasser, H. (1979): Produktion und Verbrauch von Methan und Wasserstoff durch Mikroorganismen in der Saar. *Deutsche gewässerkundl. Mitt.* **23**, 1–6.

Zaske, S. K., Dockins, W. S., and McFeters, G. A. (1980): Cell envelope damage in *Escherichia coli* caused by short-term stress in water. *Appl. Environ. Microbiol.* **40**, 386–390.

Zavarzin, G. A. (1972): Litotrofn'ie mikroorganismi. *Moskau Isdatel'stvo Nauka*, 323 pp.

Zhukova, A. I. (1963): On the quantitative significance of microorganisms in nutrition of aquatic invertebrates. In: Oppenheimer, C. H., *Symposium on marine microbiology*. Springfield, Ill., Thomas, 699–710.

Zimmermann, R. (1977): Estimation of bacterial number and biomass by epifluorescence microscopy and scanning electron microscopy. In: Rheinheimer, G., Microbial ecology of a brackish water environment. *Ecological Studies* **25**, 103–120. Berlin, Heidelberg, New York, Springer.

ZoBell, C. E. (1938): Studies on the bacterial flora of marine bottom sediments. *Sed. Petrol.* **8**, 10–18.

— (1942): Bacteria of the marine world. *Scient. Monthly* **55**, 320–330.

— (1946a): *Marine microbiology. A monograph on hydrobacteriology*. Waltham, Mass., USA, Chronica Botanica Co., 240 pp.

— (1946b): Studies on redox potentials of marine sediments. *Bull. Amer. Ass. Petrol. Gas Geol.* **30**, 447–513.

— (1963): Domain of the marine microbiologist. In: Oppenheimer, C. H., *Symposium on marine microbiology*. Springfield, Ill., Thomas, 3–24.

— (1964a): Hydrostatic pressure as a factor affecting the activities of marine microbes. In: *Recent researches in the fields of hydrosphere and atmosphere and nuclear geochemistry*, 83–116. Tokyo, Maruzen Co. Ltd.

— (1964b): The occurrence, effects and fate of oil polluting the sea. *Internat. Conf. Water Poll. Res. London 1962*, 85–118. London, Pergamon Press.

— and Beckwith, J. D. (1944): The deterioration of rubber products by microorganisms. *J. Amer. Water Assoc.* **33**, 439–453.

— and Budge, K. M. (1965): Nitrate reduction by marine bacteria at increased hydrostatic pressure. *Limnol. Oceanogr.* **10**, 207–214.

— and Cobet, A. B. (1964): Filament formation by *Escherichia coli* at increased pressure. *J. Bact.* **87**, 710–719.

— and Feltham, C. B. (1938): Bacteria as food of certain marine invertebrates. *J. marin. Res*, **7**, 312–327.

— and Johnson, F. H. (1949): The influence of hydrostatic pressure on the growth and viability of terrestrial and marine bacteria. *J. Bact.* **57**, 179–189.

— and Landon, W. A. (1937): Bacterial nutrition of the Californian mussel. *Proc. Soc. Exp. Biol. Med.* **36**, 607–609.

— and Morita, R. Y. (1959): *Deep-sea bacteria. Galathea Rep., Copenhagen* **1**, 139–154.

— and Oppenheimer, C. H. (1950): Some effects of hydrostatic pressure on the multiplication and morphology of marine bacteria. *J. Bact.* **60**, 771–781.

— and Upham, H. C. (1944): A list of marine bacteria including descriptions of sixty new species. *Bull. Scripps Inst. Oceanogr.* **5**, 239–292.

Index of Names of Genera

Subject Index